D0169341

Form 178 rev. 01-07

TUNDRA

Ecosystem

TUNDRA

Peter D. Moore

Illustrations by Richard Garratt

Facts On File
An imprint of Infobase Publishing

TUNDRA

Facts On File, Inc.
An imprint of Infobase Publishing
132 West 31st Street
New York NY 10001

ISBN-10: 0-8160-5933-0
ISBN-13: 978-0-8160-5933-1

Library of Congress Cataloging-in-Publication Data
Moore, Peter D.
    Tundra / Peter D. Moore; illustrations by Richard Garratt.
        p. cm.—(Ecosystem)
    Includes bibliographical references and index.
    ISBN 0-8160-5933-0
        1. Tundra ecology—Juvenile literature. 2. Tundras—Juvenile literature. I. Garratt, Richard.
    II. Title. QH541.5.T8M665 2008
    577.5'86—dc22   2006038355

Facts On File books are available at special discounts when purchased in bulk quantities for businesses, associations, institutions, or sales promotions. Please call our Special Sales Department in New York at (212) 967-8800 or (800) 322-8755.

You can find Facts On File on the World Wide Web at http://www.factsonfile.com

Text design by Erika K. Arroyo
Illustrations by Richard Garratt
Photo research by Elizabeth H. Oakes

Printed in the United States of America

Bang FOF 10 9 8 7 6 5 4 3 2 1

This book is printed on acid-free paper.

To Amelia, Amanda, Madeleine, and Michael.
Your hands may be very small but in them lies the future.

✦ ✦ ✦ ✦ ✦

Contents

Preface   xi

Acknowledgments   xiii

Introduction   xv
   What Is Tundra?   xv
   Where Is the Tundra?   xvi
   Why Is the Tundra Important?   xviii

1 ✦ Climate of the Tundra   **1**
   Patterns of Tundra Distribution   1
      Latitude   1
   Global Climate Patterns   6
      Albedo   8
   The Atmosphere and Climate   10
   The Oceans and Climate   11
      The Oceanic Conveyor Belt   13
   Arctic Climate   14
      A Year in the Arctic   16
   Aurora   16
   Antarctic Climate   17
   Microclimate   18
      Aspect and Microclimate   19
      Humidity   20
   Alpine Climate   21
      Lapse Rate   22
      Timberlines   22
   Conclusions   26

2 ✦ Geology of the Tundra   **28**
   The Rock Cycle   28
   Plate Tectonics   29
   Tectonic History of the Tundra Lands   33
   Polar Rocks   35
   Alpine Rocks   36

Snow and Ice   37

Ice Sheets, Ice Shelves, and Sea Ice   38

Nunataks   38

Ice Caps and Glaciers   40

Crevasses   44

Glacial Debris and Its Deposition   44

Sorting by Water and Wind   49

Particle Size   50

Geology and Rock Weathering   52

Acidity and pH   54

Soil Formation and Maturation   54

The Soil Profile of a Podzol   56

Periglacial Features   60

Conclusions   62

3 ✦ Types of Tundra   64

Tundra Forms   64

Vegetation Structure   64

Polar Desert   65

Dwarf Shrub Tundra   66

Tall Shrub Tundra   71

Forest Tundra   71

Coastal Tundra   74

Tundra Wetlands   74

Antarctic Tundra   78

Alpine Tundra   79

Tropical Alpine Habitats   82

Conclusions   85

4 ✦ The Tundra Ecosystem   86

What Is an Ecosystem?   86

Energy   87

Primary Productivity in the Tundra   88

Resource Allocation   89

Food Webs in the Tundra   91

Element Cycling in the Tundra   93

Nitrogen Fixation   97

Decomposition in the Tundra   98

Nitrification   98

Ecosystem Development and Stability   99

The Concept of Stability   100

Productivity and Population Cycles in the Tundra   101

Alpine Tundra Ecosystems   102

Conclusions   102

5 ✦ Biology of the Tundra   104

Wrapping up in the Freezer   104

Hair   105

Cold Stress 109
Plant Forms in the Tundra 110
   Plant Life-forms 112
   Biological Spectrum 113
Body Size in the Tundra 113
Color in the Tundra 115
   Why Be an Evergreen? 116
Drought in the Tundra 119
   Transpiration, Water Potential, and Wilt 120
Waterlogging and Oxygen Shortage 121
Reproduction and Dispersal in the Tundra 123
   r- and K-selection 125
   Pollination 126
Grazing in the Tundra 129
Behavioral Adaptations: Hibernation or Migration? 130
Conclusions 134

6 ✦ Biodiversity of the Tundra 136

The Meaning of Biodiversity 136
Patterns of Biodiversity 137
Biodiversity in Succession 139
   Succession 140
Microbes in the Tundra 141
   Cyanobacteria 142
Tundra Lichens 143
   Lichens 143
Tundra Mosses and Liverworts 145
The Tundra Soil Fauna 147
Vascular Plants in the Tundra 149
Tundra Arthropods 152
Birds of the Northern Tundra 154
   Ross's Gull 156
Alpine Tundra Birds 161
   Condors 163
Birds of the Southern Tundra 164
Polar Tundra Mammals 165
Alpine Tundra Mammals 170
Conclusions 174

7 ✦ Geological and Biological History of the Tundra 176

Glacial History of the Earth 176
The Pleistocene Glaciations 178
   Uniformitarianism 179
   Radiocarbon Dating 180
   Oxygen Isotopes 181
Causes of Glaciation 183
   Glaciation and Sea Level Change 183

Stratigraphy and Recent History of Tundra   186
    Pollen Analysis   188
The End of the Ice Age   190
After the Ice   192
Conclusions   194

8 ✦ People in the Tundra   196

The Emergence of Modern Humans   196
Early Tundra People and the Megafauna   197
    Mammoths   198
    Early Tundra Art   200
Modern Tundra People   202
Arctic Exploration   203
Antarctic Exploration   206
Alpine Tundra Exploration   208
Conclusions   209

9 ✦ The Value of the Tundra   211

The Problem of Evaluation   211
Hunting and Trapping   212
Mineral Reserves   214
Ecotourism and Recreation   216
Tundra and the Carbon Cycle   219
    Carbon Cycle and the Greenhouse Effect   220
Conclusions   221

10 ✦ The Future of the Tundra   222

The Threat of Climate Change   222
Biological Responses to Climate Change   224
Ozone Holes   226
Pollution of the Tundra   226
    History of the Ozone Holes   227
Tundra Conservation   229
    Albatrosses   230
Conclusions   232

11 ✦ General Conclusions   234

Glossary   237

Further Reading   247

Web Sites   249

Index   251

Preface

Increasingly, scientists, environmentalists, engineers, and land-use planners are coming to understand the living planet in a more interdisciplinary way. The boundaries between traditional disciplines have become blurred as ideas, methods, and findings from one discipline inform and influence those in another. This cross-fertilization is vital if professionals are going to evaluate and tackle the environmental challenges the world faces at the beginning of the 21st century.

There is also a need for the new generation of adults, currently students in high schools and colleges, to appreciate the interconnections between human actions and environmental responses if they are going to make informed decisions later, whether as concerned citizens or as interested professionals. Providing this balanced interdisciplinary overview—for students and for general readers as well as professionals requiring an introduction to Earth's major environments—is the main aim of the Ecosystem set of volumes.

The Earth is a patchwork of environments. The equatorial regions have warm seas with rich assemblages of corals and marine life, while the land is covered by tall forests, humid and fecund, and containing perhaps half of all Earth's living species. Beyond are the dry tropical woodlands and grassland, and then the deserts, where plants and animals face the rigors of heat and drought. The grasslands and forests of the temperate zone grow because of the increasing moisture in these higher latitudes, but grade into coniferous forests and eventually scrub tundra as the colder conditions of the polar regions become increasingly severe. The complexity of diverse landscapes and seascapes can, nevertheless, be simplified by considering them as the great global ecosystems that make up our patchwork planet. Each global ecosystem, or biome, is an assemblage of plants, animals, and microbes adapted to the prevailing climate and the associated physical, chemical, and biological conditions.

The six volumes in the set—*Deserts, Revised Edition;* *Tundra; Oceans, Revised Edition; Tropical Forests; Temperate* *Forests, Revised Edition;* and *Wetlands, Revised Edition*— between them span the breadth of land-based and aquatic ecosystems on Earth. Each volume considers a specific global ecosystem from many viewpoints: geographical, geological, climatic, biological, historical, and economic. Such broad coverage is vital if people are to move closer to understanding how the various ecosystems came to be, how they are changing, and, if they are being modified in ways that seem detrimental to humankind and the wider world, what might be done about it.

Many factors are responsible for the creation of Earth's living mosaic. Climate varies greatly between Tropics and poles, depending on the input of solar energy and the movements of atmospheric air masses and ocean currents. The general trend of climate from equator to poles has resulted in a zoned pattern of vegetation types, together with their associated animals. Climate is also strongly affected by the interaction between oceans and landmasses, resulting in ecosystem patterns from east to west across continents. During the course of geological time even the distribution of the continents has altered, so the patterns of life currently found on Earth are the outcome of dynamic processes and constant change. The Ecosystem set examines the great ecosystems of the world as they have developed during this long history of climatic change, continental wandering, and the recent meteoric growth of human populations.

Each of the great global ecosystems has its own story to tell: its characteristic geographical distribution; its pattern of energy flow and nutrient cycling; its distinctive soils or bottom sediments, vegetation cover, and animal inhabitants; and its own history of interaction with humanity. The books in the Ecosystem set are structured so that the different global ecosystems can be analyzed and compared, and the relevant information relating to any specific topic can be quickly located and extracted.

The study of global ecosystems involves an examination of the conditions that support the planet's diversity. But environmental conditions are currently changing rapidly. Human beings have eroded many of the great global ecosystems as they have reclaimed land for agriculture and urban settlement and built roads that cut ecosystems into ever smaller units. The fragmentation of Earth's ecosystems is proving to be a serious problem, especially during times of rapid climate change, itself the outcome of intensive industrial activities on the surface of the planet. The next generation of ecologists will have to deal with the control of global climate and also the conservation and protection of the residue of Earth's biodiversity. The starting point in approaching these problems is to understand how the great ecosystems of the world function, and how the species of animals and plants within them interact to form stable and productive assemblages. If these great natural systems are to survive, then humanity needs to develop greater respect and concern for them, and this can best be achieved by understanding better the remarkable properties of our patchwork planet. Such is the aim of the Ecosystem set.

Acknowledgments

I should like to record my gratitude to the editorial staff at Facts On File for their untiring support, assistance, and encouragement during the preparation of this book. Frank K. Darmstadt, executive editor, has been a constant source of advice and information and has been meticulous in checking the text and coordinating the final assembling of materials. My gratitude also extends to the production department. I should also like to thank Richard Garratt for his excellent illustrations and Elizabeth Oakes for her perceptive selection of photographs. Particular thanks are due to my wife, who has displayed a remarkable degree of patience and support during the writing of this book, together with much needed critical appraisal. I must also acknowledge the contribution of many generations of students in the Life Sciences Department of the University of London, King's College, who have been a constant source of stimulation, critical comment, and new ideas. I also acknowledge a considerable debt to my colleagues in teaching and research at King's College, especially those who have accompanied me on field courses and research visits to many parts of the world. Their work, together with that of countless other dedicated ecologists, underlies the science presented in this book.

Introduction

The world is no longer rich in wilderness. In the course of a few thousand years the human species has penetrated to all parts of the globe and has everywhere left an indelible mark in the form of habitat destruction and pollution. But of all the world's biomes the tundra remains the closest to its original, pristine condition. Tundra may be regarded as the last true wilderness. This book examines the structure and the distribution of tundra ecosystems, including both the polar regions and the high mountain tops of the world. It considers the development of the tundra during the course of geological history and looks at the impact of people on this wild ecosystem. By understanding how the tundra has come into being and how it functions, future generations will be in a better position to protect its unique features.

Tundra is structured in a way that will make relevant information both interesting and accessible to the student. Each chapter deals with a major feature of the tundra environment, including physical aspects, such as climate, geology, and geography, and biological aspects, including the adaptations of organisms to extreme conditions. The student is encouraged to integrate these living and nonliving components by viewing the tundra as an interactive ecosystem, operating like a delicately balanced machine. In order to appreciate the problems facing the tundra, the question of environmental change is finally emphasized, looking at the possible consequences of climate change and increasing pollution in this remote and wild biome. Throughout the book diagrams, photographic illustrations, and explanatory sidebars are used to ensure that the factual material is presented with clarity and maximum impact.

The tundra regions of the world have the lowest density of human populations, with the possible exception of some deserts. It is the one biome that has resisted most attempts at agricultural exploitation and human settlement. Most people, even those who spend much time in traveling the world, have never visited tundra habitats except perhaps in the depths of winter when mountaintops offer recreational opportunities, such as skiing and snowboarding. Consequently, the tundra is a mystery to the vast majority of people, a land of ice and snow that is known only through television documentaries or adventure novels. It is not surprising, therefore, that there are many misconceptions about the tundra.

The tundra is cold. This is true as far as average annual temperature is concerned, and it is certainly true with respect to the long, dark winters. But in summer temperatures can become high, and the long summer days become favorable to many plants and animals. The tundra is low in biological diversity. Again, this is true when compared to tropical forests, but the range of creatures found in the tundra is still considerable, and many of the plants and animals encountered there are found nowhere else on Earth. The tundra is wet in summer and snowy in winter. This is the immediate impression of the landscape, but the actual quantity of precipitation (rain and snow) is actually very low. The input of water from the atmosphere is similar to that of a desert, but the low temperature and consequent low evaporation mean that water or snow is often present. The tundra is a wasteland. From the point of view of human settlement and farming, this is clearly the case, but there are many inconspicuous ways in which the tundra serves humanity. The balance of gases in the Earth's atmosphere and therefore global climate are strongly affected by what goes on in the tundra. The tundra has many hidden values.

■ WHAT IS TUNDRA?

The word *tundra* is derived from the Finnish language. The rounded, treeless hilltops of northern Finland are called *tunturi,* and the word has become modified to *tundra.* The application of the word *tundra* has also become extended to cover all treeless landscapes found in generally cold places, such as the regions bordering upon the polar ice or oceans, and the summits of high mountains, even in the Tropics.

The absence of trees in the tundra, which is its most distinctive feature when first observed, is not confined to this biome. Deserts and grasslands have few trees because of drought conditions that restrict plant growth. The scrub vegetation of the chaparral is maintained free of trees because it is subject to frequent fires. But the restricting factors for tree growth in the tundra are the combination of low winter temperatures and high winds that carry abrasive particles of ice, blasting and destroying all growing parts of plants that raise themselves high above the ground. All the vegetation found in the tundra hugs the soil closely. Cushions of moss, flat plates of lichens on rocks, small domes of herbaceous plants, and low-lying stretches of dwarf shrubs form the general vegetation cover of the tundra. The only trees present are hardly recognizable as such, being reduced to contorted, dwarf forms that lie close to the ground.

Beneath this cover of vegetation lies a sheltered and protected zone, where pockets of air remain still and unaffected by the wind above the plant canopy. Temperatures rise during the day, humidity stays high, and these two factors favor the many insects and other small invertebrates that manage to eke out a living, hidden from the bleak tundra landscape. Small mammals and birds can also find a home in the low cover of vegetation, feeding on leaves, berries, and insects. Some of the birds travel great distances to take advantage of the productivity of the tundra together with the long days of summer in the polar regions. These extended days enable them to feed their young more effectively and produce bigger families, so a surprising number of birds choose to breed in the tundra. Some larger mammal herbivores, including caribou and musk ox, graze on the summer productivity. Both birds and mammals in turn may form a food source for the tundra predators, gyr falcon, arctic fox, wolf, and polar bear. Together the plants and animals form the distinctive living community of the tundra.

The landscape of the tundra is not flat and uniform. There are many microhabitats to be found within the tundra, including peaty wetlands in the lower regions and raised ridges of shingle and rock in the more elevated locations. Craggy cliffs and high mountains are found in some of the polar tundra regions, providing impressive vistas with permanent snow and ice. This varied landscape offers opportunities for a range of specialized plants and animals to find a home within the variety of microhabitats on offer.

The soils of the tundra are also distinctive in their form. With the coming of winter they freeze, while in summer they thaw once again. When temperatures in the tundra region are especially low, or when the summer melt season is particularly short, the soil may not thaw out completely. Only the surface layers thaw, leaving the deeper layers of the soil permanently frozen, which is called the *permafrost*. Animal life in the soil is thus confined to the upper layers, where conditions become warm enough for active life and growth in summer. Even the hardy fungi and bacteria in the soil are restricted in their activity, spending the winter in a dormant state and becoming active in the summer as they decompose organic remains.

■ WHERE IS THE TUNDRA?

There are three main regions of the Earth that support the tundra biome: the Arctic region, together with some adjacent areas outside the Arctic Circle that can be termed *subarctic;* some coastal locations on the continent of Antarctica, together with neighboring islands in the southern ocean, strictly *subantarctic* islands; and mountain summits throughout the world where conditions at high altitude are too severe for tree growth. These regions are shown on the map.

The Arctic and Antarctic tundra regions have much in common and can jointly be termed *polar tundra*. The mountaintop tundra, on the other hand, has many distinctive climatic features and contains some plants and animals that are not found in the polar regions. These montane sites can be classed as *alpine tundra*.

The two types of tundra share a generally cold climate, often with only a short summer season that is suitable for plant growth and animal activity. They are both subjected to ice and snow in winter, much of which may persist through the summer. Wind is an important factor in both types of tundra, restricting the height to which plants are able to grow. The vegetation of both alpine and polar tundra, therefore, has a superficial similarity. Some plant species are found in both types of tundra, while others are restricted to either the alpine or the polar tundra.

Some mountains are found within the Arctic and Antarctic Circles, while other mountains are found close to the equator. The polar mountains provide a home for a similar range of plants and animals to those found in lowland polar tundra, while the mountains of equatorial regions usually carry a very different assemblage of species. Polar and equatorial mountains also differ considerably in their respective climates, despite the fact that both experience a cold overall average temperature. Daily fluctuations in temperature are much greater on an equatorial mountain, but seasonal variation is far less. Precipitation is usually higher on mountains than is the case in the lowland polar regions. Mountains are also

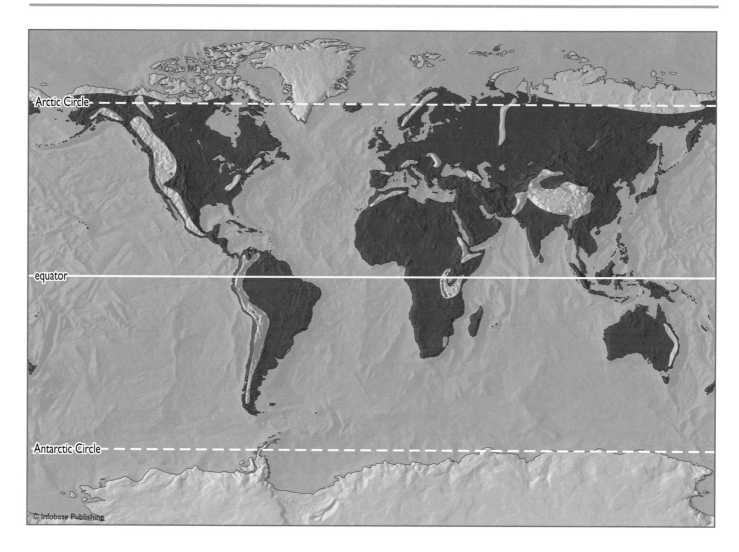

The regions of the world in which tundra ecosystems are found. Much of Greenland and most of Antarctica are covered by ice and so bear no vegetation. Tundra in regions outside the Arctic and Antarctic Circles is confined to high-altitude sites. This type of mountain tundra is called alpine tundra to distinguish it from polar tundra.

often widely separated from one another, and the intervening lowlands may be impassable for animals with poor dispersal ability. So the plants and animals of mountains may find themselves isolated, and this can lead to evolutionary separation and development much like that found on oceanic islands.

Mountain landscapes usually display a greater variety of topographical features than flatlands. High peaks are interspersed with deep valleys, providing a wide range of habitats and a great variety of locations that differ in their local climate, or *microclimate*. This diversity of habitat and microclimate can lead to a great range of animals and plants that have special requirements and make spe-

cific demands of their environment. When this is coupled with the spatial isolation of mountains, which stimulates evolutionary divergence, it is not surprising to find that there are some very distinctive organisms found on mountains, some of them very restricted in their global distribution.

Tundra, therefore, is a globally widespread and varied habitat. It is the product of extreme climatic conditions, and it contains some organisms that are highly adapted in order to cope with the stress of extreme cold. In general, tundra is not a rich biome; it does not contain a very high diversity of species, but it is the home of some very specialized species that are found in no other type of habitat. The structure, biochemistry, and behavior of tundra organisms have long attracted the interest of biologists and ecologists, who are eager to discover how these creatures survive under such stress. The presence of whole communities of these highly specialized creatures has made the tundra a living laboratory in which scientists can study the ways in which nature deals with the challenge of an extreme environment.

WHY IS THE TUNDRA IMPORTANT?

It is evident that the tundra has much interest for scientists, whether geographers who study the topography of ice-created landforms, biologists who examine cold tolerance among plants and animals, or soil scientists who investigate the effect of freeze-thaw cycles on soil structure. But the tundra is also important to people who may not be directly involved in scientific investigations. In the future the importance of the tundra may well become increasingly appreciated as the many valuable features of this biome become more widely known. This book is designed to assist in this process.

The highly adapted plants and animals of the tundra contain within their cells genetic information that determines their hardiness. There are genes, for example, that provide plants with the ability to resist freezing or to survive even when cells become frozen. There are genes within polar bears that enable them to live a healthy and vigorous life despite the fact that their diet (mainly seals) contains extremely high levels of fat. Understanding how either of these two biochemical systems operates could be of great value in agriculture and human medicine, respectively. The animals and plants of the tundra, therefore, can be regarded as an enormous untapped resource of information that may one day enable people to improve their lives.

During the history of the Earth the tundra has experienced changing fortunes. There have been long periods when this biome has been absent or found only on mountaintops. There have also been times when the tundra has spread and expanded over continents, the ice ages. It is vital that the climatic mechanisms underlying these changes be understood. There is an apparent pattern to these events that requires detailed study so that the future behavior of the tundra under ever-changing conditions can be predicted. Geologists collect evidence for past changes in the distribution of tundra in order to detect patterns and understand the processes involved in these global climatic changes.

At the present day the climate of the polar regions appears to be changing more rapidly than that of other parts of the world. It is not yet understood why this should be the case, but changing conditions in the Arctic and Antarctic may provide an early warning of what might be expected elsewhere. There are many questions that need to be answered about the role of the polar regions in global climate change, and this work is advancing rapidly. One of these questions concerns the development of the so-called *ozone hole,* in which changes in atmospheric composition, particularly in the gas ozone, can lead to potentially dangerous rays reaching the surface of the Earth. Knowledge of this phenomenon has improved greatly in recent years. Remedial actions seem to be having some effect, although there are considerable year-to-year variations and there remain many unanswered questions.

Tundra is one of Earth's few remaining wilderness areas, and a consequence of this is that there is very little pollution resulting from human activities. But even here pesticides manage to penetrate. The persistent pesticide DDT, for example, has been found in the flesh of Antarctic penguins, which provides a salutary warning of just how widely dispersed are the chemicals people introduce into their environment. The relative purity of the tundra ecosystems makes them particularly useful in monitoring the movement of pesticides and their continued presence in the global environment.

The tundra is becoming increasingly accessible to people as tourism develops and the popularity of adventure holidays and ecotourism grows. Undisturbed habitats will be increasingly exposed to stress from people, whether they come to admire the scenery and wildlife or to exploit the mineral reserves found in many tundra areas. Antarctica is currently protected by international treaty, but the increasing scarcity of oil reserves and the continued demand for minerals will undoubtedly bring increased demand for exploitation even in this last wilderness.

It is this combination of potentials and problems that makes the tundra a biome that demands attention. Despite its wide geographical distribution, the tundra has proved hard to explore and understand. The severe conditions that make the tundra so distinctive have also made exploration and scientific research particularly difficult. There is much that is yet to be discovered about the tundra, but such discovery depends upon our appreciating and understanding what is currently known about this remote and unique biome. This is the task of the current book.

Climate of the Tundra

The tundra biome, unlike most other biomes, is not restricted to one particular region of the Earth or to any distinct zones north and south of the equator. The reason for this, as explained in the introduction, is that the tundra is found both in polar regions and at high altitudes on the world's mountains, where it survives above the level of forest growth. Mountains are scattered around the world, so tundra ecosystems may occur even in the Tropics, close to the equator, as long as the mountain is high enough. In order to discover what factors control the formation of tundra, it is necessary to analyze in detail the precise distribution of this biome.

■ PATTERNS OF TUNDRA DISTRIBUTION

The map on page xvii shows the broad pattern of tundra distribution over the Earth's land surface. As explained in the introduction, there are two types of tundra present on this map: One is the polar tundra, restricted to the high latitudes (see sidebar), and the other is the alpine tundra, found only on mountains but that can occur in any latitude, both high and low, depending simply on the presence of mountains.

Polar tundra is found mainly in the Northern Hemisphere, in the region called the Arctic. The word *Arctic* refers to the northern constellation of stars called the Great Bear (*Arktos* is Greek for "bear"). The northern distribution limit of the tundra, as shown in the diagram on page 3, is defined by the Arctic Ocean, and the presence of an ice sheet over Greenland also restricts its occurrence on that large island. To the south it is bordered by the boreal forest, or *taiga*, where coniferous trees and birches become the dominant feature of the landscape. The Arctic tundra covers approximately 29,000 square miles (7.5 million ha), which represents 5.5 percent of the total land surface of the Earth. It is divided into three blocks: North America with

Latitude

The Earth is a sphere that rotates around an axis. This axis passes from the North Pole through the center of the Earth to the South Pole. The equator is an imaginary line that passes around the Earth and is equidistant from both poles. In cross section, therefore, the equator forms a right angle to the Earth's axis, as shown in the map on page 2. The equator is regarded as having an angle of zero degrees (0°), which means that the angle of the North Pole is 90°N and the South Pole is 90°S. Between these extremes it is possible to denote the position north or south of any point on the Earth's surface by reference to the angle at which it lies in relation to the equator. This is called its *latitude*. Lines of latitude run around the Earth north and south of the equator, ranging from 0° to 90°. Those with high values (close to the poles) are termed the *high latitudes,* and those close to the equator have low values and are called *low latitudes.* Polar tundra is found in the high latitudes. These lines are, of course, purely conceptual but are an important means of defining any location on the surface of the Earth. A second coordinate is required to fix a precise position, and this is given by its *longitude.* Lines of longitude run from pole to pole over the Earth's surface and are numbered from 0° to 359°, with the base line (0°) running through Greenwich in London, England. This choice of the zero line of longitude is a consequence of history, but it has proved convenient because the 180° line (the International Date Line) runs through the Pacific Ocean rather than over land surface, where changing date with location on either side of a line would be extremely confusing for the local inhabitants.

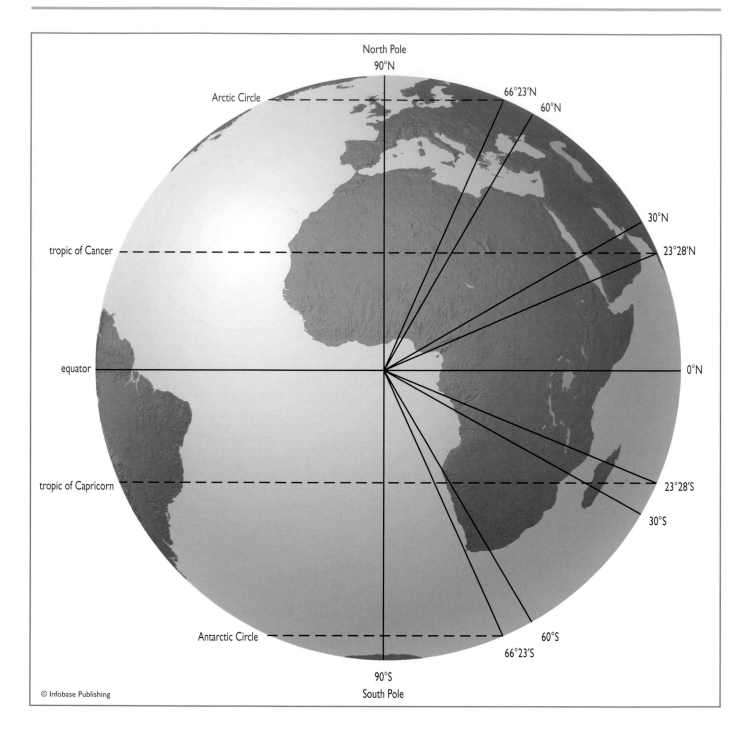

11,000 square miles (2.8 million ha), Eurasia with 10,000 square miles (2.6 million ha), and Greenland and Iceland with 8,000 square miles (2.1 million ha).

Much of Alaska is covered by tundra, extending through the Aleutian Islands, which form a chain out into the North Pacific Ocean. The northern part of Canada has a tundra cover running from the Northwest Territories around the southern shore of Hudson Bay and into northern Quebec and Labrador. Extending out into the Arctic Ocean are numerous islands, including Baffin Island and Ellesmere Island, all clothed with tundra. Only the coastal fringe of

Latitude is measured as the angle at the center of the Earth for each location on the surface, taking the equator as 0° and the poles as 90° north and south. The latitudes of the tropics of Cancer and Capricorn together with the Arctic and Antarctic Circles are shown.

Greenland is free from a permanent cover of ice, leaving little room for tundra, but the more southerly island of Iceland in the North Atlantic Ocean has a covering of tundra. The mainland of Europe has only a fringe of polar tundra in the north of Scandinavia, mainly in northern Norway, but the ice-free areas of the islands of Svalbard in the Arctic Ocean bear tundra. This biome is also found along the fringe of the Arctic Ocean in northern Russia, together with the islands

lying off that coast. It also covers the eastern parts of Siberia, where the tundra reaches the Bering Sea, separating Russia from Alaska. Effectively, the polar tundra forms a circle around the Arctic Ocean; it has a *circumpolar* distribution.

Polar tundra is also found in the Southern Hemisphere but is much less abundant. The main difference in the polar regions of the Southern Hemisphere is the presence of a continental landmass, Antarctica. Unlike the North Pole, which is situated over the Arctic Ocean, the South Pole lies within the landmass. The main effect of the presence of this

The Arctic tundra regions showing the extent of the Arctic polar tundra and the position of the magnetic North Pole. Greenland is covered by an ice sheet, so tundra vegetation occurs only around its coasts. Much of the Arctic Ocean is covered by permanent sea ice, which extends yet farther as fragmented pack ice, especially in winter.

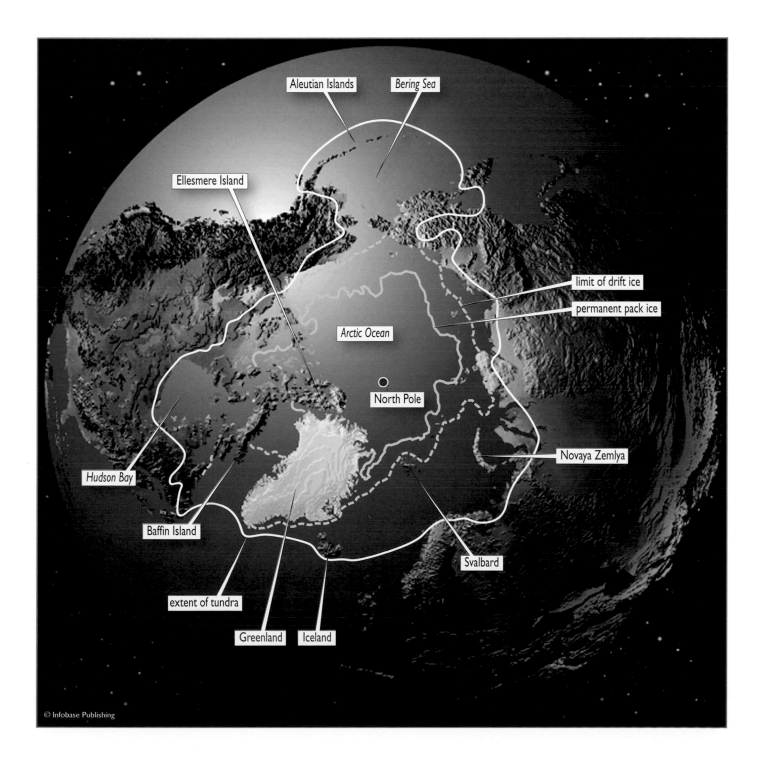

continent is the development of an ice sheet, similar to that of Greenland in the Northern Hemisphere, but much bigger. The ice sheet covers almost all Antarctica, so there is virtually no vegetation present except for small areas where rocks emerge from the ice along the coastal fringe (see illustration). On the western side of Antarctica, as shown on the map below, lies a peninsula of land that stretches up toward

(above) The edge of an ice sheet, where the ice cliff meets floating pack ice   *(University of Kiel Institute of Polar Ecology)*

(below) Antarctica, showing the location of the magnetic South Pole. The entire continent, except for a few coastal areas and islands, is covered by the world's largest ice sheet. This ice rests upon bedrock in many areas, but in some regions the ice rests upon the ocean floor (shown here in gray). There are three ice shelves, extensions of the ice sheet as floating layers over ocean bays (shown here in white). The confluence of the three oceans forms the Southern Ocean, which surrounds Antarctica.

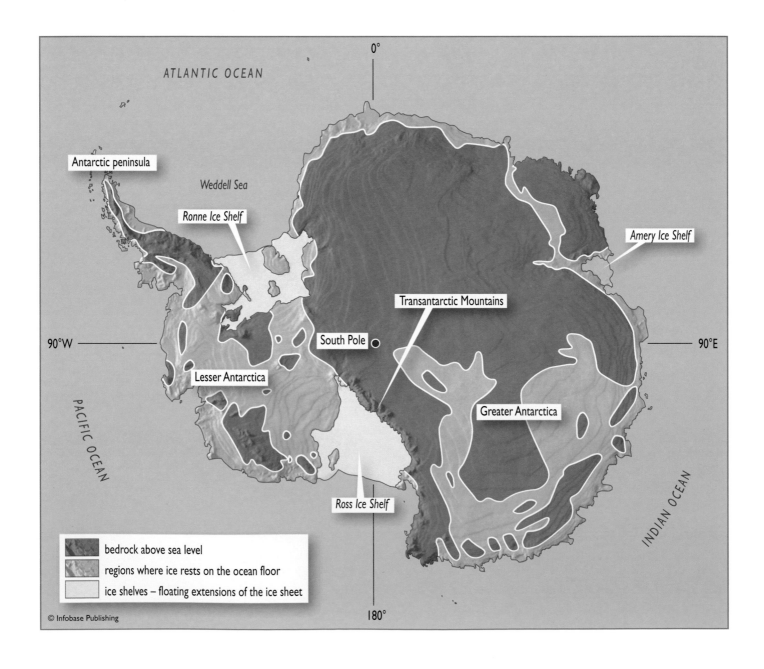

ATLANTIC OCEAN

0°

Antarctic peninsula

Weddell Sea

Ronne Ice Shelf

Amery Ice Shelf

Transantarctic Mountains

90°W

South Pole

90°E

Lesser Antarctica

PACIFIC OCEAN

Greater Antarctica

Ross Ice Shelf

INDIAN OCEAN

bedrock above sea level
regions where ice rests on the ocean floor
ice shelves – floating extensions of the ice sheet

© Infobase Publishing

180°

An alpine meadow in Switzerland shows the colorful flowers for which such habitats are famous. *(Peter D. Moore)*

South America. This is called the Antarctic Peninsula, and this, together with a number of islands lying off its coast, is referred to as Maritime Antarctica. This region extends farther north than most of the Antarctic landmass and so penetrates into relatively lower latitudes. Maritime Antactica is also surrounded by the sea, and these two factors, lower latitude and oceanic conditions, result in a slightly warmer climate, reducing ice cover and permitting the development of tundra vegetation. Most of the very limited tundra found in the Southern Hemisphere lies within the Maritime Antarctic.

The polar tundra is thus restricted overall to the higher latitudes. Most tundra in the Northern Hemisphere lies north of 70°N latitude. In the Southern Hemisphere the effect of the Antarctic ice sheet chills the southern ocean. Tundra is found on islands as far north as 55°S, but its limit is still clearly associated with relatively high latitude. This is not the case for alpine tundra. Alpine tundra is found

wherever mountains are high enough to result in restricted forest growth. When conditions become too cold or wind strengths too high for the survival of trees, tundra is able to thrive. Precisely how high a mountain needs to be in order to provide the right conditions for tundra varies with latitude.

The mountain ranges of the world are shown in the map on page xvii. As can be seen, they are quite widely scattered, but where mountains occur they often form groups or long chains. The reasons for this will be investigated later (see "Tectonic History of the Tundra Lands," pages 33–35). In North and South America the main mountain systems lie on the western sides of the continents and form roughly north-south running ridges. The mountain ranges of Alaska and the Mackenzie Mountains of Canada merge in the southern part of their range to create the Rocky Mountains, the spine of western North America. The Cascade Mountains of Washington and Oregon and the Sierra Nevada of California are essentially coastal branches and parallel systems alongside the main Rocky Mountain ridge. The range continues south into Mexico as the Sierra Madre, which run through Central America and eventually link up with the Andes Mountains that line the west coast of South America. At the southern tip of that continent the Andes pass through Patagonia and end in the extremity of Tierra del Fuego, where the alpine tundra and polar tundra meet.

In Europe an extensive mountain chain runs approximately from northeast to southwest on the western fringe of Scandinavia, mainly in Norway. The mountains of Scotland in the British Isles are a southerly outlier of this system. The remaining mountain chains of Europe are east-west in their orientation, a fact that has had considerable implications for the biogeography of that continent (see "Glacial History of the Earth," pages 176–178). In the west of Europe, the Pyrenees Mountains separate France from the Iberian Peninsula (Spain and Portugal). To the north of Italy lie the Alps, where the highest mountains in Europe are found (see illustration on page 5). Farther east, the Carpathian Mountains form a crescent on the western fringes of the Black Sea.

North Africa also has a mountain range, the Atlas Mountains, that lie in an east-west arrangement, running through Morocco and Algeria. The remaining African mountains are concentrated in the eastern regions, running south from the Ethiopian highlands. The Great Rift Valley systems of Kenya, Tanzania, Uganda, and Congo are rich in high mountains, many of them volcanic, whose chains run across the equator. In southern Africa the Drakensberg Mountains lie along the eastern side of the continent.

In Asia the north-south running Ural Mountains separate this continent from Europe and extend from the Arctic toward the Caspian Sea. It is in Asia that the highest mountain in the world, Mount Everest (29,029 feet; 8,848 m) is located, set within the Himalayas. This massive mountain block runs from Afghanistan in the west, where the mountains are called the Hindu Kush, through northern Pakistan, India, Tibet, and Nepal, ending in eastern China. They extend northward around the Gobi Desert into Mongolia. In eastern Russia lie the Altai Mountains, the Sayan Mountains, and the Transbaikalian Mountains, reaching over 11,000 feet (3,400 m). Australia is not rich in high mountains but has the north-south orientated Great Dividing Range along its eastern coast.

Alpine tundra is, therefore, much more widely dispersed than polar tundra, but separating the two is not always possible because some high mountains lie within or adjacent to the polar regions. In North America, for example, the Brooks Range of mountains lies along the southern edge of the polar tundra in Alaska, and the Laurentian Mountains are situated within the polar tundra of Labrador. In Europe the mountains of Norway merge with polar tundra at their northern extremity, as do the Ural Mountains of Asia. In eastern Asia the mountains of Siberia also merge with the tundra of the Arctic. Antarctica has a ridge of mountains running along the length of the Antarctic Peninsula and extending around the edge of the Ross Ice Shelf and the Ross Sea. In areas such as these, the polar and alpine tundras meet and are indistinguishable from one another.

■ GLOBAL CLIMATE PATTERNS

Understanding the distribution pattern of tundra demands a close study of climate patterns over the Earth. Why are the Arctic and Antarctic colder than the Tropics? And why is tundra found at lower latitudes in the Southern Hemisphere than in the Northern Hemisphere? To answer these questions, it is necessary to consider the distribution of energy around the Earth, and this requires information about the movements of air masses in the atmosphere and water masses in the oceans.

The first point to examine is the way in which energy arrives at the surface of the planet. Energy on Earth ultimately comes almost entirely from the Sun. There are some exceptions to this statement: The hot interior of the Earth releases some *geothermal energy,* natural radioactive decay also produces some energy, and the gravitational pull of the Moon creates tidal energy. But the vast bulk of the energy arriving on Earth is in the form of solar radiant energy. Energy comes in a variety of forms, which physicists differentiate in terms of *wavelength.* Light itself, for example, consists of a spectrum of energy of different wavelengths, from the very short violet end to the longer-wavelength red end. This accounts only for the visible forms of energy; there are many shorter-wavelength forms, such as ultraviolet and even shorter gamma rays, and also longer-wavelength forms, such as infrared, heat energy that can be felt but not seen. The Sun emits energy over a very broad spectrum, but most of the solar radiation that reaches the surface of the Earth lies between 0.000001 and 0.00016 inches (0.00000025 and 0.0004 cm). Such minute numbers are difficult to work with, so the measurements used by scientists are metric, and the wavelength is expressed in micrometers (1 μm = 1 millionth of a meter). The Earth receives energy mainly in the 0.2 μm to 4.0 μm wavebands. The Earth's atmosphere does not absorb energy within this range, so it passes directly to the surface of the planet. Shorter wavelengths, however, such as some ultraviolet radiation, are absorbed by certain components of the atmosphere (see "Ozone Holes," page 226). In addition, dust in the atmosphere together with cloud cover can result in the absorption or reflection of the some of the incoming energy.

When solar energy strikes the surface of the Earth, much of it is likely to be absorbed. An object placed in sunlight becomes warm as a result of the absorption of energy, and as its temperature rises it radiates some energy back into the atmosphere. In a similar way a land surface heated

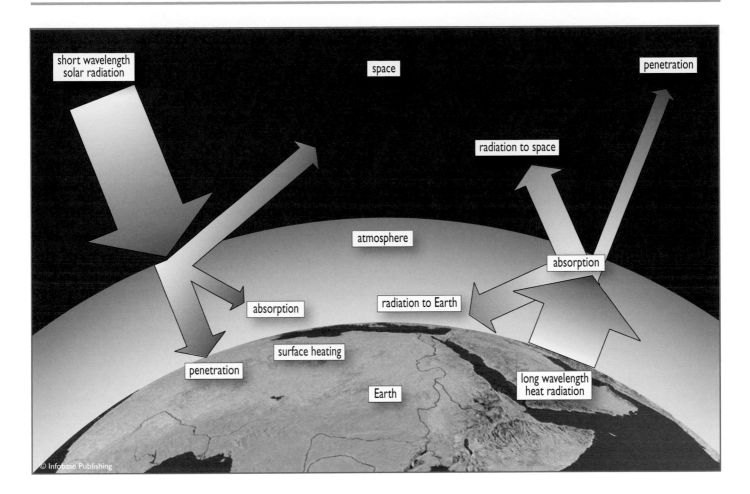

The greenhouse effect. Solar energy arriving at the Earth's atmosphere consists mainly of short wavelength radiation. Some of this is reflected, a little is absorbed by the atmosphere, but much of the energy passes through to the Earth's surface. Here it heats the land and water, and energy is then radiated back in the form of long wavelength heat. Some of the radiated heat penetrates the atmosphere and is lost to space, but much is absorbed by the atmosphere, which then radiates heat in all directions, including some back to Earth. Certain gases in the atmosphere, the so-called greenhouse gases, are particularly efficient at trapping heat and act as a thermal blanket around the Earth.

by sunlight radiates heat energy, and heat energy has a longer wavelength than light energy. Some of the heat energy radiated by land surfaces is taken up by the atmosphere, because the atmosphere is less transparent to longer-wavelength heat than it is to light. By absorbing the reradiated heat energy in this way, the atmosphere acts like a thermal blanket around the Earth, retaining heat that would otherwise be rapidly lost, as shown in the diagram above. This absorption, known as the *greenhouse effect,* is due to the presence of certain gases in the atmosphere, including

water vapor, carbon dioxide, oxides of nitrogen, and several others, which have become known as *greenhouse gases.* It is called the greenhouse effect because the entire Earth system is acting like an enormous greenhouse. Light passes through glass panes of a greenhouse without being absorbed to any great extent by the glass. The glass does not itself become hot. When light strikes the soil within the greenhouse it causes the soil to become hot, and the reradiated heat does not pass out through the glass roof because long-wave heat energy does not easily penetrate glass but remains within the greenhouse, warming the air it contains. Without the greenhouse effect the Earth would rapidly cool as soon as the Sun set in the evening. Nights would become unbearably cold, and life on the planet would be much more difficult. As will be discussed later (see sidebar "Carbon Cycle and the Greenhouse Effect," page 220), changes in the intensity of the greenhouse effect in the atmosphere have a profound influence on global climate.

There is a complication in this process because the amount of energy absorbed by a surface varies. A black surface absorbs sunlight and becomes heated much more effectively than a white surface, and a dull, matte surface absorbs more efficiently than a shiny surface. The efficiency of a surface in absorbing light and becoming heated in the process is dependent on how much of the incoming energy is reflected, and the reflectivity of a surface is expressed as its *albedo* (see sidebar below).

As can be seen from the figures in the sidebar, different types of land surfaces and vegetation have a strong effect on how much of the Sun's energy is absorbed. The figures for the surface of water also show that albedo depends quite strongly on the angle of the Sun. These aquatic data apply to a water body with a calm and smooth surface. Waves can create very complex patterns of reflection. These variations in albedo, depending on the nature of the surface, have an important impact on the world's climate. Oceans and land surfaces, forests and deserts, ice sheets and wetlands all have different capacities to absorb and reflect incoming radiation, which influences the degree to which they become warm in the sunshine and also the way they radiate heat back into the atmosphere.

The angle at which radiation strikes the Earth's surface also influences absorption and reflection. The diagram shows the way in which the Earth's curvature results in energy being received at different angles depending on latitude. The Sun is overhead at noon at the equator twice a year, when the length of day and night is equal over the whole planet. The Earth's tilt on its axis of 23.5°, however, means that this situation is not maintained. The Sun is overhead at the tropic of Cancer (66.5°N) during the Northern Hemisphere summer and at the tropic of Capricorn (66.5°S) during the Southern Hemisphere summer. Throughout the year, however, the angle of the Sun within the low-latitude Tropics is high when compared with that of the higher latitudes. In the polar regions the angle of the Sun above the horizon is never high. This difference in the solar angle between the poles and the Tropics means that energy is spread over a wide area in the high latitudes, whereas it is concentrated in a smaller area in the low latitudes. The heating potential of radiation in the polar regions is thus lower than that found in the Tropics, as shown in the diagram on page 9.

Seasonal differences in day length are also more pronounced in the high latitudes. In the Tropics there is some variation in day length depending on the position of the overhead Sun at noon, but the variation is very small. Days are roughly equal to nights in their length at all times of the year. At higher latitudes, however, there are increasingly wide disparities in day length with season. At 60°N (the latitude

Albedo

When light falls upon a surface, some of the energy is reflected and some is absorbed. The absorbed energy is converted into heat and results in a rise in the temperature of the absorbing surface. What proportion of the light is absorbed and what proportion reflected varies with the nature of the surface. A dark-colored surface absorbs a greater proportion of the light energy and reflects less than a light-colored surface; it therefore becomes warmer more rapidly. A dull, matte-textured surface absorbs energy more effectively than a shiny one. A mirror reflects most of the light that falls upon it, which is why mirrors are usually cold to the touch. The angle at which light falls upon a surface also affects the proportion of light absorbed and reflected; light arriving vertically is reflected less than light arriving from a low angle.

Scientists express the degree of reflectivity of a surface as its *albedo*. It is stated as the amount of light reflected divided by the total incident light. In order of efficiency of reflectivity, the following list gives examples of the albedo of various surfaces.

SURFACE	PROPORTION REFLECTED	ALBEDO
Snow	75 to 95%	0.75–0.95
Sand	35 to 45%	0.35–0.45
Concrete	17 to 27%	0.17–0.27
Deciduous forest canopy	10 to 20%	0.10–0.20
Road surface	5 to 17%	0.05–0.17
Coniferous forest canopy	5 to 15%	0.05–0.15
Water (overhead Sun)	5%	0.05
Water (low Sun)	100%	1.00
Overall average for Earth	39%	0.39

(above) Light from the Sun arrives at the Earth as parallel beams, but the light heats different extents of surface area depending upon the angle at which it arrives. Polar regions receive less intense energy than the tropical areas. The tilt of the Earth upon its axis means that the position of the overhead noonday Sun varies with season.

at high latitude. There are two means by which the Earth's energy can be redistributed in this way, the atmosphere and the oceans.

(below) The circulation of the atmosphere occurs in a series of cells, creating areas of low pressure near the equator and in the region of the polar fronts, and areas of high pressure close to the tropics of Cancer and Capricorn and at the North and South Poles.

of Anchorage, Alaska, the southern tip of Greenland, and Oslo, Norway) the days of summer are extremely long, while the days of winter are extremely short. Beyond the Arctic Circle (66.5°N), there is at least one day in the year when the Sun does not set. This difference in summer and winter day length results in a strong seasonal difference in temperature in the high latitudes. The length of the summer season also becomes shorter close to the poles. Within 450 miles (725 km) of the North Pole, for example, at Cape Morris Jessup in northern Greenland (about 84°N), the length of time when conditions are suitable for plant growth may be as little as three or four weeks.

When the three factors of solar angle, area of energy dispersal, and seasonal differences are put together, it is evident that the polar regions will have a smaller input of energy per unit area of Earth's surface than the Tropics. Lower energy levels result in lower temperatures overall, which means that snowfall and persistent snow cover become more likely in the high latitudes, and this has a feedback effect on albedo, resulting in even more of the incoming energy being lost by reflection.

The average temperature at the equator is 79°F (26°C), which varies very little with season, while at 80°N the January average is −22°F (−30°C) and the July average is 37°F (5°C). At this high latitude the Earth is actually losing more energy than it is receiving, so the Arctic can be regarded as an *energy sink*. But the Arctic is not becoming colder year by year, so the energy it loses must be replaced by energy from other parts of the planet. There must be energy movement from areas with an abundant supply, such as the Tropics, to the energy sinks

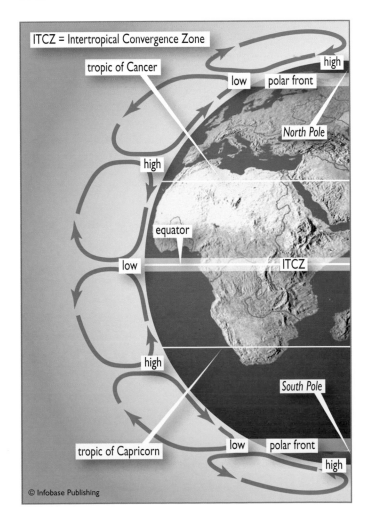

■ THE ATMOSPHERE AND CLIMATE

Heating in the Tropics and cooling at the poles results in a global energy imbalance, and this in turn gives rise to atmospheric turbulence. Warm air has a lower density than cold air, so there is a tendency for cool, dense air to move toward the equator, pushing the warm, low-density air upward. Air masses approach the equatorial regions from both north and south of the equator, creating an *intertropical convergence zone* where the two air masses meet. The warm, moisture-laden air rises and cools in the process (see sidebar "Lapse Rate," page 22). Cooler air is less able to hold moisture, so water condenses into raindrops, and the equatorial regions receive high levels of precipitation, resulting in the development of the tropical rain forests. In the upper atmosphere, around 10 miles (16 km) above the ground, the air that has been forced upward moves out toward the poles, as shown in the diagram on page 9, and since the air has now cooled and become denser, it falls toward the Earth's surface once more, creating a high pressure belt between about 25° and 30° north and south of the equator. These falling air masses have lost their moisture over the equatorial regions, so they cause dry conditions when they arrive back at the Earth's surface. These belts are where the world's deserts are mainly found.

The hot, dry air of the desert belts may be deflected back toward the equator, or they may travel toward the poles. If they take the latter route, then these air masses meet cold, polar air moving in the opposite direction. Where the two air masses collide they create a boundary zone, called the *polar front,* and this region is characterized by unstable weather conditions. The precise position of the polar front is very variable in space and time, but the mix of air masses results in the cyclonic depressions and accompanying precipitation typical of latitudes roughly between 40° and 60° north and south of the equator. The polar front is also associated with a strong wind, the jet stream, which moves from west to east at an altitude of around 35,000 feet (10,000 m) at the boundary between the lower part of the atmosphere,

The atmosphere is divided into a series of layers, becoming less dense with altitude. The lights of the aurora occur in the thermosphere, between 60 and 260 miles (100 and 420 km) above the Earth's surface. This is the layer of the atmosphere in which satellites orbit the Earth.

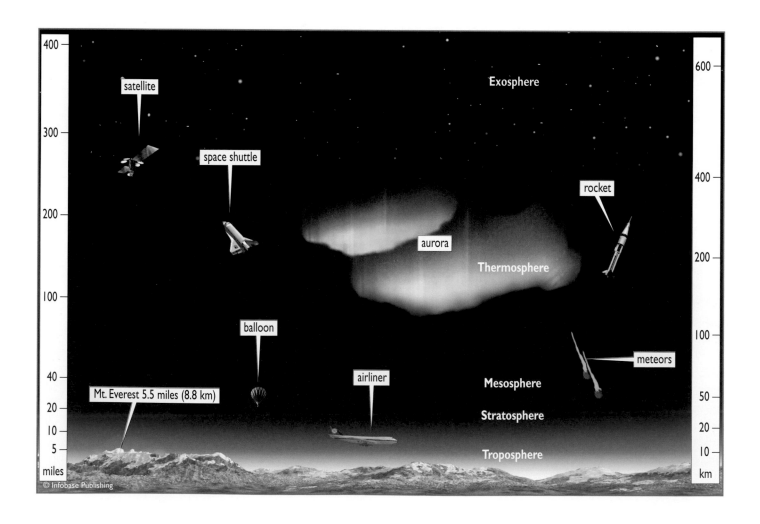

the *troposphere,* and the upper part of the atmosphere, the *stratosphere.* This boundary, as shown in the diagram on page 10, is called the *tropopause.* The jet stream causes cyclones (low-pressure systems) and anticyclones (high-pressure systems) to move around the world in an easterly direction. The latitude of the polar front and the jet stream varies with the season. In summer in North America the polar front and jet stream lie roughly along the Canadian border, dipping south to Washington, D.C., in the east. In winter it runs approximately from Los Angeles through northern Texas to South Carolina.

In the Southern Hemisphere the polar front encircles Antarctica and lies within the Southern Ocean. For much of its length, where the Southern Ocean meets the South Atlantic Ocean and the Indian Ocean, it lies approximately along 50°S

The oceanic currents play an important part in the redistribution of energy around the world. In the North Atlantic warm tropical waters from the Caribbean move up the western edge of Europe and enter the Arctic Ocean. This penetration of warmth into the Arctic is much less strong in the Pacific, so the north coast of Alaska is much colder than the north of Scandinavia.

latitude. In the region where the Southern Ocean meets the South Pacific, however, the southern polar front is located along latitude 60°S, as shown in the diagram on page 17.

Some of the air forced upward in the turbulent regions around the polar fronts moves on toward the poles at high altitude. Here the air becomes very cold and dense, so it begins to lose altitude, and its descent creates another high pressure zone over the poles, as shown in the diagram on page 9. As in the high-pressure zone of the outer Tropics and sub-Tropics, the outcome is the production of low-precipitation, desert conditions. But in the polar regions the lack of precipitation does not result in drought at the surface of the Earth because the very low temperature at ground level restricts the degree of evaporation taking place, so water may accumulate over low-lying parts of the landscape even though water input from the atmosphere is very restricted.

■ THE OCEANS AND CLIMATE

The atmosphere is not the only determinant of global patterns of climate; the oceans also have a profound influence. Water has certain physical properties that affect climate. A

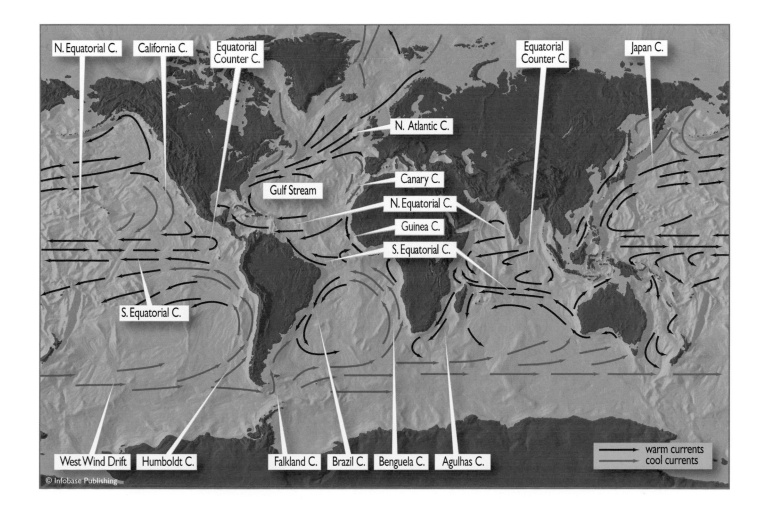

N. Equatorial C.   California C.   Equatorial Counter C.   Equatorial Counter C.   Japan C.

N. Atlantic C.

Gulf Stream   Canary C.   N. Equatorial C.   Guinea C.   S. Equatorial C.

S. Equatorial C.

West Wind Drift   Humboldt C.   Falkland C.   Brazil C.   Benguela C.   Agulhas C.

warm currents
cool currents

© Infobase Publishing

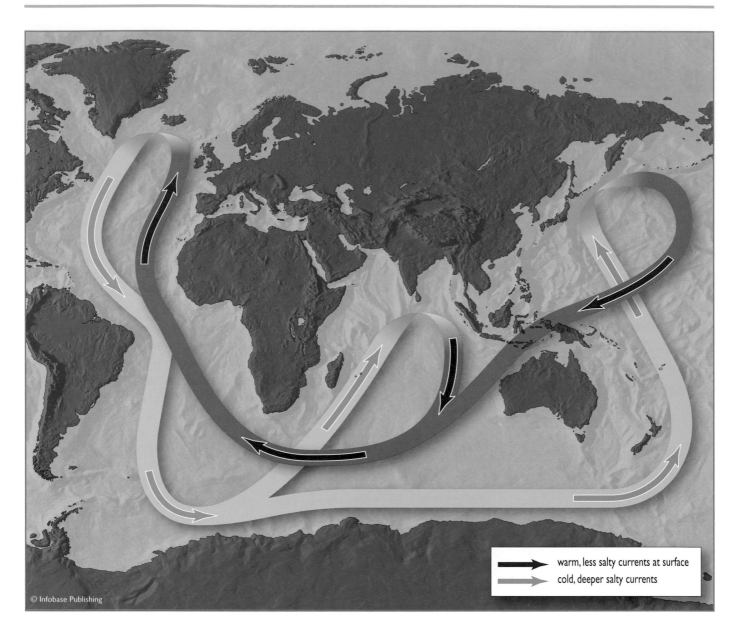

© Infobase Publishing

| warm, less salty currents at surface |
| cold, deeper salty currents |

The global circulation of the Earth's oceanic waters is called the oceanic conveyor belt. Warm, low-salinity, and low-density water moves along the surface of the ocean. On arrival in the North Atlantic Ocean the water cools and forms a deep-water current that flows in the opposite direction.

sample of water both gains and loses heat more slowly than an equivalent weight of rock. Areas of land surrounded by water are, therefore, protected to some extent from extreme variations in temperature; land close to the ocean is kept cool in summer and relatively warm in winter. The proximity of the sea, therefore, has a considerable influence on the climate of a region. A site close to the sea is said to have an *oceanic* climate, whereas one situated far from the ocean's influence has a *continental* climate. For example, the city of

Archangel, at a latitude of 65°N in western Russia, lies on the edge of the Barents Sea and has an average annual temperature of 33°F (0.4°C) and an average temperature for its coldest month of 5°F (−15°C). But in the continental east of Russia, Verkhoyansk, in Siberia, which lies on the same latitude, has an average annual temperature of 3°F (−16°C), and the average for its coldest month is as low as −58°F (−50°C). The oceanic regions of Russia are kept warmer by the proximity of the sea. Oceanic regions also receive more precipitation because the sea is a source of atmospheric moisture. Thus, Archangel receives 18 inches (47 cm), while Verkhoyansk has only five inches (13 cm) of precipitation each year.

The oceans also act as channels for the redistribution of energy around the world. Surface water moves around and between the great oceans in the form of currents, as shown

The Oceanic Conveyor Belt

In addition to the surface currents in the oceans (see the diagram on page 11), which are driven mainly by winds in the atmosphere, there is a global circulation of waters driven by the changing density of seawater, as shown in the accompanying diagram on page 12. The surface water of the oceans tends to be warmer, less salty, and hence less dense than the deep water. When these warm surface waters move into the high latitudes, they give up some of their heat to their surroundings and consequently cool, become denser, and sink. This chilling of surface water occurs most strongly in the North Atlantic Ocean, where the warm tropical surface waters of the Gulf Stream pass northward between Iceland and the British Isles and meet the cold waters of the Arctic Ocean. As the warm waters lose their heat to the waters of the Arctic Ocean, they cool, sink, and begin to make their way back south through the Atlantic and eventually into the Southern Ocean surrounding Antarctica. Passing eastward into the Pacific Ocean, these cold, deep waters eventually surface either in the Indian Ocean or in the North Pacific. There they become warm and drift westward once more around the Cape of Good Hope in South Africa and into the Atlantic to begin their circulation over again.

This thermohaline circulation of water through the oceans of the world plays an important part in the redistribution of energy between the Tropics and the high latitudes, particularly the North Atlantic. These movements of water have strongly influenced world climate in the past. A failure in the North Atlantic system 12,000 years ago temporarily plunged the Earth into an ice age.

de Coriolis (1792–1843) in 1835 and has come to be called the *Coriolis effect*. In the North Pacific Ocean and the North Atlantic Ocean there are west-east currents bringing warm tropical water into the higher latitudes, but in the Atlantic Ocean there is a gap between Iceland and the British Isles that allows the continued movement of warm water (called the Gulf Stream) north past Scandinavia and into the Arctic Ocean. The penetration of warm water into the Arctic Ocean is far less from the Pacific basin. The northeast-directed waters in the Pacific either circulate around southern Alaska and return south or pass down the west coast of California and Mexico and back into the Tropics.

The difference in oceanic circulation patterns in the Atlantic and Pacific Oceans has a very considerable impact on the climate of the tundra. The penetration of warm water from the North Atlantic into the Arctic Ocean results in mild and ice-free conditions persisting through the summer months around the northern coast of Scandinavia. On the north coast of Alaska at the same latitude of 70°N, on the other hand, there is little direct injection of warm water into the adjacent part of the Arctic Ocean, and conditions remain very cold even in summer. Even here, however, there is some effect from the input of warm water into the Arctic Ocean. The Gulf Stream that enters the Arctic Ocean between Iceland and Scandinavia initiates a circulation of waters around the North Pole in an anticlockwise direction when viewed from above. This movement, combined with the increased energy input of the summer months, ensures that pack ice around the coastal regions of the Arctic Ocean partially breaks up during the summer. The center of the Arctic Ocean, however, is permanently covered by pack ice.

Oceanic circulation occurs in deep water as well as at the surface. There is a global movement of water around the Earth that is driven by changes in the density of seawater, itself a product of salinity and temperature. Warm water with relatively low levels of salt is less dense than cold water with higher salt concentrations. As water moves around the world it changes in its temperature and its salinity. This circulation pattern is called *thermohaline* circulation (*thermo-* refers to the water temperature, and *-haline* relates to its saltiness), or the oceanic conveyor belt. The overall pattern of global thermohaline circulation is shown in the diagram on page 12 and is explained in the sidebar at left.

The thermohaline circulation of water in the oceans is an important means of exchanging energy between different parts of the world. Without this energy exchange there would be a much stronger gradient of temperature between the Tropics and the poles. There have been times in the Earth's history when this conveyor belt has slowed down or even ceased altogether in certain parts of its cycle, and this has led to severe climatic change. When the conveyor belt faltered around 12,700 years ago in the North Atlantic, for example, the Earth began to move into a new ice age. The normal cir-

in the diagram on page 11. In general the pattern follows that of the winds and is driven by them. In the Southern Hemisphere the main surface circulatory pattern of the oceans is in the form of anticlockwise motions, whereas in the Northern Hemisphere the movements are clockwise. The reason for this is the spin of the Earth on its axis. Any free moving object in the Northern Hemisphere, whether in the oceans or in the atmosphere, tends to be deflected to the right, hence motions become clockwise. In the Southern Hemisphere free objects are deflected to the left, so the motions become anticlockwise in direction. This tendency was first described by a French engineer, Gaspard Gustave

culation pattern was restored within a few centuries, however, and the climate began to warm once again (see "The End of the Ice Age," pages 190–192). This illustrates the critical role played by oceanic circulation in controlling the climate of the entire Earth.

■ ARCTIC CLIMATE

Both the Arctic and the Antarctic lie in positions where they receive low energy input from the Sun yet still radiate heat back into space. They lose energy faster than they gain it, so they are energy sinks. The circulation of the atmosphere and the oceans, however, restores some of the balance so that these high latitude regions are not becoming constantly colder over long periods of time. The Arctic consists of an ocean almost completely surrounded by land, whereas the Antarctic consists of a large continental land mass, so the climates of the Arctic and Antarctic differ. The Arctic Ocean receives a sufficient input of warm water from the Gulf Stream of the North Atlantic Ocean to ensure that some ice-free conditions exist in summer even as far north as 80°N. This means that communities of living things can survive even at high latitudes in the Arctic.

The word *Arctic* is used in a number of different ways. In common speech it is often used in a very general sense to describe very cold conditions, "arctic conditions." It is used by biogeographers to describe certain types of distribution patterns, called *arctic-alpine,* applied to organisms that are found in both polar and alpine tundra sites. More specifically, the word *Arctic* can refer to all parts of the Earth lying beyond the Arctic Circle (66.5°N). Climatologists, however, define the Arctic as those regions around the North Pole where the mean temperature of the warmest month does not exceed 50°F (10°C) and the mean temperature of the coldest month is below freezing. The region that falls within this definition of Arctic coincides closely with the tundra biome in these northern regions. These climatic conditions also define the *timberline,* the northern limit of tree growth. South of this line some climatologists define a further zone, the *subarctic.* This region has a climate in which a maximum of four months in the year has a mean monthly temperature exceeding 50°F (10°C), and the mean temperature of the coldest month is below freezing. In this climate tree growth is possible, so the subarctic bears taiga, or boreal forest vegetation, rather than tundra. Many land areas that lie north of the Arctic Circle are subarctic rather than Arctic in their climate and bear coniferous forest vegetation, as shown in the diagram on this page.

Some northern areas of ocean can also be regarded as subarctic rather than Arctic in their climate. These include the northern Bering Sea and Bering Strait, through which the Arctic Ocean communicates with the Pacific Ocean,

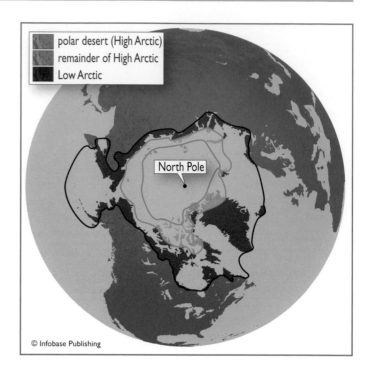

polar desert (High Arctic)
remainder of High Arctic
Low Arctic

North Pole

© Infobase Publishing

The Arctic tundra can be divided into the High Arctic and the Low Arctic. The High Arctic consists mainly of polar desert, ice, and lands with low, sparse vegetation. The Low Arctic has a more complete vegetation cover but no true trees. The southern boundary of the Low Arctic is formed by the northern limit of tree growth, and this corresponds closely to the 50°F (10°C) July mean isotherm (not shown on the diagram).

together with the seas around Greenland and Iceland and the Norwegian Sea that joins the Atlantic Ocean to the Arctic Ocean.

The Arctic can be subdivided on the basis of its climate into the High Arctic and the Low Arctic. The High Arctic consists of numerous islands lying in the maritime Arctic. These include the Svalbard, northern Novaya Zemlya, Franz Joseph Land, Severnaya Zemlya, the Canadian Queen Elizabeth Islands and Ellesmere Island, and the northern part of Greenland. The climate of the High Arctic is extremely cold. The mean monthly temperature lies below freezing for at least nine months of the year. The snow-free summer period lasts usually less than six weeks, so the effective growing period for plants does not exceed two months. All except the surface one to three feet (20 cm to 1 m) of the soil remains frozen throughout the year. Even at a depth of 100 feet (30 m) the temperature is permanently below 14°F (–10°C). This is called the *permafrost* (see the map on page 58). The High Arctic is sometimes referred to as *polar desert* or *polar semidesert* because of the very low level of precipitation in the high pressure conditions at these high latitudes (see page 9). Precipitation is mainly in the form of snow in the High Arctic and may amount to only about four inches (10 cm) per year. The most northerly parts of Arctic Canada have only two inches (5 cm) of precipitation per year. The very cold conditions, however, mean that the evaporation rate is very low, so drought is experienced only where there is very good soil drainage, such as on sand and shingle ridges.

Where there is permanent ice cover, as in the case of the Greenland ice sheet, the climate is even more severe, with the mean monthly temperature remaining below freezing throughout the year. Under such conditions there is no vegetation. The shiny white surface of the ice has a very high albedo, so almost all of the incident energy is reflected and very little is absorbed. The middle of the Greenland ice sheet has a mean annual temperature of –18°F (–28°C), and even in summer the temperature rarely rises above freezing. Ice sheets are maintained by the arrival of new material in the form of snowfall, replacing any loss from melting. In the case of the center of the Greenland ice sheet, however, precipitation is very low. The ice mass survives because wastage by melting is also extremely slow. Precipitation is higher only where the ice sheet lies close to the sea and the climate is more oceanic.

The Low Arctic consists of the remaining regions of the tundra, mainly the northern parts of Siberia, Alaska, Canada, coastal Greenland, and Iceland. The Low Arctic has a climate that is cold but not quite as extreme as that of the High Arctic. The mean monthly temperature is below freezing for between seven and nine months of the year, and the snow-free period is between three and four months in length. This means that plants may be able to grow for as long as five months during the summer. As in the case of the High Arctic, the soil remains permanently frozen below the surface layer, with the temperature at 100 feet (30 m) below the surface remaining at about 25°F (–4°C). Precipitation is generally still extremely low, at about eight inches (20 cm) per annum, but the ground is generally wet during the summer because of the low evaporation rate. In particularly oceanic regions, such as the coastal tundra around the Bering Sea, precipitation may be as high as 30 inches (76 cm).

One result of the low precipitation in the Arctic tundra is that snow cover is generally thin. Vegetation is low-lying and uniform, so this does not influence the shallow covering of snow. Wind speeds, on the other hand, can be very high, and this can lead to drifting and the development of deeper snow patches in the lee of rocks and promontories. The shallow snow cover has important implications for the temperature of the soil and the penetration of frost. Snow acts as an insulating blanket over the surface of soil, so a poor covering of snow leads to deeper frost penetration. This, in turn, may lead to the development of distinctive features in the topography, such as the ice mounds of palsa mires (see "Tundra Wetlands," pages 74–78). On the other hand, the lack of snow means that vegetation is quickly able to take advantage of rising light levels and increasing day length in the spring and begin its growth more rapidly. Early growth then attracts grazing animals, so the shallow snow cover has extensive implications for the ecology of the tundra.

The wind is itself an important climatic factor in polar tundra. High winds are frequent, and these bring with them a *wind chill* effect. Wind constantly removes the warm layer of air around an object or an organism, so heat energy is taken from the object faster than in still air. From the point of view of a living organism, the conditions feel much colder when there is a wind blowing. Human beings subjected to a wind of 50 feet per second (15 m/sec) in air at 32°F (0°C) experience the same degree of cold that they would in still air at 14°F (–10°C). So conditions in the tundra are even more unpleasant for living creatures than is apparent simply from the temperature data. An additional factor that makes things even worse is the way in which ice crystals are often suspended in the moving air, acting as an abrasive agent in the wind. Very few plants can survive exposure to ice-laden winds that can strip the surface cells from a leaf.

Although overall temperatures are low, and the growing season is short in the Arctic, this does not mean that conditions are eternally cold. There are occasions when the air temperature can be quite high. In the Taymyr Peninsula in the north of Siberia, for example, July temperatures often exceed 68°F (20°C) during the day, and temperatures as high as 84°F (29°C) have been recorded. But the temperature is strongly affected by local conditions of topography (see "Microclimate," pages 18–21), and this means that the summer climate is very variable over relatively short distances. This has a considerable effect on vegetation. The climate also depends on the distance to the ocean, especially if it is influenced by the warm waters of the Gulf Stream (see "The Oceans and Climate," pages 11–14). The cooling effect of the ocean during the summer is also an important factor in determining the climate of the tundra. If the summer temperature rises far enough, then trees can survive and tundra vegetation is lost. This is why all polar tundra habitats lie relatively close to the sea, because the more continental climates found deep within land masses have high summer temperatures and become clothed with forest. High mountain tops are an exception, of course, but the lowland tundra of the polar regions is almost always found within 300 miles (500 km) of the sea or other large body of water.

So the polar tundra is oceanic in its distribution, and the proximity of the ocean keeps summer temperatures cool and winter temperatures less intensely cold than those at locations deeper in the continental interiors. The seasonal variation in temperature, therefore, is smaller than that found in the continental regions of the subarctic lands. In such locations the summer temperatures are higher (leading to the development of forests), but the winter temperatures

A Year in the Arctic

An observer in the coastal region of the Arctic tundra, whether in Alaska, Baffin Island, southern Greenland, or coastal Siberia, would find great changes in the landscape during the course of a year. At the March equinox (when the midday Sun is overhead at the equator), day and night in the Arctic are equal in length, the Sun spending 12 hours above the horizon. The temperature, however, is very cold, not rising above freezing all day. Pack ice persists over the ocean, covered in a thin blanket of snow. During April the midday Sun is sufficiently strong to melt the snow in places. Between May and June the pack ice begins to break up, and the Sun no longer dips below the horizon at night. Some days are cloudless and bright, while others are overcast, the cloud cover bringing light flurries of snow or sleet. By August the weather can be quite warm, and a fine drizzle of rain occasionally falls. But by the end of September the frosts have returned, and the sunlight is becoming shorter in duration and noticeably less strong. In October ice begins to form over the ocean, and soon the pack ice cover is complete once more. The ground by November is hard and frozen, and much of the year's snow arrives at this time of year. In the middle of winter the skies are often clear and starry, lit by the flashing lights of the aurora borealis. The cold is not as intense as that experienced in the winters of the continental interiors, such as southern Canada or central Siberia, but the persistence of cold, dark conditions for several months, with hardly a glimpse of the Sun, can be depressing. Storms may occur, bringing high winds and sometimes driving light snow or ice crystals, and these are most frequent in the tundra regions around the North Atlantic and the Bering Sea. The observer who persists in a vigil through the long Arctic winter is undoubtedly relieved by the coming of spring and the short summer season once more.

ies so greatly between summer and winter. All places north of the Arctic Circle have at least one 24-hour day during the summer and at least one 24-hour night in the winter. At the North Pole the Sun rises for six months and then sets for six months. But the angle of the Sun in the sky is always low, and this, when combined with the short summer night, results in relatively little variation in temperature during the course of a tundra summer day. In the case of a 24-hour day, the midnight Sun lies low upon the horizon, gradually becoming elevated to its maximum height at noon but still failing to reach a high angle. The temperature thus rises and falls through the day with the Sun's angle but varies relatively little. Similarly, in winter the Sun may rise a little way above the horizon around midday and then set again before the temperature has been strongly influenced by its presence (see sidebar). So both in summer and winter the diurnal fluctuation in temperature in the Arctic tundra is relatively small. As will be seen, this is in marked contrast to the alpine tundra climate.

■ AURORA

One of the most impressive sights in the Arctic tundra night is the aurora borealis, or northern lights. An observer sees curtains and patches of light swirling across the sky in a range of colors, mainly green and red. Sometimes there are explosive bursts of light that twist and pulsate, and these periods of intense activity can last for an hour or more. The auroral displays take place in a vast oval ring about 5,000 miles (8,000 km) in diameter, centered upon the magnetic pole. From the ground only a small part of the total display is visible, perhaps just a 435-mile (700-km) section of the entire ring. The ring extends farthest from the pole on the night side of the planet, reaching as far south as latitude 67°. On the day side the aurora is situated at about latitude 75°. Particularly strong displays can sometimes be seen as far south as the northern United States and England.

The light itself is generated at great heights above the Earth in a region of the atmosphere known as the *thermosphere,* at between 60 and 250 miles (100 to 400 km). This is the kind of altitude where satellites orbit the Earth (see page 10). The aurora is named after the Roman goddess of the dawn, and it occurs in both the Northern and the Southern Hemispheres. In the Arctic it is called the *aurora borealis,* and in the Antarctic, the *aurora australis. Boreas* is Greek for the "north wind," and *Auster* is Latin for "south wind."

The aurora is caused by charged particles emitted by the Sun. The outer region of the Sun, the corona, is so hot that it constantly emits gases containing these charged particles, mainly protons and electrons. The stream, called the *solar wind,* passes into space at a great velocity of around 187 miles per second (300 km/sec). When the solar wind

are even lower than those found in the tundra. The "Cold Pole" of the world is located at Verkhoyansk, in Siberia, well south of the Arctic, within the taiga biome. The winter temperature there may fall as low as –90°F (–67.8°C), but the summer temperature can be as high as 100°F (37°C).

Seasonal climatic variation in the Arctic tundra is thus less than might be expected, given that the length of day var-

strikes the upper part of the Earth's atmosphere, it interacts with the Earth's magnetic field. Captured charged particles move along the lines of the magnetic field and emit energy in the form of light. The process is very similar to what takes place within neon tubelights. Some excited, charged oxygen atoms emit a green light, and this occurs in the lower part of the aurora, at about 60 miles (100 km) high. At higher altitudes, above 155 miles (250 km), other oxygen atoms emit red light. Sometimes violet light produced by nitrogen atoms can also be observed.

The Sun is not constant in its production of solar wind; there are times when its output is much greater than others. Observers on Earth can determine when the solar wind output is greatest because of an increased density of sunspots on the surface of the Sun. Sunspots appear dark because they are slightly cooler areas where there is intense magnetic activity and a strong output of charged particles. Sunspot density follows a very regular cycle, peaking every 11 years. During such a peak in sunspot activity, the auroral displays are at their most spectacular, and this enhanced performance can last for up to two years, being most prominent in the spring and fall. The last activity peak was in the year 2000.

■ ANTARCTIC CLIMATE

The Antarctic differs from the Arctic in the presence of a large continental landmass covered with an ice sheet. The ice sheet covers more than 97 percent of the entire continent and even extends in a permanent cover over some ocean regions in the form of ice shelves. As a result there is very little land that is free of ice on the Antarctic mainland and that can truly be called tundra. The ice sheet has a great impact on the climate of the region, just as the Greenland ice sheet affects the climate in its vicinity in the Northern Hemisphere. The high albedo of the ice ensures that the surface does not become warmed by the Sun, and the climate remains cold throughout the year. In the heart of the continent the mean monthly temperature varies between –22°F (–30°C) in summer (that is, December and January in the Southern Hemisphere) and –85°F (–65°C) in the middle of winter (May and June). The center of the Antarctic continent lies far from the influence of the ocean, resulting in much colder conditions than those found in the Arctic. The coldest conditions ever recorded on Earth occurred on July 21, 1983, at Russia's Vostok Research Station on the Antarctic ice sheet. The temperature was –128.6°F (–89.2°C). The site lies at a latitude of 78°S, well short of the South Pole, but it is at a greater altitude (11,475 feet; 3,500 m) than the pole itself, which may account for the colder conditions. The summit of the ice sheet has an altitude of 13,115 feet (4,000 m), which may prove to be

even colder, but as yet no meteorological station has been established at that location. The South Pole has a mean annual temperature of –58°F (–50°C), which is much colder than the mean annual temperature at the North Pole, a relatively warm 0°F (–18°C).

The temperature rises above freezing for just two or three months of the year only on the Antarctic Peninsula (see the diagram on page 4). In summer the temperature here may rise as high as 59°F (15°C). So the Antarctic Peninsula is the nearest equivalent to the Arctic tundra found on the southern continent.

The Antarctic, like the Arctic, has a permanent high-pressure system lying over its center (see the diagram on page 9), which results from the patterns of atmospheric circulation over the planet. The Southern Ocean, which surrounds the Antarctic continent, is the site of much atmospheric turbulence. Warm air from the lower latitudes meets cold air streaming from the vast Antarctic ice sheet, and the outcome is a series of low-pressure systems that constantly circle the continent, passing from west to east. The Southern Ocean is also turbulent because cold water from the melting ice meets warmer waters from

The Antarctic and subantarctic islands, showing the position of the polar front

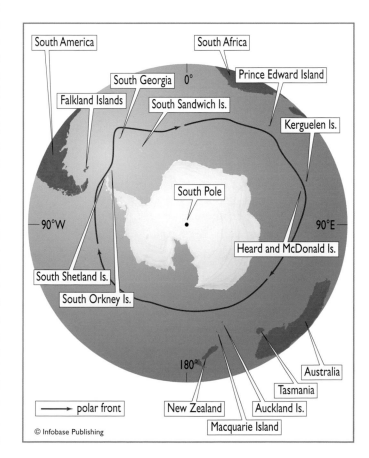

the north, leading to the formation of a polar front, just like that found in the North Atlantic. The two water bodies do not mix easily but create a sharp divide in which temperature and salinity changes abruptly. A strong ocean current constantly passes around the Southern Ocean in a clockwise direction, as viewed from above the South Pole, carrying 4.8 million cubic feet (135 million m³) of water from west to east every second. This Southern Ocean is the major link between all the other oceans of the world and plays a vital role in the global conveyor, the thermohaline ocean circulation system (see sidebar "The Oceanic Conveyor Belt," page 13).

It is not surprising, therefore, that this region of the world, with its atmospheric and oceanic turbulence, should experience some of the most violent weather conditions on Earth. When sailors first encountered these latitudes they named them the Roaring Forties, the Furious Fifties, and the Screaming Sixties! Very high winds and wild seas have always been a major deterrent to those seeking to explore these southern regions. Cape Denison, on the side of Antarctica that faces New Zealand, is often battered by winds in excess of 70 miles per hour (110 km/hr) during the Antarctic summer. Conditions are calmer in the winter, when winds drop to around 30 miles per hour (50 km/hr). In storm conditions the winds can rise to twice the speed of hurricane force, 185 miles per hour (300 km/hr). Some of the most destructive winds of Antarctica are called *katabatic winds*. These descend from the ice sheet toward the coast, becoming colder, denser, and faster as they advance. Because of these winds some parts of coastal Antarctica experience hurricane-force winds on average every three days.

As in the Arctic, the Antarctic high pressure and low temperature result in dry conditions. The air is so cold that it can hold virtually no moisture. Snowfall over the high plateau of the ice sheet is only two inches (5 cm) per year, making this region one of the world's driest places, similar in its precipitation to the Sahara in Africa. When all these climatic stresses of cold, wind, and drought are taken into consideration, it can be seen that Antarctica is one of the most inhospitable places on Earth. It is not surprising that so few plants and animals can survive there.

The Southern Ocean has many islands, and those lying south of the polar front (see "The Atmosphere and Climate," pages 10–11) are called subantarctic islands. These include South Georgia, the South Orkney Islands, the South Shetland Islands, and numerous others, as shown on the map on page 17. The climate of these islands is cold but not as severe as that of the Antarctic mainland, so they are important sites for tundra vegetation and wildlife. Islands to the north of the polar front, such as the Falkland Islands off Patagonia in South America and the Auckland Islands to the south of New Zealand, have a cool, temperate climate with much rain and snowfall, but they cannot be regarded as part of the tundra biome. The Auckland Islands are warm enough to bear some forest vegetation, consisting of rata trees (*Metrosideros robusta*). It is the most southerly forest in the world (51°S).

■ MICROCLIMATE

The climatic features considered so far are largely based upon average conditions over relatively large areas. This is sometimes called *macroclimate, macro-* meaning large. Within the macroclimate of any region there is a great deal of local variation depending on the physical features, or *topography,* of the land. The local climate of a small area is called a *microclimate, micro-* meaning small. The microclimate is often as important to a living organism as the macroclimate because animals and plants are small compared to landscapes, so it is the microclimate of their immediate surroundings that determines whether they are able to survive.

A ridge of land projecting above the general level of a landscape experiences a different microclimate from the sheltered hollows between ridges. An exposed ridge receives the full blast of the wind, reducing the temperature by constantly removing any air that may have been warmed by contact with the ground. The wind also removes snow in winter, and this also affects the temperature of the underlying soil. Snow consists of finely branched crystals of frozen water, and when it accumulates it traps air within its porous structure. Air is an excellent insulator, so any residual warmth in the soil, gained from the Sun during the summer, is retained longer. When snow is removed by wind, the soil is deprived of any insulation, and frost penetrates more rapidly and more deeply. On the other hand, a snow-free ridge receives the light of the Sun as soon as it rises above the horizon in the spring, so it will become warmed more quickly than sheltered hollows containing late-lying snow. The hollow may have been protected from some of the effects of temperature during the winter, but the summer season is made shorter by late melting of the snow cover. Ridges are also drier than hollows because water drains quickly from them. A ridge is thus exposed to more sunlight and drains more freely than a hollow, so it may suffer drought during the summer.

Slopes combine some of the characteristics of both ridges and hollows. Their microclimate depends upon whether they face the prevailing wind or are sheltered from it. Their orientation in relation to the Sun is also important. The compass direction a slope faces is called its *aspect* (see sidebar on facing page).

Polar tundra landscapes may be uniform and relatively flat, or they may be rugged and rocky. In all types of tundra landscape, however, the resident plants and animals are strongly influenced by microclimate. The oppos-

Aspect and Microclimate

The angle of the Sun above the horizon is never large in the polar regions. A flat surface of ground, therefore, receives sunlight at an oblique angle, and this means that much of the energy is reflected rather than absorbed. Absorption of light energy is most efficient if the surface is at right angles to the incident rays. This can happen, even if the Sun is very low in the sky, if the surface is steeply sloping. A cliff face that is oriented toward the low Sun is well placed to receive the incoming energy and is much more effective in its energy absorption than a flat area of ground. But the absorption efficiency is also dependent upon the direction the cliff faces, called its *aspect*. In the height of the Arctic summer the Sun moves around the horizon through the course of the day, rising to its maximum elevation at noon and falling to its lowest at midnight. Under these conditions all cliff faces, whatever their aspect, receive some direct sunlight. But during the early and late parts of the summer the Sun is above the horizon for a limited period and appears only in the southern part of the sky. A cliff facing south, therefore, receives direct sunlight, while one facing north receives none; it is in permanent shade.

Sloping ground is similarly affected by aspect, as shown in the diagram below. The sloping side of a ridge in Arctic tundra that faces south experiences a longer season of warmth and higher soil temperatures than a slope that faces north. From the point of view of a herbaceous plant, a small shrub, a beetle, or a moth, the south-facing aspect has much to recommend it because of its warmer conditions. But the down side is that such a situation may become dry in summer because precipitation is low and sunlight may be quite intense. A moss that is sensitive to drought may find itself unable to survive on a south-facing aspect and becomes confined to north-facing slopes. Steep slopes are more efficient at trapping sunlight when the Sun is low on the horizon, so they are more easily warmed. But steep slopes are more likely to shed their thin snow cover in winter, which allows the frost to penetrate more deeply. Once again, there are advantages and disadvantages for a plant or animal occupying such a situation.

The angle of the Sun in the polar tundra regions is always low, so the steepness of slopes and the direction in which they face (their aspect) are important in determining their microclimates. In the Arctic (shown here) a south-facing slope or the south-facing side of a boulder receive more solar energy and become warmer than a north-facing side.

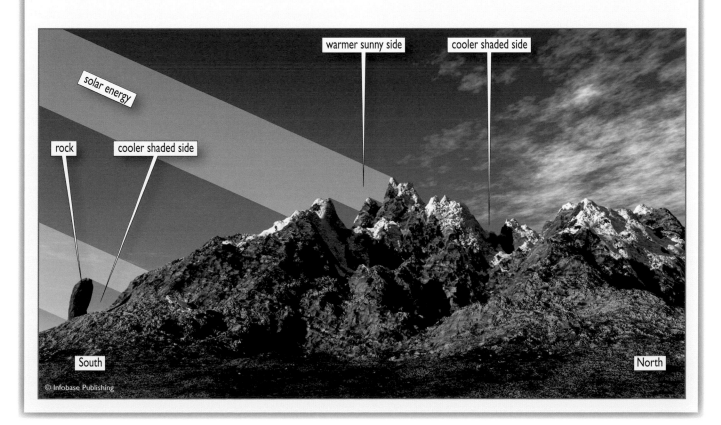

South

North

© Infobase Publishing

Humidity

Humidity is an expression of the quantity of water vapor in the air. This can be measured in absolute terms, such as the mass of water vapor in a given volume of air. Humidity may also be measured in terms of *vapor pressure*. All of the gases in the atmosphere (largely nitrogen and oxygen) contribute to the air pressure, and water vapor also plays a part. The pressure of water vapor in the atmosphere is an expression of the air's humidity, or wetness. The amount of water vapor that can be contained within a packet of air depends upon the temperature of that air. Air at low temperature is able to hold less water than air at high temperature. So if air cools some of the water vapor condenses into water droplets that may fall as rain. The temperature at which the air becomes saturated with water and condensation occurs is called the *dew point*. By definition, therefore, any packet of air at a temperature above its dew point is not fully saturated with water. The difference between the pressure of water vapor present in this air and the vapor pressure needed to saturate it is called the *vapor pressure deficit*. This is a measure of the air's capacity to absorb more water vapor. If the vapor pressure deficit is large, water in the environment is more likely to evaporate to fill the gap. An alternative way of expressing this deficit is the *relative humidity* of the air. To compute this term, the quantity of water present in the air is calculated as a percentage of the amount of water needed to saturate the air under the given conditions of temperature and pressure. A high relative humidity means that the air is close to saturation and is therefore unable to take up much more water vapor.

ing faces of rocky bays and headlands with steep cliffs may experience great differences in their microclimatic conditions, depending on aspect, and their vegetation and animal life vary accordingly. A single rock lying upon the ground may have different types of plants growing upon its surfaces depending upon their aspect. Many invertebrate animals will likewise choose to live in the shade of a rock or even beneath it, where the Sun does not penetrate. In such a shaded position water is more effectively retained, and the humidity of the air is kept at a high level (see sidebar above). Many invertebrates are more comfortable in conditions of high relative humidity because

they are unable to control loss of water from their bodies and are therefore prone to desiccation. At high relative humidity water evaporates more slowly, and the animal is placed under less stress.

The growth of plants on the surface of the ground also affects microclimate. A cushion of stems and leaves produced by a saxifrage plant, for example, or by a small tuft of moss on the surface of a rock, create their own miniature climates. Air is held still within the mass of plant tissue and is not chilled by the wind or dried by the removal of its moisture. The microclimate created by this low-lying canopy of leaves is therefore warmer and has a higher humidity than the air above the plant canopy. This permits the survival of small animals anxious to escape the cold or the drought conditions in the polar air. The microclimate within a plant cushion also protects the lower parts of the plant from frost and drought, so that buds containing new shoots and flowers can develop there.

Snow patches can also create different microclimates in their vicinity. They are formed in sheltered locations, often with an aspect facing away from the spring sunshine. Away from the direct warming rays of the Sun, snow patches may persist long after the general cover of snow in the tundra has melted, and this means that the ground beneath the snow patch, together with any plant and animal life on or within the soil, is kept covered. In other words, the growing or active summer season is made shorter beneath the snow patch. Snow patches gradually shrink during the course of the summer. Many disappear totally, but those protected from direct sunlight may partially persist throughout the summer. The ground surrounding a snow patch experiences varying microclimatic conditions depending upon when the snow retreats and the soil is exposed to full light. The vegetation around snow patches is often arranged in a series of concentric rings, the outermost being released first from the snow cover and having the longest growing season. Locations close to the snow patch in midsummer, when the melting has ceased, are exposed for the shortest period, and only plants capable of coping with long periods of snow burial and very short growing times can survive here. There are some mosses that manage to grow under such conditions, but very few.

Snow patches also have a strong influence on soil temperature in their vicinity. Soil temperatures close to snow patches are lower because the melting snow releases very cold water into the surrounding soils. There is often a gradual rise in soil temperature with distance from a snow patch, and this affects plant root growth, seed germination, and also the activity of microbes and invertebrates in the soil. In a study of mountain soils on the Himalayan mountain of Pir Panjal, scientists found that in summer the soil at eight inches depth (20 cm) and 33 feet (10 m) from a snow patch had a temperature of 36.5°F (2.5°C). At a distance of 66 feet

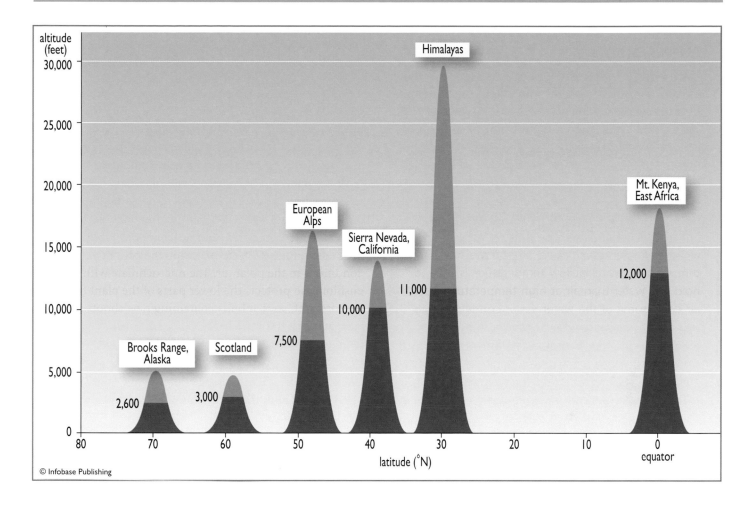

A selection of Northern Hemisphere mountains bearing alpine tundra. The closer a mountain is to the equator, the higher the altitude at which the tundra biome begins.

(20 m) the temperature at the same soil depth was 48°F (9°C). This difference is very significant for all organisms in an environment where the growing season is so short. So snow patches effectively prolong the winter conditions in their immediate environment.

Microclimate, therefore, is very important in its impact upon living organisms. Indeed, the microclimate of a location often exerts as great an influence on the survival of a species as the overall climate, or macroclimate. Those organisms that are rooted to the ground, or very small, or that lack mobility are particularly affected by their local microclimates. Larger, more mobile animals can move from one area to another and choose their microclimate to suit their needs. Some species, such as the polar bear, are capable of digging a den in the snow and effectively creating their own microclimates when seeking to escape the worst of the Arctic weather.

■ ALPINE CLIMATE

The overall climate of alpine tundra has much in common with polar tundra. Average annual temperature is low, and winter is often accompanied by high winds with suspended ice particles. But there are also many differences between the two types of tundra climate.

The general coldness of the alpine tundra is a result of altitude rather than latitude. The polar tundra is confined to high latitudes where the energy input from the Sun is relatively small, but this is not the case with the alpine tundra because it may be found in almost any latitude. High altitudes are generally cold, but for very different reasons from the high latitudes. The rate at which temperature falls with altitude is called the *lapse rate,* and this varies with a number of factors (see sidebar on page 22).

Falling temperature with altitude means that mountains are always cooler environments than the lowlands. Mountains in the high latitudes are particularly cold because the climate there is usually cool even at sea level, so the decline in temperature with altitude starts from a low point. Mountains in low latitudes, however, such as Mount Kenya, close to the equator, are relatively warm at their

Lapse Rate

Solar energy passes through the atmosphere and is absorbed by the surface of the ground. As the ground becomes warmer, some of the heat is transferred to the air in contact with it, so the atmosphere becomes heated at its base. A fixed amount of air, call it a "packet," warmed in this way expands, which means that the molecules of gas move around faster and separate from one another, so they occupy a larger volume. The packet of air thus becomes less dense, and when it is less dense than the air immediately above, it is displaced by the cooler, denser air and is forced upward. As it rises, the warmer air is subjected to less pressure because the weight of air above it is smaller. When it experiences lower pressure, the packet of air expands again, but this expansion results in the molecules of the gases losing energy as they have to push other molecules out of the way. As they lose energy the molecules slow down, and the temperature of the gas falls. This is called *adiabatic cooling*. The word *adiabatic* means that there is no energy exchange between the gas and its surroundings. Adiabatic cooling, therefore, is caused by the expansion of atmospheric gases with altitude, and the outcome is a gradient of falling temperature with height above sea level. The steepness of this gradient, or the rate at which temperature falls with altitude, is called the *lapse rate*. If the air is not saturated with water, that is, if it is relatively warm and its relative humidity is less than 100 percent (see sidebar "Humidity," page 20) then the term used for the fall in temperature with altitude is *dry adiabatic lapse rate* (DALR). At higher altitudes the air may become so cool that the water vapor it contains is sufficient to saturate it. At this stage condensation occurs, and water droplets form, resulting in rain. The air now has a relative humidity of 100 percent, and further cooling with altitude then follows the *saturated adiabatic lapse rate* (SALR), which is usually less steep than the DALR. This is because the condensation of water releases latent heat. Lapse rates vary with many factors, but on average the DALR is about 5.4°F with every 1,000 feet in altitude (9.8°C per 1,000 m), while the SALR is about 3°F with every 1,000 feet (5.5°C per 1,000 m).

Timberlines

The timberline can be defined as the border between forest vegetation and the treeless tundra environment. The boundary may be clear and distinct, tall trees abruptly ceasing and a marked line of dwarf shrub vegetation replacing them, but more frequently the boundary is ragged. The closed forest may be replaced by isolated patches of trees scattered among the tundra vegetation, especially in protected sites where the microclimate may be suitable for some tree growth. Some ecologists distinguish between a *forest limit,* where uninterrupted forest comes to an end, and the *tree limit,* beyond which there is no tree growth at all. Tree species, such as pines (*Pinus* species) and spruces (*Picea* species) may even grow beyond the tree limit in the form of stunted, dwarf specimens mixed among the low-growing willow and birch scrub of the tundra. This zone of dwarf trees is given the name *krummholz,* a German word meaning "stunted forest." The timberline is thus often a blurred boundary, or *ecotone,* between the forest and the tundra.

Climate is a major controlling factor in determining the position of the timberline at any location, and temperature during the growing season is particularly important. The position of the timberline around the Arctic tundra corresponds quite closely with the 50°F (10°C) mean temperature of the warmest month. Trees need warmth in summer so that their new shoots can grow and so that their roots can extend in the soil to take up water and nutrient elements. Temperature also affects the production of seeds, so reproduction may fail in prolonged cold conditions. Even if seeds arrive from far away, they may not germinate or establish themselves if the summer is not warm enough. So there is evidently a critical set of temperature conditions that must be achieved over extended periods if the tree is to survive, and this is reflected by the 50°F isotherm.

Although temperature is generally the most important factor controlling timberlines, precipitation can also play a part, as can topography, wind, soil conditions, and grazing intensity, particularly by domestic stock. Steep cliffs are difficult habitats for trees to invade, so the timberline may be held at a low level if the topography becomes excessively steep. Similarly, very thin soils, especially if they are unstable or subject to avalanches, may limit the growth of trees and the position of the timberline. Wind has a chilling effect and often restricts plant growth. Young trees are particularly susceptible to wind chill, and the krummholz form often results from severe wind activity above the tree limit.

base, so tundra habitats are confined to very high altitudes. The timberline, the altitude where forest ends and alpine tundra begins (see sidebar on page 22), is thus higher in mountains that lie closer to the equator, as shown in the diagram. In the mountains of Alaska (70°N) and Scotland (60°N), the timberline lies at about 2,600 feet (792 m) and 3,000 feet (914 m), respectively. In the high Alps of central Europe (48°N), trees can reach altitudes of 7,500 feet (2,286 m), while in the Sierra Nevada of California (40°N) the timberline lies closer to 10,000 feet (3,048 m). At 30°N in the Himalayas of Nepal and Tibet, the timberline is at about 11,000 feet (3,353 m), and on Mount Kenya, situated almost on the equator, trees extend up to 12,000 feet (3,658 m). As a broad generalization, the timberline rises by about 360 feet (110 m) for every 1° of latitude closer to the equator.

Temperature is an important factor in determining the altitude of the timberline (see "Tundra Conservation," pages 229–232), but there are other factors that can influence it, including the effects of wind and the impact of human land use, especially the grazing of domestic animals (see "Grazing in the Tundra," pages 129–130).

High altitude can thus create the condition of low temperature that is a major feature of all tundra habitats, but the fact that mountains occur at different latitudes results in certain other conditions being very variable, such as day length, for example. A mountain lying within the Arctic or Antarctic Circles experiences the same seasonal variation in day length as any other part of the polar tundra. Days are long, occasionally perpetual, in the summer, and short or even absent in the winter. But closer to the equator, as latitude decreases, variation in day length with season becomes less marked. Within the Tropics, the day and night sequence changes very little from 12-hour days and 12-hour nights whatever the time of year, so the sequence of day and night is, on the whole, less marked in alpine tundra, depending on latitude.

Day length variation is not the only factor that changes with latitude. In low latitudes the angle of the Sun above the horizon at noon is much higher than in the polar regions, even in polar midsummer. In the Tropics the midday Sun is close to an overhead position at all times of the year, and this applies to mountaintops just as much as to the lowlands. The incident energy from the overhead Sun

---

The annual cycle of temperature in polar and alpine sites, expressed as the mean monthly temperature. The mountain site (Colorado) has less annual temperature variation than the two polar tundra sites (Manitoba and Northwest Territories).

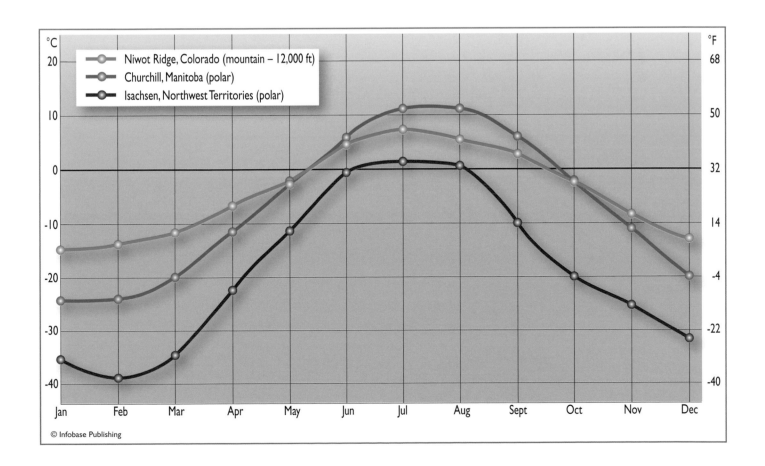

© Infobase Publishing

is much greater than from the Sun when it is close to the horizon, so daytime in the alpine tundra can be hot in the low latitudes and the nighttime temperature very cold. In other words, the diurnal (daily) fluctuation in temperature is much greater in low latitude alpine tundra than in the polar tundra or the high latitude alpine tundra. It is not an exaggeration to say that the alpine tundra of tropical mountains experience summer every day and winter every night, the amplitude of daily fluctuation often exceeding 27°F (15°C). The thin atmosphere at high altitude means that its insulating properties are much poorer than in the lowlands, so the mountaintops radiate heat rapidly once the Sun has set. Temperature also increases very rapidly when the Sun rises in the morning; a change of 18°F (10°C) in half an hour is not unusual. Mountains in the temperate latitudes also experience wide fluctuations in temperature between day and night but not as extreme as those found in tropical mountains. At Mount Whitney in California (14,496 feet; 4,418 m), the highest mountain in the United States outside Alaska, a day in early September may have a maximum of 62°F (17°C), falling to 22°F (–6°C) during the night.

Although there are such wide ranges in temperature through a 24-hour period in alpine tundra sites, when overall average temperatures are considered, the alpine and polar tundra appear more similar. In the diagram on page 23 the average monthly temperatures over a year are shown for three sites. Niwot Ridge, Colorado, is an alpine tundra site lying at an altitude of 12,000 feet (3,657 m). The other two sites are in the lowland polar tundra of Canada, at Isachsen, Northwest Territories, and Churchill, Manitoba. Niwot Ridge and Churchill have quite similar graphs, except that

the annual amplitude of temperature variation is greater at the polar tundra site. The difference between summer and winter average temperatures is thus more pronounced in polar tundra than in alpine tundra. This is best explained by the strong diurnal variation in temperature at the alpine site. Cold night temperatures have the effect of damping down the summer monthly averages, and warm day temperatures in winter prevent the winter monthly means from falling too low. As a result, temperature fluctuations during the course of a year are lower than in the polar regions.

Altitude affects temperature because of the influence of lapse rate (see sidebar "Lapse Rate," page 22), and this means that the warmth needed for plants to begin their growth arrives later in the year at higher altitudes. In general, the higher the location, the shorter the growing season and the later it begins. In the northwest Himalaya, for example, the growing season at 10,000 feet (3,050 m) begins in early April and ends in early October, lasting about 24 weeks. At an altitude of 15,000 feet (4,570 m), the growing season is delayed until early June and is over by early September, lasting just 13 weeks. At the very high altitude of 20,000 feet (6,100 m), plant growth cannot commence until late June and comes

The effect of prevailing wind and precipitation on the vegetation zonation of the Sierra Nevada in California. Westerly winds from the warm Pacific Ocean bring high precipitation to the west-facing slopes of the mountain range. The eastern slopes lie in a "rain shadow" and receive little precipitation because the descending air becomes warmer, so water does not condense. Under these conditions forest extends to higher altitude in the east than in the west.

to an end in early August, so only six weeks are available in which in which organisms must complete growth and reproduction. Needless to say, very few plants and animals are able to survive under such extreme conditions.

The upper altitudinal limit for the development of alpine tundra vegetation is determined by the permanent presence of snow. This is sometimes called the *snow line.* In fact, there is very rarely a distinct line between permanent snow and a cover of vegetation with associated snow patches; more often there is a transition zone. Much depends on the topography, the aspect, and the influence of prevailing winds in moving snow from one place to another. Snow arrives as precipitation, so the arid mountain ranges, such as the northern Himalaya, may have a higher snow line than might be expected.

Mountainous areas can induce some unusual weather conditions. On high mountains the air above permanent snow and glaciers becomes chilled, making it denser, and cold air masses slip down the sides of the mountains and into the valleys between them. The cold air forms a dense layer close to the ground, forcing the warmer air upward. The result is a stratified, or layered, arrangement of air, with the cold air close to the ground and warmer air up above, which is the opposite of the normal condition in which air is cooler with altitude (see sidebar "Lapse rate," page 22). This situation is called a *temperature inversion,* and it results in quite stable atmospheric conditions. When air close to the ground becomes hot, it induces convection currents and atmospheric turbulence, often resulting in stormy weather in mountain areas, but the temperature inversion reduces convection and creates a period of relative stability.

Winds in mountain regions can also be complicated in their behavior. Mountain ranges may receive incoming wind from a specific direction, a *prevailing wind.* If this wind arrives from the ocean it may bear a heavy burden of water vapor, and this affects the precipitation received by the mountains. A good example of this is the Sierra Nevada of California, as shown in the diagram on page 24. Winds mainly arrive from the west, having passed over the Pacific Ocean and picked up water vapor on the way. As the warm, water-laden winds are pushed up the slopes of the Sierra Nevada, they become cooler, and water condenses as rainfall in summer or snowfall in winter. The precipitation is greatest on the exposed, west-facing slopes and usually in the middle to upper elevations. By the time the air has reached the summit of the mountains much of its water content has been shed. This effect is particularly noticeable in very high mountains, such as the Himalaya. When air has risen to a height of 24,000 feet (8,000 m) it contains only 1 percent of the water present at sea level, so very little snow falls at such a high altitude. The Sierra Nevada, however, is much lower than this, hardly exceeding 14,000 feet (4,270 m), but

much of the water content has been lost by the time the air mass crosses the ridge. On the eastern side of the Sierras, the air begins to descend, and as it does so it becomes warmer because it comes under increasing atmospheric pressure. This is called *adiabatic warming* and is the opposite of the adiabatic cooling that takes place on the western side of the mountains. The dry adiabatic lapse rate is approximately 5°F warming for every 1,000 feet fall in altitude (1°C for every 100 m) (see sidebar "Lapse Rate," page 22). So warm dry air flows down the mountain slopes on the protected eastern side. Warm winds on the lee side of mountains are given the name *föhn winds* (pronounced "fern"), derived from a German term that was first coined in the central European Alps. One of the effects of the different conditions on the eastern side of the mountains is that the timberline is higher by as much as 1,000 feet (300 m) in the warmer, drier, less windy conditions than it is on the western side, as can be seen in the diagram. Lands that lie on the lee side of mountain ranges are often relatively dry because the air coming from the mountains has a low relative humidity. Such lands are said to lie in a *rain shadow.* The dry lands to the east of the Sierra Nevada are a fine example of the rain shadow effect, as is the Gobi Desert in central Asia, which lies in the rain shadow of the Himalaya.

Apart from in the very highest of mountains, precipitation generally increases with greater altitude. In the Rocky Mountains to the west of Boulder, Colorado, for example, at an altitude of 5,250 feet (1,603 m) the average annual precipitation is 15.5 inches (40 cm). At an altitude of 8,460 feet (2,580 m) precipitation is 21.3 inches (54 cm), and at 12,300 feet (3,750 m) it reaches 25.2 inches (64 cm). So alpine tundra may receive more precipitation than polar tundra. The annual precipitation at Fort Yukon in Alaska, for comparison, is 6.8 inches (17 cm). Alpine tundra, therefore, is generally not the cold desert of the polar tundra, and this represents one of the major climatic differences between the two types. There are, however, exceptions to this rule. In the northwest Himalaya, the high mountains border onto the arid lands of Afghanistan. Water-laden winds come from the south from the Indian Ocean during the summer monsoon season, but the water has condensed and precipitated by the time the air masses arrive in the northwest of the mountains. The annual precipitation here may be no more than three inches (8 cm). Strong winds and intense sunlight add to the desiccating conditions even at high altitudes in these mountains.

Aspect, the direction in which a slope faces, is clearly important in mountains because it influences the exposure to wind and rainfall as well as to sunlight. Aspect also has an effect on the temperature of a site, in the same way as in the polar tundra. Slopes that face in the general direction of the equator (south-facing in the Northern Hemisphere and north-facing in the Southern Hemisphere) receive more sunlight and become warmer and drier as a consequence.

This favors plant species that are light-demanding and are capable of coping with drought. Species that require moist, shady conditions, on the other hand, are found on slopes that face in the opposite direction.

One additional factor that needs to be considered is the low air pressure at high altitude. Atmospheric pressure results from the downward force exerted by the weight of atmospheric gases. At higher altitude there is less atmosphere pressing down from above, hence less atmospheric pressure. The pressure of the atmosphere decreases exponentially with altitude, that is, in a logarithmic rather than a linear manner. The main way in which this is important to living organisms is that oxygen is much less dense, and therefore respiration can become difficult. The pressure exerted by the oxygen component of the atmosphere at an altitude of 18,000 feet (5,500 m), for example, is only half that found at sea level. Strenuous activity becomes very difficult for animals at such low oxygen levels, and most mammals, even wild sheep and ibex, are largely restricted in their distribution to altitudes below 19,000 feet (5,800 m).

Mountains thus create their own climates quite independently of the general climate of the regions in which they lie. High mountains can support tundra vegetation on or near their summits because of their generally low temperatures. But there are some features of alpine tundra habitats that differentiate their climate from that of the polar tundra of the high latitudes, and these features are most evident in the high tropical mountains. It is not surprising, therefore, that the alpine tundra of tropical mountains is so very different from that of polar tundra and from the alpine tundra of higher-latitude mountains.

■ CONCLUSIONS

The tundra biome is widely distributed over the land surface of the Earth. It is found in two main forms, polar tundra and alpine tundra. Polar tundra, as its name implies, is located in the high latitudes and forms a ring around both the North Pole and the South Pole. Polar tundra is not found at the poles themselves because the North Pole lies on floating ice over the Arctic Ocean, and the South Pole is situated deep within the Antarctic ice sheet. The edges of these frozen areas, however, including islands and coastal regions, bear the main concentrations of the polar tundra. The southern edge of the Arctic tundra is determined by the occurrence of boreal forest (taiga), and the limits of the forest are in turn defined by climate, in particular the degree of summer warmth.

Alpine tundra is widely scattered and is not limited by latitude, but it is limited by altitude. As in the case of polar tundra, the lower limit of alpine tundra on mountains is determined by forest growth, which is again constrained by the need for summer warmth. The altitude of this timberline varies with latitude, being at low elevation in the high latitudes and higher in the low latitudes. Temperature is the most important factor, and this depends on lapse rate, the steepness of the decline in temperature with altitude.

Both types of tundra, therefore, survive because the overall temperature of their environment is too cold for forest growth. Polar tundra is cold because of the curvature of the Earth and the low angle of the Sun at high latitudes. The presence of snow also means that the Earth's surface in this region has a high reflectivity, or albedo. North of the Arctic Circle and south of the Antarctic Circle there are times in summer when the Sun does not set and times in winter when it never rises. This means that there are considerable temperature differences between seasons. The polar regions radiate more energy than they receive from the Sun, and they would become even colder if they did not receive energy from the lower latitudes, brought to them by the oceans. This is particularly true of the Arctic, which receives warm water from the North Atlantic as the Gulf Stream brings in water from the Tropics.

Alpine tundra, especially when found in low latitude sites, differs from polar tundra in its general lack of day length variations. Near the equator there is little change in day length with season, and the angle of the Sun in the sky also varies little between summer and winter. As a result, alpine tundra generally has less temperature variation between seasons than polar tundra. The high angle of the Sun at noon and the thin atmosphere at high altitude result in temperatures rising steeply during the day and falling rapidly at night. The diurnal fluctuation in temperature in alpine tundra may be as great as, or greater than, the variation in seasonal averages.

In both types of tundra microclimate is important in creating a considerable range of local climatic variation. The direction a slope or a cliff faces (its aspect) can cause large differences in the local climate, both in temperature and humidity. The presence of a rock or the persistence of a patch of snow into the summer can create a set of conditions that determine what plants and animals can survive in that location.

Polar tundra is situated in regions of high pressure, where descending air masses bring little precipitation. The High Arctic and the Antarctic ice sheet are effectively as dry as deserts as far as precipitation is concerned. But low temperature means that evaporation is also low, and most of the water received is retained as ice. Montane sites, where temperature decreases with altitude, create conditions that encourage precipitation. As rising air is cooled, water vapor condenses to rain drops or snowflakes, so the side of a mountain exposed to wind often receives a good supply of precipitation. Very high mountain peaks, however, are so cold that the atmosphere is unable to hold much water vapor, and they develop arid climates. On the sheltered, or lee, side of a mountain chain, the descending air becomes warmer and retains its moisture more effectively, creating a dry rain shadow.

The two types of tundra thus have some features in common, such as cold overall climate with high winds, but they also differ in such features as day length, daily and seasonal temperature variation, and precipitation patterns. Both the similarities and the differences are important in creating a range of environments where plants and animals struggle to survive.

2

Geology of the Tundra

Geology is the study of rocks, their formation, and their breakdown. Geologists define rock as any mass of mineral matter found in the Earth's crust that has formed naturally. Most rocks are composed of hard materials, solid stony masses built up by the fusion of different minerals. But the geologist does not limit the word *rock* to hard materials; rocks can also be soft. Deep desert sands, the layers of clay sediments in a lake bed, and the detritus left behind when a glacier melts are all rocks according to the geological definition. Only those materials constructed by human activity, such as concrete and plastics, lie beyond the confines of the word *rock.*

Neither the Earth nor the rocks that compose its crust are static; they are constantly changing. Rocks are forming even today as sand, silt, clay, and the bodies of microscopic animals and plants sediment to the floor of the ocean and become compacted. Rocks formed in this way are called *sedimentary* rocks. Rocks are being modified by the intense pressures and high temperatures they experience within the Earth's crust. These are called *metamorphic* rocks. Slate is an example of a metamorphic rock that began its existence as a sedimentary deposit but has been subjected to pressures and temperatures that have compressed and changed its constitution. Metamorphic rocks are often produced in the vicinity of volcanic activity or at great depth within the crust. *Igneous* rocks are the third major rock type, and these come into being when molten rock, or magma, from the Earth's core breaks up through the crust and solidifies as it cools. All of these rock types are found within the polar tundra regions and also in the high mountains of the world.

■ THE ROCK CYCLE

The interior of the Earth consists of molten rock, or magma, and convection currents are set up within this fluid core that push upward into the solid crust at the Earth's surface. Sometimes these rising currents in the magma force their way into the crust but fail to reach the surface before the

rock cools and solidifies. Occasionally the magma bursts through the crust and erupts in the form of a volcano, and the molten rock solidifies on the surface of the Earth as masses of lava. Solidification of the magma involves many chemical changes in its composition, depending in part upon the location (above or below ground) where it finally cools. Magma contains about 14 percent of dissolved gases, mainly water vapor and carbon dioxide, and these may be discharged into the atmosphere. The chemical constituents of the magma, called *minerals,* have different melting points, so they may separate out during the cooling process, depending upon the rate at which the temperature falls. Consequently, there are different kinds of igneous rock that vary chemically according to the conditions under which they were formed.

Igneous rocks that have failed to reach the surface and are buried deep in the crust are protected from the process of rock breakdown, or *weathering,* suffered by surface rocks, but eventually even these may be exposed to the atmosphere as the crust above them is worn away. Weathering involves the chemical and physical breakdown of the rocks into their component minerals or even into smaller chemical units. Living organisms also play a part in this decomposition of the rocks, and the process is the basis of soil formation (see "Geology and Rock Weathering," pages 52–54). Solid rock is converted by weathering into smaller particles, some of which may be soluble in water, and they are transported by water draining over and through the soils in which they lie. This transport of particles is called *erosion,* and the movement of dissolved materials is called *leaching.* As a result of these two processes, the rocks exposed at the Earth's surface are constantly being worn away, and their components are transported down to the oceans, where they may remain suspended for a while but eventually sediment to the bottom.

The sedimentation of particles derived from land-based rocks is accompanied in the oceans by the deposition of tiny remains of planktonic organisms that have spent their lives in the surface layers of the ocean. Creatures such as foraminifera and coccolithophorids, single-celled organisms belonging to the kingdom Protoctista, accumulate calcium carbonate (lime) from the sea water and build cases for themselves that survive and sink into the ocean sediments after the death of the living creature. The diatoms (also members of the kingdom Protoctista) build cases out of silicon dioxide, and this relatively inert material, with the same composition as sand, also joins the other particles in the steady rain of fine materials that constantly accumulate in the sediments of the ocean floor.

The rock cycle. Volcanic activity creates igneous rocks derived from the molten magma of the Earth's interior. Pressure and heat modify the physical and chemical nature of adjacent materials, forming metamorphic rocks. Uplifted landmasses are weathered and eroded by the atmosphere and climate, and the transported materials are sedimented in the oceans, over time creating sedimentary rocks.

These sediments build up in the course of time and are converted into harder rocks by the pressure from above. The sedimentary rocks produced in this way may eventually be forced beneath the crust to enter the magma once more. Alternatively, they may be crushed, folded, and buckled by movements in the Earth's crust, raised above the surface of the ocean, and exposed to the forces of weathering and erosion once again. In this way the rocks of the Earth are involved in a constant cycle that takes many millions of years to complete, as can be seen from the diagram below.

■ PLATE TECTONICS

The rock cycle describes the movement of materials between the Earth's crust and the Earth's core, and this exchange involves other processes that play an important part in the patterning of continental land masses over the

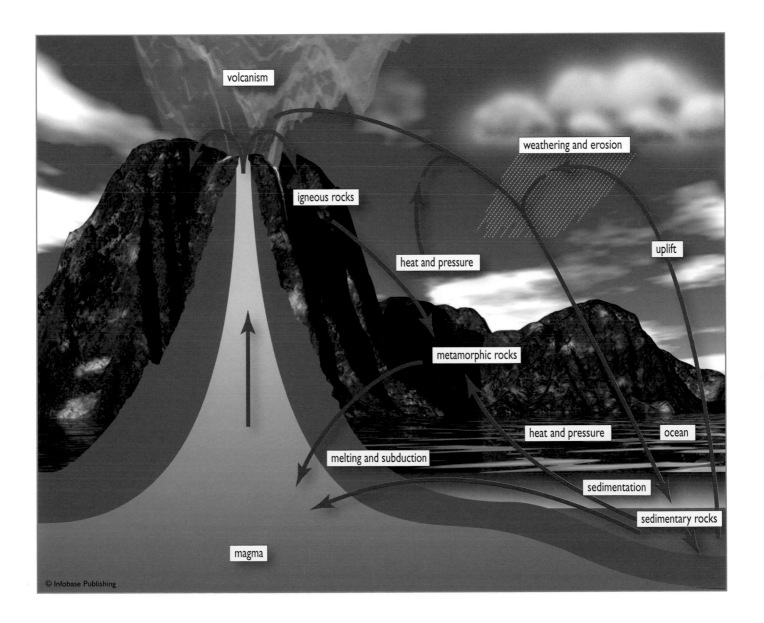

volcanism

weathering and erosion

igneous rocks

uplift

heat and pressure

metamorphic rocks

heat and pressure

ocean

melting and subduction

sedimentation

sedimentary rocks

magma

© Infobase Publishing

face of the Earth. The location of the land masses, the continents of the world, is itself a major factor in determining the existence of the tundra biome.

Geologists have long known about the workings of the rock cycle, but until the last 50 years they were unaware of the remarkable mobility of the Earth's crust. They pondered over such questions as the origin of the world's oceans and the difficulties of interpreting some of the fossils found in unlikely places, such as tropical animals found in London, England, and fossil forests in the heart of Antarctica. Past climates must have behaved in a very strange and unpredictable way. The answer to these problems came from a very unlikely person, a professor of astronomy and meteorology in Berlin, Germany, who had never studied geology. His name was Alfred Wegener (1880–1930). He noticed that the east coast of South America and the west coast of Africa are mirror images and would fit together very comfortably if the South Atlantic Ocean closed up. The same is true of eastern North America and the west coast of Europe, especially if one considers the edges of the continental shelves rather than the current coastlines. He concluded that there was once a single "supercontinent" that has subsequently split into a number of components, and these have gradually drifted apart. His first publication on the subject came in 1912, but the outbreak of World War I and the limited circulation of his German publication meant that it was not noticed by the scientific community. His subsequent publications, including a book on the subject, were received with ridicule. Scientists throughout the world considered his views eccentric and pointed out the lack of any mechanism to explain the wandering continents. Wegener died in an accident in Greenland in 1930, his ideas still universally rejected.

Geological evidence supporting a close association between continents in the past continued to accumulate over the years. The rocks of the Andes Mountains in the west of South America, for example, are also found in the Antarctic Peninsula and in New Zealand. Rocks from eastern Australia

The crustal plates of the Earth. Plate boundaries are the locations of intense tectonic activity. Some boundaries are constructive, the lines along which new crustal material is being formed; other boundaries are destructive, the places where crustal material plunges back into the Earth's interior, called subduction zones. Volcanic activity is particularly strong along the boundaries of subduction zones, shown here by red shading.

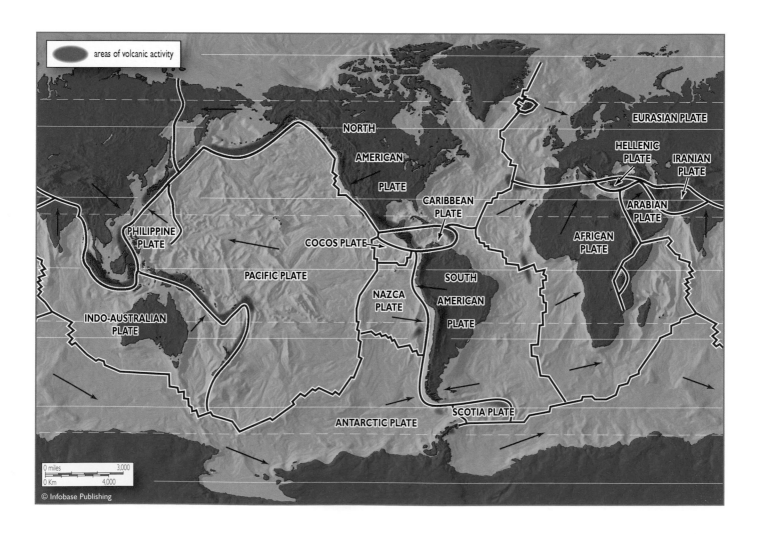

also continue in a band across Antarctica. Fossils of a seed fern were found in southern South America, southern Africa, Antarctica, Madagascar, India, and Australia, which added weight to the hypothesis that these land masses had once been much closer together or even joined in one supercontinent. Then geophysicists discovered a new technique that changed many opinions and attitudes. It was called *paleomagnetism,* and it is based on the observation that when rocks are first formed they are strongly affected by the Earth's magnetic field. Any particles that can be magnetized, such as iron compounds, become aligned with the magnetic field as the rock solidifies, and they subsequently retain this orientation of magnetization. By studying rocks of different ages in Australia, geologists were able to work out that this continent had migrated around the South Pole and then northward into its present position over the past 500 million years. The only alternative explanation was that the South Pole itself had wandered during this time, but further studies in many parts of the world during the 1950s and 1960s confirmed that the continents had indeed been on the move. Wegener was at last proved right.

Proving that continents, such as Europe and North America, are moving apart is not easy because the rate of movement is very slow. Geologists also needed to find a mechanism that would account for the movement, and they found this at the bottom of the sea. The gradual separation of Europe and North America, for example, can be explained only if there is a dividing line out in the Atlantic that is pushing apart the ocean floor as well as the two continents. This is found in the Mid-Atlantic Ridge, a line of submarine mountains rich in active volcanoes that runs the entire length of the Atlantic Ocean and even emerges above the surface on the island of Iceland. Along this line magma is rising to the surface and solidifying into new crustal rocks that are pushed either east or west, forcing the continents apart. Lines of separation of this kind are now known to occur around the world, dividing the Earth's crust into a series of plates and forcing them apart, as shown in the diagram on page 30. As two plates are pushed apart, an ocean basin may be formed between them, as in the case of the Atlantic Ocean, which is continuing to enlarge by between two and four inches per year (5 and 10 cm per year).

If new crustal material is constantly being created, then there are two possible consequences. Either the new material will force older parts of the crust to buckle and fold, forming new mountain chains, or there must be places where the crust is being destroyed or is slipping down into the Earth to become magma once again. In fact, both of these options occur. Moving plates may collide, and the force of the collision can cause crumpling and upthrusting of the crust, creating high mountains, the future home of the alpine tundra. This type of mountain-building collision occurs when both of the plates involved bear continental land masses. India,

for example, is part of the same plate as Australia, as can be seen from the diagram, and when Australia drifted northward, India also moved and eventually collided with the Eurasian plate. The result was the creation of the Himalaya Mountains along the collision boundary. A similar collision between the African plate and the Eurasian plate resulted in the building of the Alps in central Europe, the Caucasus Mountains of eastern Europe, and the Alborz Mountains of Iran, and the ripples caused by that collision even created hills in southern England.

The crust beneath the oceans is often only about four miles (6.5 km) thick, much thinner than the continental crust, which may be up to 20 miles (32 km) thick and is composed of older rocks. When an oceanic plate collides with a continental plate, it usually results in the oceanic crust sliding underneath the continental one, where it is pushed back down to rejoin the magma. This is called a *subduction zone,* as shown on the diagram on page 32. The existence of subduction zones ensures that there is a balance between the production of new crust at the ocean ridges and its destruction by burial. The oceanic crust is heavy, so on collision it tilts downward and begins its descent back into the magma below. As it does so, it drags some of the lighter continental plate materials with it and creates a trench in the zone of subduction. As this mix of rock, often saturated with water, descends into the Earth it begins to melt in the increasing heat, and the lighter, more volatile constituents form an unstable mass of material that forces its way back up to the surface. The consequence is that subduction zones are often accompanied by intense volcanic activity. One of the most active volcanic regions of the world lies along the junction of the Indo-Australian plate with the southeastern edge of the Eurasian plate in the vicinity of Java. The famous island of Krakatoa, which erupted and exploded in 1883, is just one of the volcanoes that have been created along this subduction zone.

Plate boundaries are often zones of stress for the Earth's crust, but their movements are not always in direct opposition to one another, as in the case of plate collisions. Sometimes plates are moving alongside but in opposite directions, and they slide past one another. This is called a *conservative plate boundary.* A good example of this type of boundary lies immediately west of the Sierra Nevada in California, where the Pacific plate is moving northward with respect to the North American plate, having a relative motion of about half an inch (1.2 cm) per year, although it has been recently accelerating. The friction between the two plates causes them to stick at times, and the pressure is then released by a sudden and catastrophic movement that causes a major earthquake. The San Francisco earthquake of 1906 was due to one of these sudden releases of pressure between the sliding plates, and it resulted in a sudden movement of 21 feet (6.4 m). A similar sudden shifting of plates took place in December 2004 beneath the sea in Southeast

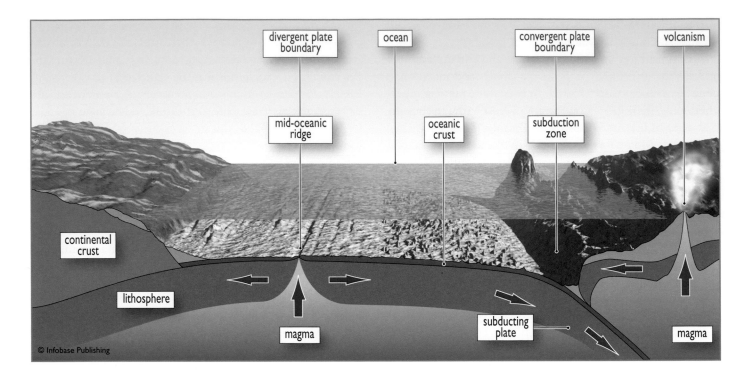

Convection currents in the Earth's molten interior cause a constant movement of the surface crust. New oceanic crust is created at a mid-oceanic ridge, forming a divergent plate boundary. The outward spreading of the oceanic crust results in a collision with continental crust at a convergent plate boundary, and the thinner oceanic crust is forced beneath the continental crust, creating a subduction zone. Melting of the submerged crust results in a high incidence of volcanism along the continental edge of the plate boundary.

Asia, and the outcome was a devastating tidal wave, or tsunami, which resulted in hundreds of thousands of deaths in the coastal regions of lands surrounding the Indian Ocean. Plate boundaries are the places where the Earth is at its most dynamic.

The polar regions and their associated tundra habitats lie on three main plates. The Arctic regions occupy the North American and the Eurasian plates, and the Antarctic has a plate of its own, as shown in the diagram on page 30. The North American and Eurasian plates are moving apart as new crustal material is generated along the Mid-Atlantic Ridge. Most of this ridge is situated in deep oceanic waters, but in the North Atlantic it comes to the surface and creates the volcanic landscape of Icelandic tundra. The spectacular combination of fire and ice results in the formation of hot springs and geysers in a tundra environment. Volcanoes, such as Mount Hekla, are active and periodically eject lava in what are known as "quiet eruptions." The lava emerges

from the volcanic vent at a temperature of about 1,500 to 2,200°F (800 to 1,200°C) and flows gently down the slopes of the volcano, often traveling for tens of miles. As the liquid rock flows over the ground surface, it gradually cools, forming a jagged and fragmented mass. Where the molten lava flows directly into the sea, however, it cools quickly into rounded blocks of *pillow lava*. The Icelandic volcanoes are mainly of the quiet type, but their activity can be spectacular, as in the creation of a new island off the south coast of Iceland in 1953. An underwater eruption elevated a volcanic cone above the surface of the sea and produced the new island of Surtsea overnight. The establishment of this totally new expanse of land within the tundra region has provided scientists with a unique opportunity to study the development of ecosystems in cold regions (see "Ecosystem Development and Stability," pages 99–100). Volcanic activity is also found where the North American plate meets the Pacific plate along the line of the Aleutian Islands and along the Kamchatka Peninsula in eastern Russia, leading southward to Japan, but the remainder of the Arctic region is relatively free of volcanism.

The Antarctic continent has its own plate, so it is not fragmented by plate boundaries. Nevertheless, this southern polar plate is rotating in a clockwise direction, as viewed from above the South Pole, causing stress along its edge. There is scattered volcanic activity around this edge, particularly at the tip of the Antarctic Peninsula and in the South Sandwich Islands, which are currently moving away from the Antarctic Peninsula. There was major volcanic activity on Deception Island during the period 1967–70, when two research stations were destroyed.

The great mountain ranges of the world and their alpine tundra habitats are also the products of plate tectonic movements. In western North America the North American plate is being crushed against the Pacific plate, creating the ridges of mountains that run parallel to the western coast, including the Cascade Mountains, the Rockies, and the Sierra Nevada. This line of mountains continues south through Central America and into the Andes range of South America, where the South American plate is moving west while the Nazca plate in the east Pacific is forcing its way east. The mountains of Central America and eastern South America are rich in volcanoes, but in North America volcanic activity is now restricted to Alaska and the Cascades.

The complex of plates that constitute northeast Africa are moving north and becoming compressed against the southward movement of Eurasia, and this has created lines of mountains through Turkey and northern Iran, where volcanoes are still active. This line continues eastward through the Hindu Kush of Afghanistan and the Himalaya Mountains, elevated by the opposing movements of India (northward) and Eurasia (southward). The Himalaya, however, are not volcanically active, but the contact between the Indo-Australian plate, the Eurasian plate, and the Pacific plate in southeast Asia has created a volcanic and tectonically unstable zone of mountains in this part of the Tropics. The mountains of East Africa are also volcanic and lie along a north-south running rift, where the crust of the Earth is tearing.

Both the polar and the alpine tundra regions of the world owe their existence and their current positions to the tectonic activity of the Earth's crust. The long-term history of the tundra, therefore, has been determined by tectonic history.

■ TECTONIC HISTORY OF THE TUNDRA LANDS

By studying the patterns of rock arrangements and their fossils in different continents and their residue of magnetic alignment, geologists can trace the ways in which land masses have moved in the past. The complicated story of past continental wandering is now more or less completely unraveled.

In Devonian times, around 380 million years ago, the continents then in existence bore no resemblance to those found today. There were three major landmasses. One consisted of the lands that would one day become North America and Europe, a second was a block of land destined to become Siberia, and the third was a massive supercontinent called Gondwana, consisting of most of the remaining land areas of the world. By the late Permian, some 260 million years ago, all the world's land masses had fused into one supercontinent, called Pangaea, as shown in the diagram on page 37. At that time the region that is now called Asia was fragmented and its component parts well separated around the edge of the supercontinent, but the other major continental masses were intact and joined together. But this single, unified world soon began to break up, forming two major land areas, one called Laurasia, which consisted of North America, Europe, and Asia, and the other, the original Gondwana, consisting of South America, Africa, India, Australia, and Antarctica. By the Early Cretaceous, some 140 million years ago, even these continental masses were fragmenting as epicontinental seas began to form within the supercontinents and the major continents that make up the modern world started to emerge, as shown in the same diagram.

In Late Cretaceous times, 90 million years ago, Gondwana had split into its component parts, with Africa, Madagascar, India, and New Zealand completely separated off, and South America and Australia linked to Antarctica only by tenuous land bridges. Even these links were broken by the Eocene, about 40 million years ago, and the world had begun to take on its familiar form. India had collided with southern Asia, creating the Himalaya Mountains in the process, and northeast Africa was about to make contact with western Asia to create the Arabian Peninsula. The bridge between North and South America was not yet in existence. The land masses that were later to bear Arctic tundra now formed an almost unbroken circle around the North Pole, and Antarctica was in its present isolated position over the South Pole.

The wandering of the continents over the face of the Earth has had important climatic consequences. Eastern North America and western Europe, for example, were far south of their current geographical positions 60 million years ago and experienced subtropical climatic conditions. At that time what is now Greenland and the Hudson Bay region of Canada would have been considerably warmer. In northern Greenland fossil leaves of the breadfruit tree (*Artocarpus dicksoni*) indicate that conditions must have verged on the tropical. Geologists have taken corings from the sediments of the Arctic Ocean and discovered fossil materials from 70 million years ago that demonstrate much warmer conditions existed even at very high latitudes. They speculate that the average annual temperature in the Arctic Ocean at that time was 60°F (15°C) and that the region was ice free throughout the year. There are two possible reasons for this Arctic warmth. First, the entire Earth at that time is likely to have been warmer. There is evidence that carbon dioxide gas was more abundant in the atmosphere than at present and that the world was in a "greenhouse" condition (see "The Atmosphere and Climate," pages 10–11). The second factor is that the Arctic Ocean, although largely surrounded by land, must have been receiving an influx of warm tropical water from the North Atlantic or North Pacific, or both. One thing is sure: The Arctic tundra as we know it did not exist at that time.

When all continents were fused into one massive landmass, oceanic waters had free access to the polar regions, so the redistribution of energy over the face of the Earth was uninterrupted. The Poles were considerably warmer than today, so there were no polar icecaps. The southern supercontinent of Gondwana experienced mild conditions with high precipitation as a result of the evaporation

of water from warm oceans and its condensation over the land. During mid-Carboniferous and Permian times, from about 330 million years ago, Gondwana was drifting toward the South Pole, and this prevented any warm tropical water from reaching this part of the world, where the Sun's energy was in low supply. As a consequence, an ice sheet began to form on the Gondwana continent. Meanwhile, the northern landmasses, including what is now North America and Europe, were close to the equator, so tropical swamp forests prevailed there, laying down deep deposits of peat that would later turn to coal. Despite the formation of a southern polar ice cap, Gondwana retained a relatively mild summer climate around its edges, as the fossil floras from Carboniferous and Permian rocks record. One particular plant fossil dominates, namely, a shrub or small tree with relatively thick woody stems and broad leaves. It has been given the name *Glossopteris,* and, together with the associated plant species, the community of that time is called the *Glossopteris* flora. The fossil remains of woody parts of *Glossopteris* have very distinct annual growth rings, indicating that the plant exhibited summer growth and winter dormancy. It is remarkable, however, that this lush vegetation survived even within 5° latitude of the contemporaneous South Pole, indicating that global temperature was

Tectonic plate movements have been taking place throughout geological history. In Permian times (A), about 260 million years ago, all the Earth's landmasses were linked in one supercontinent. This split into two by Cretaceous times (B), about 140 million years ago, and the breakup continued through the Cretaceous period until the continents had begun to assume their familiar shapes by the late Cretaceous (C), 90 million years ago. By the Eocene period (D), 40 million years ago, India had collided with Eurasia, and South America had broken its links with Antarctica but had not yet formed a bridge to link it with North America. The North Atlantic Ocean had formed, however, separating the Old World from the New World.

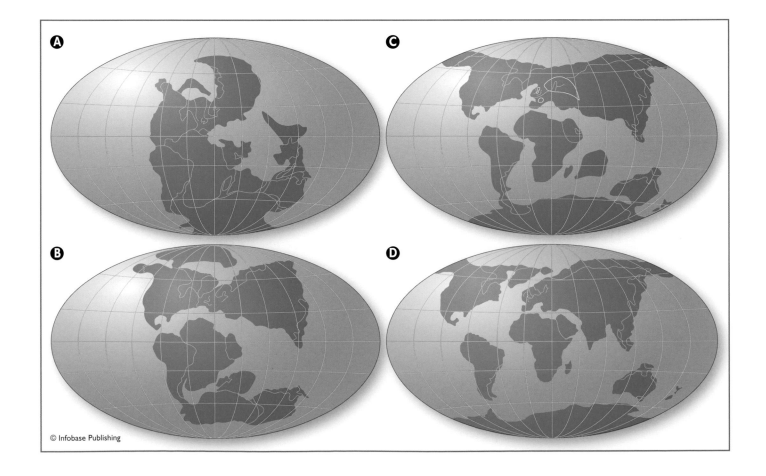

warmer at that time. The *Glossopteris* flora has become important in the reconstruction of tectonic processes because its fossil remains have been found in the rocks of Australia, Antarctica, South America, and southern Africa. The fossils thus provide important evidence for the former close contact between these continents.

The breakup of Gondwana during Cretaceous times was accompanied by the development of extensive forests in what is now Antarctica. Geologists have found the remains of tree trunks fossilized in their original positions, and this has enabled them to reconstruct the ancient forests. The trees were widely spaced, about 10 to 16 feet (3 to 5 m) apart and held their foliage in a vertical rather than horizontal plane. This would have allowed them to catch the maximum sunlight that arrived at a very low angle. Among the trees present was a species that closely resembles the modern maidenhair tree (*Ginkgo biloba*). Late Cretaceous times (99 to 65 million years ago) saw a great diversification of flowering plants, and these made their way through the fragmenting supercontinent on migration routes that passed from South America, through Antarctica and South Africa, to Australia. Among these plants, which persist in all of these continents apart from Antarctica, was the southern beech (*Nothofagus* species) and members of the family Proteaceae. The migratory routes became disrupted as the supercontinent broke up, and by about 23 million years ago the Drake Passage had formed between Antarctica and South America. Antarctica was finally isolated. Over the past 100 million years Antarctica has been subjected to varying climate, and its vegetation has changed in response to this. It is clear, however, that the Antarctic tundra ecosystem has evolved only in the geologically recent past. The relatively recent history of tundra vegetation will be considered in detail later (see "Stratigraphy and Recent History of Tundra," pages 186–190).

◼ POLAR ROCKS

Piecing together the tectonic history of the continents has helped geologists understand the current pattern of land masses over the surface of the Earth, but the rocks themselves can provide further evidence about the formation and development of the tundra regions. The land masses that migrated over the face of the planet are composed of rocks of different ages and origins, and these supply information about the history of the region as well as giving clues about the current vegetation and landforms.

The rocks surrounding the Arctic Ocean, as shown in the diagram on page 36, are particularly ancient, dating back to a time beyond 542 million years, known as the Precambrian. The Precambrian was once thought to represent the early stage in the Earth's history when life did not exist. Microscopic fossils do occur in these rocks, however, and these have enabled geologists to divide the Precambrian into three eons, the Hadean (before 3,800 million years old), the Archean (from 3,800 to 2,500 million years ago), and the Proterozoic (from 2,500 to 542 million years ago). Some of the rocks of northern Canada date from the Archaean eon. The Archaean was a time of crustal development for the Earth, so these rocks represent the early solidification and spreading of the surface crust. During the Proterozoic this crust had begun to be modified by erosion, redeposition, and metamorphism.

The Precambrian rocks surrounding the Arctic Ocean are arranged in three distinct regions, called "shields:" the Canadian and Greenland Shield, the Baltic, or Fennoscandian, Shield, and the Russian Angara Shield. Of the three, the Canadian Shield is perhaps the most intensively studied. Here ancient Archaean rocks form a basal core, and around its edge lie Proterozoic rocks that have been modified by mountain-building events, the heating and cooling of rocks as they metamorphosed, and the erosion of materials, sometimes exposing the underlying Archean layers. These Proterozoic rocks often consist of gneiss (pronounced "nice") that has been so strongly modified that it is impossible to determine the nature of the original rock. The shield continues through Greenland, but there the ice sheet makes detailed geological survey extremely difficult. Along the coast of Greenland, however, the ancient rocks are exposed, and in the southwest of Greenland are some of the world's oldest rocks, dating back 3,800 million years. The shield rocks are generally poor in the elements required for plant growth and are acidic in their reaction, so they produce relatively poor soils. The fact that they are constantly being shattered and ground down by ice action, however, means that their meager content of elements is constantly being released into the soil.

The following period, the Cambrian, lasted from 542 million years ago until 488 million years before the present. At that time the Canadian Shield of Precambrian rocks formed the core of the continent, and what is now the Northwest Territories of Canada lay beneath the waves. Sediments that were later to form Cambrian sandstone, limestone, and shale accumulated in these seas, and they now form the basal bedrock of the tundra in northwestern Canada. Many of these ancient rocks, however, are hidden beneath more recent deposits.

Antarctica also lies on a shield of Precambrian rocks, but, like Greenland, most of its central region is so deeply covered by ice that its geology is difficult to survey. This shield links up with similar Precambrian rocks in Australia, southern Africa, Madagascar, and India, all part of the Gondwana supercontinent. Some parts of the Antarctic Shield have become metamorphosed as a result of tectonic activity. For example, the Antarctic Peninsula that leads

northward toward South America was formed by crustal plates colliding in Mesozoic times (250 to 65 million years ago). The Transantarctic Mountains, running along the eastern edge of the Ross Sea, are composed of a light yellow-colored sandstone, called the Beacon Sandstone, of Devonian to Jurassic age, overlying Cambrian limestone. All of these beds of sedimentary rocks are fossil-bearing. Rocks dating from the Cretaceous period and from the later Cenozoic era are present in Antarctica and have provided evidence of the forests that once occupied this continent (see "Tectonic History of the Tundra Lands," pages 33–35).

The lands now occupied by polar tundra, therefore, are composed of a wide range of different rock types that vary both in their age, their fossil content, and their chemistry. It is their chemical constitution that is most influential in affecting the nature of soils and vegetation in the tundra (see "Geology and Rock Weathering," pages 52–54).

■ ALPINE ROCKS

The geological process of mountain building is called *orogeny*. As has been described (see "Plate Tectonics," pages 29–33), many mountains owe their creation to the movements and collisions of crustal plates. The physical habitats that would later become occupied by alpine tundra thus came into existence.

In the Devonian period (416 to 359 million years ago) the North American and the European plates were in collision, and this led to an uplift of rocks in both regions. On the North American plate it resulted in the formation of the Appalachian Mountains of eastern North America, and on the European plate it created the long ridge of Scandinavian mountains and the Caledonian Mountains of Scotland. In North America this mountain-building period is called the Acadian Orogeny, and in Europe the Caledonian Orogeny. This was one of the continental collisions that eventually led to the formation of the supercontinent of Pangaea.

The location of Precambrian shields in the Northern Hemisphere. These ancient rock platforms are a common component of the geology of the Arctic.

From Jurassic times (200 million years ago) the Pacific plate increased its pressure on the western sides of North and South America, creating a strong subduction zone and generating volcanic activity and folding that gave rise to the western mountain chains of the Cascades, the Sierra Nevada, the Rocky Mountains, and the Andes of South America. During the Miocene Epoch (23 to 5 million years ago) the relative motions of the Pacific and North American plates changed from one of direct collision to a sliding sideways motion. This created the highly faulted and earthquake-prone conditions that now characterize California. In the last 3 million years the Sierra Nevada have risen about 6,500 feet (2,000 m), and they are still rising. In 1872 near the town of Lone Pine, California, a strong earthquake created cliffs of more than 20 feet (6.5 m) in height, showing that the process of uplift in these mountains is still taking place.

The Miocene epoch was also a time of particularly active mountain-building in Europe and Asia. At that time the African plate was crushing against Europe, and the outcome was the Alpine Orogeny, which resulted in the buckling of the rocks in what now form northern Italy, Switzerland, France, and Austria, creating the Alps. The slow but crushing impact of this orogeny was felt as far north as southern England, where a massive dome of chalk was forced upward over what is now the English Channel between England and France. Subsequent erosion has removed most of this mountain mass but has left remnants in the form of the chalk hills called the Downs, together with the white cliffs of Dover. The Himalayan Orogeny also took place at this time, resulting from the collision of India with southern Asia.

The creation of mountains by volcanic action and by crushing and folding as a result of plate collision means that the rock types found in alpine regions are extremely variable. Some mountain ranges consist largely of igneous rocks resulting from past volcanic activity, while others are formed from metamorphic rocks that were modified because of their proximity to volcanic processes and the degree of heat they suffered as a consequence. Some mountains, however, may simply consist of sedimentary rocks that have been twisted, contorted, and thrust upward by the intense pressure of one continental plate upon another. The varied geology of mountains results in a wide variety of soil types formed by the decay and erosion of the rocks, and the nature of the rock underlying alpine tundra thus has a strong impact on the type of vegetation and animal life that is able to survive there.

Whatever rocks underlie the tundra, however, the severe cold of the climate has a profound effect on the development of landscapes, soils, and vegetation. The landscape structure of the tundra may be founded upon rocks, but it is shaped by snow and ice.

■ SNOW AND ICE

Snow forms as water droplets freeze in the atmosphere, assuming a wide range of complex crystalline forms. In the polar tundra precipitation is extremely low, while in alpine

tundra it is more variable. At very high altitude the air may be so cold that it can hold very little water vapor, and precipitation becomes scarce. In both situations, however, the low temperature results in much of the precipitation falling as snow, and slow rates of melting and evaporation frequently lead to its accumulation on the ground. Wind can lead to drifting and redistribution of snow so that it is stripped from exposed ridges and accumulates in hollows and valleys. When snow accumulates faster than it melts or is eroded, then it can persist from year to year, building up and eventually turning to ice.

Snow crystals, with their complex and intricate forms, become altered within a matter of days after they settle. They may fragment into pieces, they may melt in the sun, or they may become compacted as additional snow accumulated from above. Compaction results in their becoming denser and harder, forming sugarlike grains that are crushed together. As they are crushed, many of the air spaces between them are eliminated, although some bubbles of gas usually

An Austrian glacier, flowing slowly down a valley from its mountain source. As it moves it carves out a round-bottomed valley. *(Peter D. Moore)*

remain, forming small trapped fragments of atmosphere within a solidifying matrix of packed snow. If the temperature remains low enough to permit their survival, the snowflakes will have changed completely over the course of two years to produce what geologists call *firn,* or "old snow." Within another three to four years the firn will become further compacted to ice. Snow that persists through the summer in the tundra landscape is called a *snow patch.* If it persists through many summers and assumes a permanent status in the landscape, then it may form the beginnings of a *glacier.* The illustration on page 37 shows a glacier in the European Alps, in Austria.

Accumulating snow is rarely pure in its composition. If the sea lies nearby, then it may contain many dissolved elements and compounds, such as sodium chloride and magnesium chloride. Dust from wind-blown soils and eroding rocks, fragments of plant tissue, pollen grains, and air-borne dust from volcanic eruptions may all find their way into the growing bank of snow. All of these fragments and dust particles are valuable to geologists who study past environments because the snow accumulates in a series of layers that correspond to the time they were laid down. Deep layers in the snow or ice represent older periods of time. The content of the snow can thus provide forensic evidence of past conditions: Gas bubbles indicate the nature and composition of past atmospheres, and plant fragments tell of former vegetation in the region. Volcanic dust particles can prove particularly important, as they can be used to date particular horizons in the deposit. Volcanic dust is clearly recognizable under a microscope because it consists of small rounded and lobed fragments of glass called *tephra.* The chemical composition of the tephra particles can be determined by microchemical techniques, and this can assist in the identification of the source volcano. Each volcano has its own chemical signature, and sometimes it is even possible to determine the precise eruption that has given rise to a particular tephra sample. Gradually, geologists are building up a database of eruptions, their chronology, and their chemistry, which means that layers of tephra in ice or in other sediments can be dated with accuracy. In this way tephra provides a time framework into which all other pieces of forensic evidence, including atmospheric composition and fossil plant material, can be placed. The technique of *tephrochronology* is now an important tool in geological studies.

Nunataks

Nunataks are mountain peaks that pierce the even surface of ice in a glacier or ice sheet. The word is derived from the Inuit language, in which it means simply mountain top, but it is used by geologists solely for the mountains tops that form lonely projections above the ice. Their bleak location means that they are subject to intense erosion as the forces of frost and wind operate on their exposed rock surfaces. This weathering of the rock leaves its mark in the form of a *trimline,* a sharp boundary between the exposed part of the nunatak and the lower parts of the mountain that are buried by ice. If climate changes and the ice retreats from the nunatak, then the former depth of the ice can be determined from the position of this trimline. Geologists can use this feature to study the extent and the volume of former glaciers and ice sheets.

Nunataks have long been a source of controversy for biogeographers because some scientists have regarded them as possible refuges for plant and animal life during an ice age. According to this hypothesis living organisms find favorable locations on a nunatak where they escape the ice and survive, possibly for thousands of years, before spreading out from their refuge when warmer conditions cause the ice to retreat. The alternative explanation for the reappearance of life after an ice age is that the plants and animals withdraw as the ice expands and move to warmer climates for the duration of the ice age, spreading back into former habitats when conditions improve for them. Most biogeographers now favor the latter explanation for the recovery of species after an ice age. Nunataks are extremely severe environments, and while some species of tundra plants and animals might be able to survive in such locations, it is very unlikely that any warmth-loving species could cope. The nunatak hypothesis for perglacial survival of organisms has now largely been abandoned.

ICE SHEETS, ICE SHELVES, AND SEA ICE

Ice accumulates as a result of the continued snowfall, but it is also lost by melting, runoff, erosion, and evaporation. The loss of ice is called *ablation.* If snowfall exceeds the combined ablation losses, then the ice mass grows, but if ablation equals or exceeds snowfall, then the ice mass remains static or shrinks. Both situations occur in nature, depending upon factors such as slope, aspect, and climate. In situations where ice builds up to form a permanent feature of the landscape, it may take a range of different forms, the most important of which are ice sheets, ice shelves, and glaciers.

An ice sheet is an extremely extensive area of permanent ice. The most commonly accepted definition is that it must occupy at least 19,300 square miles (50,000 km²). There are currently only two expanses of ice on the planet that merit the term ice sheet, one in the Northern Hemisphere, lying over Greenland, and the other in the Southern Hemisphere, covering most of Antarctica. The Greenland ice sheet is more than 650,000 square miles (1,683,000 km²) in extent. The weight of its ice load is so great that it has depressed the Earth's crust, forming a hollow basin 1,000 feet (300 m) below sea level at its center. The Antarctic ice sheet is even bigger, measuring 5.5 million square miles (14 million km²) in extent, making it twice the size of Australia and bigger than the whole of Canada. About 85 percent of the world's freshwater is locked up in the Antarctic ice sheet, which in places achieves a depth of 2.5 miles (4 km). As in the case of the Greenland ice sheet, this massive weight of ice has depressed much of the land surface of Antarctica below sea level, pressing the Earth's crust deeper into the fluid layer of mantle beneath until the weight from above achieves equilibrium with the pressure from below. This equilibrium is called *isostasy*. Changes in the ice loading at different points of the land surface, as when the Earth enters an ice age, cause isostatic adjustments, leading to alterations in the relative land-sea level (see sidebar "Glaciation and Sea Level Change," page 183).

Both the Greenland and the Antarctic ice sheets are interrupted by mountain peaks that project above the ice surface. These are called *nunataks* (see sidebar on page 38).

Cross section of Antarctica from west to east through the Ross Ice Shelf. There are two major sections of ice, both resting upon rock, much of which lies below the level of the Southern Ocean. The ice shelf itself covers a marine bay bounded on its eastern edge by the Transantarctic Mountains. Compare this section with the map on page 4. Note the exaggeration of the vertical scale.

The main nunataks in Antarctica are on the western side, associated with the Transantarctic Mountains, but there are also some in the coastal regions on the eastern side of the continent. In Greenland the nunataks are largely confined to the coastal region, for example Mont Forel, which reaches 11,000 feet (3360 m) near the eastern coast.

Ice sheets have not always been present on the Earth (see "Glacial History of the Earth," pages 176–178) but have periodically appeared, spread, then shrunk and disappeared once more. An ice sheet grows when the supply of precipitation exceeds ablation, and this may occur when the climate becomes colder (reducing ablation rates) or when the supply of snow increases. This is said to provide the ice sheet with a positive *mass balance*, leading to increased ice volume. There are three main theories concerning the growth of ice sheets. In mountainous areas with high precipitation the mountains may form a focus for a buildup of ice that subsequently spreads out from its place of origin into surrounding regions, including lowlands. Alternatively, ice may suddenly begin to accumulate simultaneously over a wide area because of climate change, leading to increased general snowfall or slower snowmelt. In this case the increasing albedo (see sidebar "Albedo," page 8) accelerates the process, forming a positive feedback to the system, and widespread ice accumulation results. This process is called the "snowblitz" theory. A third possibility is that ice forms over the sea and then invades the land, forming an ice sheet. It is quite possible that all of these different processes have been involved in the development of ice sheets in various places in the past.

Where modern ice sheets make contact with the sea they often extend out to form floating expanses of permanent ice. These features are called *ice shelves*. Antarctica has three major ice shelves, the two largest, the Ross Ice Shelf and the Ronne Ice Shelf, being on the western side. The Ronne Ice Shelf lies in a bay of the Weddell Sea, partially enclosed by the Antarctic Peninsula. The Amery Ice Shelf is relatively small in comparison to the other two major

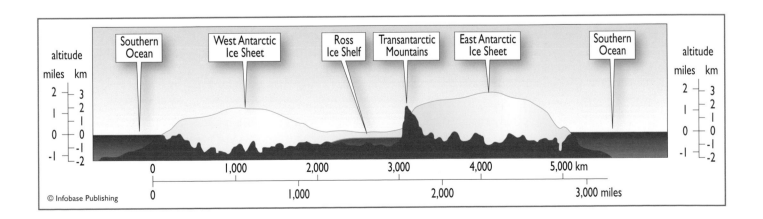

© Infobase Publishing

Antarctic ice shelves and lies in a bay on the eastern side of the continent.

The Ross Ice Shelf, as shown in the diagram on page 39, lies in the major inlet of the Ross Sea, bordered by the Transantarctic Mountains and opening into that part of the Southern Ocean that abuts the Pacific Ocean. It covers 190,000 square miles (500,000 km²), which is about twice the area of France, and it floats upon the Ross Sea. Where the ice shelf ends and gives way to the open ocean it forms a massive ice cliff, varying in height between 30 and 200 feet (10 to 60 m). The ice cliff runs for 500 miles (800 km) and at its greatest extent is 620 miles (1,000 km) from the shore. The shelf of ice is being fed by the continental ice sheet of Antarctica, so it is constantly moving seaward at a rate of about 0.7 miles (1 km) per year. At its seaward edge, however, it is collapsing to form floating icebergs, a process called *calving*. These icebergs can be enormous. One was recorded in March 2000 with an area of 4,250 square miles (11,000 km²), measuring 183 by 23 miles (295 by 37 km).

Large icebergs of this type are usually of the *tabular* form. Tabular bergs are flat-topped and are essentially blocks of the floating ice sheet that have become detached. The ice from which they were originally formed came from the land-based ice sheet, so they consist of frozen freshwater. This means that they have a relatively low density, but their great mass still causes them to lie low in the water. It is indeed true that the bulk of an iceberg (about 90 percent) lies below the surface. Hydrologists have calculated that a large tabular iceberg can weigh hundreds of millions of tons, enough to provide a city of a million people with freshwater for about three years. Icebergs can be extremely attractive, especially those with a deep blue color, which is characteristic of ice derived from great depth. The compression of the ice over many thousands of years, resulting from the continuous accumulation of more ice from above, causes any remaining gas bubbles to be crushed, and the outcome is a very clear blue color in the ice. Despite their beauty, however, icebergs are extremely dangerous to shipping, mainly because so much of their bulk lies out of sight below the water. Perhaps the most famous disaster involving icebergs is the wreck of the *Titanic* on its maiden voyage from Southampton, England, to New York in 1912, when it struck an iceberg and sank with the loss of more than 1,500 lives.

Much of the Arctic Ocean is also occupied by a permanent ice shelf, attached to the land in northern Greenland, Ellesmere Island, and some of the most northern Queen Elizabeth Islands. As in the case of the Antarctic ice shelves, the Arctic ice is constantly calving icebergs around its edge.

In both the Arctic and the Antarctic the winter brings an extension of sea ice around coasts and ice shelves. As conditions become colder and the days shorter, thin plates of ice begin to form on the surface of the sea. Ice crystals develop on the calm water and extend into thin sheets with sufficient strength to support snow that falls on top. Even when the ice layer is four inches (10 cm) thick, it is still easily broken by the motion of the sea and is split into rounded leaflike plates resembling water lily leaves, sometimes called "pancake ice." Pancakes gradually fuse together, forming ice floes, which thicken from below as new ice is added from the sea. Unlike the ice of ice shelves, this sea ice is not purely fresh but is about 0.6 percent salt, just 10 percent of the salt concentration of seawater. As the seawater freezes, some of the salt is retained in the ice in the form of tiny packets of brine. Over the course of time these bubbles of brine move downward through the ice under the influence of gravity, leaving the upper layers of one-year-old sea ice fresh enough to drink, with a salt concentration of only about 0.1 percent. This is an important source of freshwater for Inuit hunters on the ice.

Sea ice in the Arctic may last for up to eight years, while Antarctic sea ice rarely lasts more than a year. Researchers have found that up to 90 percent of Arctic sea ice is older than one year, while only 10 percent of Antarctic sea ice lasts beyond one year. Older sea ice is thicker and stronger than young ice, so ships have difficulty making a passage through it. In the Antarctic approximately 7,700,000 square miles (20 million km²) of sea ice form during the winter, more than doubling the total area of the continent. In the Arctic the winter sea ice is somewhat less extensive, covering an area of 5,400,000 square miles (14 million km²). Some sea ice survives through the summer in both of the polar regions, amounting to 1,500,000 square miles (4 million km²) in the Antarctic and 2,700,000 square miles (7 million km²) in the Arctic. Sea ice is an important feature of polar tundra areas for many of the animals that inhabit these regions.

ICE CAPS AND GLACIERS

Isolated domes of ice smaller than 19,300 square miles (50,000 km²), the required size for the term *ice sheet* to be applied, are called *ice caps*. Ice caps may cover only a few square miles, but like ice sheets they completely cover the underlying rock with a deep and permanent layer of ice. Nunataks may pierce their surface, breaking up the dome of ice. They are common features of polar regions, particularly on plateau areas at elevated altitude. Some of the Arctic islands, such as Svalbard, for example, have their own ice caps. The Svalbard ice cap, although not large enough to be termed an ice sheet, is approximately the size of Connecticut, so ice caps are very substantial features of the tundra landscape. Iceland also has its own ice cap.

As in the case of an ice sheet, the accumulation of snow on the top of an ice cap causes pressure and forces the accumulating ice to move outward from its source. Lobes of ice

The Matterhorn in Switzerland has been carved by glacial erosion, stripping its various faces of rock and leaving a pyramidal structure. *(Peter D. Moore)*

move out from the center of the ice cap, working their way downslope through valleys. Where these tongues of ice meet the sea, they may extend out to form areas of floating ice that fragment to create icebergs.

Ice caps are rare outside the polar regions and are replaced in alpine tundra by *glaciers*. These are essentially miniature ice caps that are localized in their occurrence and development. They may cover several square miles of landscape, especially when they occupy high-altitude plateaus, or they may extend for tens of miles in a strip running down a valley with mountain peaks on either side. Some of the world's longest glaciers occur in Alaska, such as the Bering Glacier that has a length of 125 miles (200 km). By contrast, the glaciers in the European Alps are relatively small, the longest being the Grosser Aletschgletscher in Switzerland with a length of 14.3 miles (23 km). In situations where a glacier is confined to a single valley, its upper limit may be formed by a half circle of steep mountains enclosing the head of ice. This

is called a *cirque*, or *corrie*, and is a typical feature of glaciated mountain terrain. If glaciers develop on different sides of a mountain peak, erosion can lead to an isolated, pyramidal peak, as in the case of the very distinctive and famous Matterhorn in Switzerland (see photograph above).

As the ice moves down a valley it grinds against the rocks at its base and along its sides, eroding the material and creating its own smooth channel through the mountains. The floor of the valley becomes rounded, while the sides of the valley become steeper as the ice crushes the lateral slopes. If the climate changes and becomes warmer, these features remain long after the glacier has disappeared. The valley left behind is called a *U-shaped valley* because of its form in cross section, as shown in the photograph on page 42.

Sometimes glaciers run down parallel valleys, gradually enlarging their channels and eroding the ridge of mountain that separates them. Eventually the ridge becomes sharpened by the grinding activity of the glaciers on each side, just as a knife is sharpened by a grindstone, and the ridge that remains is called an *arête*.

When the valleys through which glaciers descend eventually open out onto flatter land, the glacier also spreads out over the landscape, forming a *piedmont glacier*.

The Melaspina Glacier in Alaska is a fine example of this type. If, on the other hand, the glaciated valley meets the sea, then, like ice sheets and ice caps, the glacier extends out into the water and fragments under the action of waves, calving icebergs.

Glaciers in temperate mountains may spend part of the year in temperatures above the freezing point of water. During such episodes ice melt takes place, especially at the lower end of the glacier. Often such glaciers, known as *warm glaciers,* develop a *snout,* or a hole at the lower end, through which the meltwater discharges into streams that flow from the glacier. Beneath the glacier a network of streams and water pockets becomes established that feeds the emergent meltwater stream. Glaciers in colder climates, where the temperature rarely rises above freezing point, are not subjected to these periods of excessive ablation. The temperature in the upper part of the ice mass may vary a little with seasonal factors, but melting does not take place. At the base of the ice in these *cold glaciers,* however, where the ice meets the bedrock, some melting may take place. Insulated from the cold air by the mass of ice above, the geothermal heat from rocks below finds its way

to the basal layers of ice. The intense pressure of ice also has the effect of lowering its melting point, so some melting does take place, and pockets or streams of water may be found deep beneath the ice. Such water pockets have even been found beneath several miles of ice in Antarctica. If these basal water bodies occur under thinner layers of ice, then light from the Sun may penetrate the ice layer and heat the darker rocks below. Some lakes in Antarctica that are permanently buried below ice have had recorded temperatures as high as 75°F (25°C) as a result of this form of solar heating. Warm glaciers are most typical of mountain regions in low latitudes, whereas cold glaciers are mainly restricted to polar regions and high-latitude mountains. Warm glaciers are found in many alpine areas of the world, such as New Zealand, the Himalaya, the European Alps,

---

The rounded presence of a U-shaped valley in a mountain area, such as this site in Austria, provides evidence of past glacial activity. The glacier has now gone, but the outcome of its erosive power is revealed by the shape of the valley.
(Peter D. Moore)

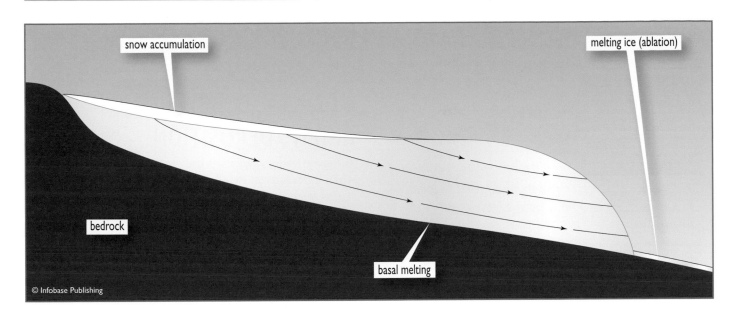

Labels on the diagram: snow accumulation, melting ice (ablation), bedrock, basal melting

© Infobase Publishing

Profile through the long axis of a valley glacier. Snow accumulates on the surface and is gradually compressed into ice, which gradually slips downhill, eventually releasing water at its snout. Some melting of ice takes place at the junction between the glacier and the warmer underlying rock, and this helps to lubricate the motion of the glacier.

the Peruvian Andes, and some of the high mountains close to the equator in East Africa.

Glaciers are mobile masses of ice. They lie upon slopes, and the ice they contain is in constant movement in response to the force of gravity. Any molecule of water landing upon the surface of the glacier (usually in a solid, snowflake form) does not remain static in its position of landing. Even when buried by additional molecules, it does not retain its location in relation to the surrounding landscape but begins to move laterally even while being buried from above. The progress of individual water molecules in a glacier is shown in the diagram. As each molecule is buried and enters deeper layers of ice, it moves downslope toward the snout of the glacier. The speed of this process varies with a number of factors, such as the rate of precipitation, which adds additional weight to the ice mass, and the degree of slope, which accelerates ice movement. Generally, the movement of a glacier is too slow to be detected by eye. A velocity of three feet (1 m) per day is fast for a glacier, but occasionally glaciers increase their speed in what is known as a *glacial surge*. In 1986 the Hubbard Glacier in southern Alaska suddenly increased its rate of movement to about 33 feet (10 m) per day, advancing across a bay and crushing trees on an island before it. The glacier finally came to rest when it collided with the far side of the bay, where it blocked off a large lake on its landward side. It seems to have been caused by a sudden increase in the supply of ice from one of its tributary glaciers as a result of exceptionally high levels of precipitation. The surging of glaciers is not common, but it is known to have taken place in Iceland, Greenland, Alaska, and several locations in Russia. The Medvezhiy Glacier of the Russian Pamirs surges in a cyclic pattern approximately every 19 years. It is possible that all surging glaciers run in cycles, but if so the cycle for most is much longer than this, so prediction is difficult. Glacial surges are very unusual in mountain glaciers away from the Arctic and subarctic, but they have been recorded in the South American Andes.

Increases in precipitation are one cause of glacial surges, increasing the mass of the ice as the snow accumulates. Earthquakes may cause landslips or avalanches that stimulate surges, but most surges are likely to be caused by conditions at the base of the ice. Changes in the production of meltwater beneath the ice can add to the lubrication of the ice mass as it slips over the bedrock, so excessive water production or the deflection of a subglacial stream can destabilize a glacier and send it into a surge. One of the first signs of a surge is the thickening of the ice mass at the head of the glacier, followed by the appearance of masses of crevasses (see sidebar on page 44) in its upper parts. The surge then advances downslope, eventually reaching the snout, where the most spectacular movements and advances can be seen. The most extensive surge ever recorded was on the island of Svalbard, where a glacier rapidly moved forward 12.5 miles (20 km) along a 19 mile (30 km) front. Since the surge actually begins at the head of the glacier, it is possible that some surges never actually arrive at the snout but are absorbed by compression of the ice before they reach that far. Once the surge is completed, the steepness of the glacier slope is reduced, so it becomes more stable for a while. The cyclic nature of glacial surges may be related to a process of increasing thickness at the glacial head, leading to an unstable steep slope of ice, followed by released pressure during the surge, and then stability until thickness at the glacial head builds up once again.

■ GLACIAL DEBRIS AND ITS DEPOSITION

The ice of valley glaciers is not clean. It may begin that way when it lands as snowfall, but even snow contains contaminants such as dust particles and tephra (see "Snow and Ice," pages 37–38). Glacial ice accumulates even greater quantities of larger debris as it moves down a valley, as shown in the photograph on page 45. Moving ice has a very high capacity for abrasion, and the rock walls of a valley down which ice has moved become scarred and scratched by the forces it exerts. As it grinds along the valley sides, it cuts into the rock, bearing away fragments and causing collapses of overhanging rock from above the level of the scouring ice. The rock particles that fall onto the ice add to its abrasive capacity because they are carried along by the ice and may become deeply embedded within it, providing additional tools for the grinding activities of the glacier. The surface of a glacier is often strewn with assorted rock debris, giving it a dirty and desolate general appearance. Clearly, this deposition of debris is possible only if the glacier is surrounded by steep mountain slopes that provide a source of fragmented rocks. Ice sheets and ice caps are free from such influences except in the vicinity of nunataks, so they have clean surfaces and are relatively devoid of debris.

The size of the rock fragments carried by a glacier is variable. Collapses of rock faces can bring large rocks onto the ice surface, while the constant grinding of rock walls produces a very fine material, sometimes referred to as *rock flour*. If vegetation grows along the sides of a glacier, it, too, along with attached soil, may fall upon the ice surface and be carried down a valley. If a glacier passes a particularly unstable area of rock, then a constant stream of eroded material may fall upon the ice surface and will be carried along in the form of a strip of fallen rock and soil stretching from the point of origin along the length of the glacier. Such strips of eroded material are called *medial moraines*. Patches or strips of rock detritus on the surface of a glacier are very apparent against the generally white background, and their dark surfaces and high capacity to absorb solar energy (see "Albedo," page 8) lead to a local warming effect. This warming may be sufficient to melt some of the ice in the vicinity of the detritus and cause the rock fragments to sink into the ice mass. On the other hand, rocks have a lower specific heat than ice, so they lose heat faster in cold conditions and may then protect ice from melting, eventually standing on raised areas of ice that has survived in their shade and protection.

Crevasses

The surface of a glacier is rarely smooth. More often it is broken into sections by the development of deep splits in the ice called *crevasses*. These splits are generally much deeper than wide, penetrating far down into the mass of ice. They are particularly dangerous for mountaineers because they often become buried beneath snow, which can form unstable and treacherous bridges over the crevasse. Often the crevasse is thus totally hidden from sight, but many are open-topped, expanding and contracting with the slow movement of the ice. Crevasses usually form where the ice is under tension, for example, where the glacier crosses a change in slope of the bedrock or bends around a corner in a valley. When the glacier passes into a part of the valley that slopes more steeply, the speed of the ice increases, and crevasses tend to form in a transverse pattern along the lines of contour as the surface ice is placed under tension. Near the snout of the glacier, crevasses often align themselves longitudinally, following the line of flow of the ice and splaying out in a fan pattern if the glacier expands into less confined space. Sometimes, however, the crevasses cross one another and produce a complex pattern in which pillars of ice stand in isolation from one another. These ice pillars, known as *séracs,* are particularly unstable and are often found where there is a severe break of slope, creating a condition called an *ice fall*.

The deep parts of crevasses and the exit tunnels of glacial snouts often shine with an impressive blue light. The reason for this is that the ice selectively absorbs the red end of the light spectrum, allowing only the blue light to penetrate. Viewing sunlight through any thickness of ice therefore gives an impression of a vivid blue world. This selective absorption of light of different wavelengths is found in all water bodies, including the ocean, where again red light is absorbed and only blue light penetrates to deeper layers.

The Seas of Ice glacier in the French Alps carries a load of rock detritus on its surface as it flows downhill. This will be deposited in a series of moraines when the glacier eventually melts. *(Peter D. Moore)*

The effect of debris on ice depends upon the local climate, aspect, and the degree of shading of the glacier by surrounding slopes.

The concentration of debris on and in the ice generally increases toward the snout of the glacier as more material accumulates and the ablation rate of the ice increases. Near the snout, especially during summer when ablation is at its maximum, the sides of the glacier may be covered with layers of rock and finer detritus that slips down the ice slopes and accumulates at the base.

An observer can usually see only the debris that is carried along on the top of the ice, but more material lies out of sight below the surface. Some of the eroded rock that lands upon a glacier from the slopes of a cirque or the valley sides becomes buried by continued snowfall and ice accumulation. Some rocks may work their way down through the ice because of their absorption of sunlight, melting surrounding ice layers. Other material may fall into crevasses and

become incorporated into deeper ice. The ice forms a matrix in which many fragments and particles are transported. If a glacier enters the sea at the end of its downslope journey, then the icebergs calved into the ocean carry the load of surface and embedded detritus out to sea. Icebergs may carry their burden of rock far over the ocean and release it gradually as they enter warmer waters and slowly decay. Iceberg-transported materials found in ocean sediments have provided geologists with much information about former climate and oceanic conditions (see "The Pleistocene Glaciations," pages 178–183).

At the base of a glacier, where the ice comes into contact with the bedrock, yet more detritus accumulates. The rock that underlies a glacier may well have been subjected to the effects of a cold climate, including frequent freezing, before the glacier developed. As a consequence, much of the bedrock is shattered and weakened so it is easily scraped up by the moving ice. Blocks of rock incorporated into the basal ice scour their way along the bottom of the glacier and their movement, together with the great pressure as a result of ice above, makes them very efficient agents of further erosion of the bedrock. The grinding also affects the transported material, rounding and polishing the rocks carried at the base of the ice. The load of mobile rock debris at the base of a glacier

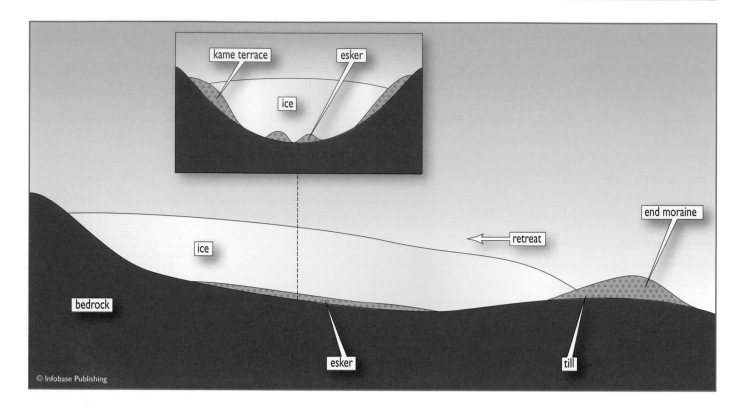

© Infobase Publishing

Cross section of a valley glacier along both its long and short axes. As the ice retreats it deposits a mass of detritus in the form of an end moraine. Rock debris accumulates along the side of the glacier, forming kame terraces, and the flow of meltwater beneath the glacier creates sinuous ridges of deposits called eskers.

is called the *glacial sole,* as in the sole of a shoe. Fine chippings of silt and clay produced in the constant rock scouring are also carried along beneath the glacier, and when they are eventually discharged in meltwater they give the water a pale milky appearance, called *glacial milk,* that can be detected in rivers for many miles downstream.

The accumulation of material beneath a glacier may become so thick and dense that it remains behind even when the glacier itself has vanished as a result of climate change. This deposit is called *lodgement till.* The word *till* refers to any unsorted material deposited by a glacier. It was once called *boulder clay* because it is a collection of rock fragments of many different sizes, ranging from tiny clay particles to massive boulders, all mixed in a random manner with no sign of layering. Geologists now prefer a modern term, *diamicton,* rather than till, but till remains the expression most frequently used for this material. The presence of till deposits in regions now far removed from glaciers provided some of the most important evidence when ideas concerning glacial history were first being proposed (see "The Pleistocene Glaciations,"

pages 178–183). Ancient till deposits become hardened and cemented in the course of time, and these fossil tills are termed *tillites.* These tillites have supplied evidence of former ice ages in the history of the Earth (see "Glacial History of the Earth," pages 176–178).

Glaciers, then, are constantly offloading some of their cargo of detritus, but deposition is greatest when a glacier is in retreat. Glacial advance and retreat is ultimately controlled by the mass balance of the glacier, and this in turn is strongly affected by climate. The mass balance of a glacier describes the relationship between ice accumulation, determined by the arrival of snow, and ablation, which is affected by a number of factors, a major one of which is temperature. If the climate becomes colder or the amount of precipitation increases, then ice accumulation also increases, and the glacier grows. In the case of a valley glacier, it extends down a valley, and ice envelops land that was formerly ice free. Glacial growth eventually slows and finally attains equilibrium with respect to the new climatic conditions as the lower end of the glacier experiences warmer temperatures, melting becomes faster, and the ablation rate increases. The mass balance of the glacier then achieves an equilibrium state. If the overall climate becomes warmer or precipitation diminishes, on the other hand, the reverse process operates. The ablation rate increases and eventually exceeds the input of new material from snow, resulting in a decrease in the total ice volume. The lower end of the glacier then begins to retreat up the valley, leaving in its wake the masses of rock, soil, and mineral sludge that it contained. Retreat continues until equilibrium with the new climatic conditions is attained once more.

The deposits of detritus around the lateral and terminal edges of a glacier are called *moraines*. Medial moraines are a special type of deposit still in the process of being transported on top of the ice, but most moraines are in the form of mounds and ridges of unsorted material left behind after the retreat of a glacier. When a glacier remains in a static, or equilibrium, state for a period and then retreats, it often leaves a very distinct ridge of rock detritus at the limit of its former extent, as shown in the diagram on page 46. This is called a *terminal moraine* or *end moraine*. The quantity of material left in this way may be sufficient to construct a massive rampart across a valley, sometimes achieving a height of 200 feet (60 m). An example of a terminal moraine in the Austrian Alps is shown in the picture below. A terminal moraine may obstruct the flow of water down the valley or out of a cirque, leading to the formation of a *glacial lake* behind it. Convict Lake in the Sierra Nevada of California is a fine example of such a glacial lake held in place by a terminal moraine. The positions of terminal moraines indicate where glaciers have come to

a standstill and then retreated in the past, so, if they can be dated accurately, they provide valuable evidence about past climate (see "The Pleistocene Glaciations," pages 178–183). At the former edges of a valley glacier ridges and terraces of detritus are left by the retreat of the ice. These are called *lateral moraines*. These are usually more difficult to detect than terminal moraines in valleys where the glacier has retreated because they are quite quickly eroded from steep valley sides or obscured by landslips from above.

One conspicuous feature of glaciated valleys is the presence of *hanging valleys* along their sides, as shown in the diagram on page 48. Tributary glaciers joining the main glacier from lateral valleys create their own U-shaped valleys, but the valley floors are often less deep than that of the main glacier valley. When the ice has gone, these tributary valleys end abruptly on the steep sides of the main valley, perched high above the main valley floor. The streams emerging

A glacier in retreat near Obergurgl, Austria, has left behind a terminal moraine of rock detritus, which is being eroded and sorted by the flow of water from the melting ice. *(Peter D. Moore)*

hanging valley

main valley floor

© Infobase Publishing

Small glaciers may form tributaries to a main glacier, and, on retreat, the junction is marked by a steep divide between the tributary and main valleys. The tributary valley (called a hanging valley) comes to an abrupt end, and its stream drops into the main valley, forming a waterfall.

from these hanging valleys then cascade over the lip of the tributary valley, forming high waterfalls as they join the main valley stream.

Lodgement till is left in the form of a blanket over the valley floor or the plain formerly occupied by a glacier, ice cap, or ice sheet. This type of till can cover whole landscapes, burying the bedrock beneath deep layers of mixed detritus that is sometimes called *drift*. On a valley floor the lodgement till may be confined and forced into movement by the flow of the ice, and this leads to some distinctive forms called *drumlins*. Drumlins, as shown in the diagram on page 49, are elongated, streamlined mounds composed largely of till but with a solid projection of bedrock at their upper end, in other words, the end that received the main force of the

ice flow. An accumulation of till has taken place beneath the glacier in the lee of the projecting rock, sheltered from the erosive power of the moving ice. Drumlins vary in size, but typically they can be about a mile (1.6 km) in length and 60 to 100 feet (18 to 30 m) in height. They rarely occur in isolation but are mainly found in groups called *drumlin fields*.

Tills often contain clues about the direction in which the ice moved in the past. Stones and rocks embedded in the matrix of silt and clay are often aligned according to the direction of flow of the ice. Geomorphologists, scientists who study the geological forms of landscapes, can use this type of evidence to trace the movements of former ice cover and can sometimes detect ice advances that have arrived from different directions at various times in the past. The rocks found within the till can also provide clues concerning its origin and direction of movement. Rocks may be transported over hundreds of miles by extensive ice sheets and then deposited in locations of very different geology. These displaced rocks are called *erratics,* and their presence indicates both the direction and the distance covered by the former ice sheet.

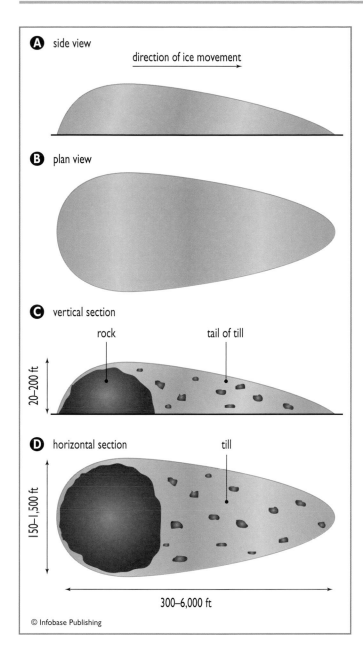

A side view

direction of ice movement

B plan view

C vertical section

rock   tail of till

20–200 ft

D horizontal section   till

150–1,500 ft

300–6,000 ft

© Infobase Publishing

The structure of a drumlin. Drumlins develop beneath glaciers and have a streamlined form created by the erosion of the ice as it flows, as shown in the side view (A) and the plan (B). In cross section, (C) and (D), it becomes evident that there is often a rock protruding at one end of the drumlin that has created a shelter in its wake for the accumulation of glacial debris, forming a long tail.

In relatively flat areas, where ice caps at the head of glaciers have finally decayed, the till they leave behind does not show any signs of the direction of flow. Here the till is deposited in a series of random mounds called *hummocky till.* Large blocks of ice called *dead ice* may sometimes become embedded in till and isolated from the main retreating glacier, especially in areas of relatively flat topography. These ice blocks gradually decay, leaving deep, steep-sided hol-

lows in the ground called *kettle holes.* As conditions become warmer the kettle holes often fill with water, forming small lakes within the till landscape. The pothole wetlands of the prairie regions of North America have developed from the infilling of kettle hole lakes.

Waterfalls, glacial lakes, and kettle hole lakes all illustrate the ways in which water becomes an important influence in transforming the landscape following glacial retreat, and many of the topographical features, the *geomorphology,* of formerly glaciated regions can be understood only by considering the influence of water.

■ SORTING BY WATER AND WIND

The melting of an ice sheet, ice cap, or glacier results in the release of large volumes of water. Even prior to glacial retreat, water drains from glaciers as part of the ablation process, and this water movement erodes materials over which and through which it travels, picking up and transporting suspended materials, taking them from their original locations, and depositing them elsewhere. The efficiency of water as a means of transporting rock fragments depends upon the speed of the water and the size of the particle. Geologists and soil scientists have devised a simple scheme for the classification of particles into groups according to their size, so terms such as gravel, sand, silt, and clay have precise scientific definitions (see sidebar on page 50).

Fast-moving waters occur where large volumes of water pass down steep gradients, and these have the capacity to move large rocks and boulders. Such water is highly energetic. Large rocks are moved mainly by being rolled along the bottom of a river or stream, and their movement will take place during periods of intense water flow, as in the spring melt period when the volume of water moving is at its greatest. The beds of glacial streams and rivers become littered with large rocks moved in this way.

Smaller stones, pebbles, rock fragments, and gravel can be moved by less energetic waters. Like larger rocks, they may roll along the bed of a stream, or they may bounce along the bottom, being periodically lifted by the motion of the water and then falling back to the stream bed. This type of motion is called *saltation.* It also occurs when wind blows across a sandy beach and causes sand grains to bounce over the ground in a series of leaps. Fine particles of rock flour, the silt and clay fraction of glacial debris, are small enough to remain suspended in the moving waters of a glacial stream and give the water its glacial milk appearance. These particles can be carried many miles, eventually settling when the water movement becomes much slower or less energetic, often in lakes farther down the course of the river.

Particle Size

Soil scientists and geologists have devised a scheme for classifying different particles of rock according to their size. In this way they can provide a precise definition for terms that could otherwise be used very loosely, such as *sand* and *clay*. Unfortunately, scientists in different parts of the world have failed to agree upon a universally accepted set of definitions, but within the United States there is some measure of agreement. A rock fragment with a diameter greater than 10 inches (25.4 cm) is called a *boulder,* while smaller particles that exceed 2.5 inches (6.4 cm) are *cobbles.* Any particle greater than 0.08 inches (0.2 cm) is called *gravel. Sand* consists of particles between 0.0008 and 0.08 inches (0.002 to 2 cm) in diameter, while *silt* is still smaller, having a diameter of 0.00008 to 0.0008 inches (0.0002 to 0.002 cm). *Clay* is the smallest of all particles, having a diameter of less than 0.00008 inches (0.0002 cm).

Clay is distinctive not only as a consequence of its small size but also because of its chemical structure. The tiny particles of clay are composed of chemical sandwiches in which layers of silicon oxides alternate with layers of aluminum oxides. The outcome of this complex crystalline structure is that the surface of the clay particle is negatively charged, and when clay particles are suspended in water the effect of these negative charges is to repel any neighboring clay particle. Clay is thus suspended in water both by its small size, which makes sedimentation slower, and also by the repulsion between the fine particles. Clay remains in suspension in water for considerably longer than all other particles and eventually settles only under very calm conditions.

The transport of materials by water is therefore linked to its energy content, and glacial streams are often at their most energetic close to the melting glacier. The area below a melting glacier, termed the glacial foreland, is often covered by a series of parallel fast-flowing streams, called braided streams, gradually converging into fewer and larger streams and eventually a single river. Large rocks and boulders do not travel very far along the stream courses, while progressively smaller rock fragments and particles travel farther. Where pools form the water loses some of its energy, and smaller particles are sedimented. The movement of water thus sorts the different particle sizes, and geologists can detect the effects of this sorting when they examine the layered, or *stratified,* deposits that are left behind. The processes of water sorting, transporting, and redeposition of rock debris that occur in and around glaciers are given the name *glaciofluvial* processes resulting in glaciofluvial landforms.

Water sorting takes place within and around an active glacier, even before glacial retreat produces excess masses of water. Wherever summer meltwater flows, glaciofluvial processes occur. Some of the ice along the side of a glacier may melt in summer because it is in contact with dark-colored rocks that absorb the sunlight and are warmed by it. Water flowing from the edge of the glacier transports lateral debris slipping down the steep icy sides and sorts it into ridges and mounds of different sized particles. The ridges of sorted material found along the edge of valley glaciers are called *kame terraces.* If a glacier melts in a series of stages, then several lines of kame terraces may develop along the valley sides, as shown in the diagram on page 46. These leave a record of the stages of glacial retreat. Kame deltas may form at the front of a glacier, where summer meltwater streams sort the detritus that falls in front of the ice. Kames can also form within crevasses deep inside the glacier, but these are often eroded and distorted by the eventual decay of the ice, so they are difficult to detect when the ice has gone.

Water flowing beneath the ice creates erosion channels and ridges of sorted and redeposited lodgement till in tunnels below the ice mass. The subglacial streams coalesce to emerge from the glacier at its snout, which is the lowest point of the glacier ice, as shown in the photograph on page 51. When the ice eventually decays, the effects of these subglacial streams can be seen in the form of elongated, winding ridges composed mainly of sorted sand and gravel called *eskers,* which are shown in the diagram on page 46. The word is derived from a Celtic word, *eiscir,* meaning ridge. The esker ridges can be as high as 100 feet (30 m) and may extend over several miles, either as complete ridges or broken into sinuous sections. They can be recognized as glaciofluvial features because the detritus from which they are formed is sorted and stratified into layers of different particle size, varying with the pattern of water flow. One surprising feature of eskers is that they sometimes lead upslope as well as down a valley. This is best explained by considering the hydrostatic pressure of water in tunnels below the ice. The force of water moving along the basal weaknesses in the pressurized ice would have been sufficient to drive the subglacial streams uphill at times, taking the load of sediments with it. Eskers can be confused with terminal moraines, but they differ in having their sediments sorted, whereas terminal moraines are composed of unsorted debris. Eskers run along the line of flow of the ice, whereas terminal moraines are formed at right angles to the ice front.

The retreating glacier leaves behind an outwash plain of glacial foreland split into strips by the braided streams.

The sediments, apart from larger boulders and rocks, consist mainly of sands and gravels because the fine particles of silt and clay have been carried away, suspended in the stream waters. These fine particles eventually settle in glacial lakes or in lake ecosystems farther down valley, and they sediment in a distinctly stratified pattern. Working from boats or from the solid winter ice cover of a glacial lake, geologists use boring equipment to obtain deep cores of the bottom sediments, and the silts and clays that form these sediments are found to consist of a series of distinct but very thin layers called *varves*. Careful microscopic analysis reveals that the thin lines are caused by an alternation of very fine clay particles with coarser silt. Each pair of lines represents a year's deposition, with the coarser material representing the summer influx of sediment borne by the spring and summer meltwater, and the fine material accumulating in winter when the lake is ice covered and very little water is flowing in. Cores of sediment from varved lakes are extremely useful to scientists studying past conditions, partly because their annual accumulation pattern provides a simple means of attaching a date to any particular layer.

Water is thus an important medium for the transport and sorting of material left behind by a retreating glacier, but wind can also play a part in the movement of rock detritus. The drying masses of rock debris left behind by a retreating glacier are subject to wind erosion, especially in an open, bleak landscape where no vegetation is present to break the force of the wind or to bind the soil particles together. High winds may move sand particles by saltation, but silt particles (see sidebar "Particle Size," page 50) can become suspended in fast-moving air and carried over considerable distances. Redistributed silty particles of this type are called *loess*. Soils derived from wind-transported loess deposits are surprisingly common over the Earth's surface, covering about 10 percent of continental land areas. The largest single area is in central China, where the loess deposits often exceed 1,000 feet (300 m) in thickness and have mainly accumulated during the past 2.5 million years, when glaciation has been particularly prevalent. Much of eastern Europe and Ukraine have extensive loess deposits, as do the Great Plains

At its lower end a glacier forms a "snout," as in the case of this Norwegian example. Some glacial ice melts at the base, where it is in contact with the relatively warm rock, resulting in subglacial streams that emerge at the snout. *(Peter D. Moore)*

of North America. Loess can be more than 100 feet (30 m) deep in parts of Kansas. In South America, Argentina and Uruguay are rich in loess soils. The silt particles of loess are often rich in lime and calcium carbonate, and they produce relatively rich agricultural soils.

Although wind transport of sand is less extensive than that of loess silts, some parts of the world have considerable deposits of sand derived from the erosion of glacial deposits. Northern Europe, especially the Netherlands, France, Belgium, and Germany, are rich in these *coversands,* and in Nebraska these deposits have formed dune fields that occupy an area of more than 20,000 square miles (50,000 km²).

The detritus carried along below, above, and within ice sheets and glaciers is thus eventually disgorged upon the ground, where it is sifted and sorted by water and by wind. The outcome is a layer of fragmented particles of rock ranging in size from boulders to clay, and it is from this raw material that the tundra soils are formed.

■ GEOLOGY AND ROCK WEATHERING

Retreating glaciers leave sediments of rock debris in their wake, but this layer of mineral fragments does not constitute a true soil. It contains three of the important components of soil, namely mineral particles, water, and an atmosphere, but it lacks the additional components that bring a soil to life. Living organisms, including plant roots, invertebrate animals, algae, fungi, and bacteria are all needed to complete the development of a true soil. Fortunately, all of these additives are available in the vicinity of a glacier, so the development of soils can begin as soon as the effects of water and wind have abated and the sediment can stabilize.

The mineral component of the soil, which consists of the various rock fragments, forms the main soil skeleton. Bedrock is constantly being degraded into smaller fragments, as shown in the photograph of a soil profile on this page. The chemical constitution of the particles depends upon the rocks from which they were derived, and this has a lasting effect upon the nature of the soil that develops. The study of tundra soils is complicated by the fact that their constitution and chemistry is not necessarily related to the underlying bedrock. The mineral components of the soil may have been derived from rocks many hundreds of miles away, where a glacier carved them from their original location. The soil particles have been crushed, ground, and fragmented during their transport, and they continue to break up even when they have come to rest. This decay of rock to produce ever smaller components is called *weathering.* Weathering takes place as a result of the activity of physical, chemical, and biological factors.

The profile of a soil developing over a bedrock of chalk. The disintegrating rock at the base is overlain by smaller rock fragments mixed with organic matter derived from the surface vegetation. *(Peter D. Moore)*

Physical weathering in tundra soils is particularly important. The constant freezing and thawing of water within the soil place considerable stresses on the mineral fragments. Water penetrates into the minute cracks and weakened layers of the particles, and when it freezes it expands, increasing its volume by about 10 percent and splitting the rock into smaller pieces. The heating of rocks in the sunlight can also lead to their physical breakdown. As the dark rocks absorb the light their temperatures rise, and this can cause cracking and breakage along planes of weakness. Rocks consist of mixtures of crystalline minerals, and these become separated from one another during the weathering process. Granite, for example, is broken down into three main minerals, feldspar, mica, and quartz. During its formation these three minerals crystallized separately from the molten magma in the cooling process, and the physical weathering of granite breaks them apart and liberates them into the developing soil. Heating splits layers from rocks, rather like the peeling of an onion, in a process called *exfoliation.*

Chemical weathering acts upon the fragmented rock particles. Water is an important medium for chemical weathering because it is an excellent solvent and contains many dissolved substances. Rainwater contains a range of dissolved materials, most importantly dissolved carbon dioxide. When dissolved in water, this gas forms a weak acid, carbonic acid, and this operates on the minerals in the soil, reacting with them to form new compounds. The atmosphere also contains small amounts of other gases, such as oxides of sulfur and nitrogen, which are derived from both natural sources and from pollution. The oceans are a source of dimethyl sulfide, and volcanic activity produces hydrogen sulfide, as do some wetlands. In the atmosphere these sulfur compounds become oxidized and then dissolve in water to produce sulfuric acid, a very much stronger acid than carbonic acid. The burning of fossil fuels in industrial processes and automobiles has added to the atmospheric load of sulfur, and this form of pollution affects even the most remote of tundra habitats in the Arctic, Antarctic, and high mountains. Nitrogen oxides are also produced by fossil fuel combustion and also by burning vegetation in forests and grasslands. In the atmosphere these oxides also dissolve in water to form nitric acid, another strong acid. When this collection of acids reacts with the mineral components of soils, the result is chemical breakdown. The feldspar mineral from granite, for example, is broken down by carbonic acid to form potassium carbonate and a relatively inert material, kaolinite. Kaolinite, or china clay, is a clay mineral (see sidebar "Particle Size," page 50) and is not broken down further. It plays an important part in controlling the behavior of certain elements in the soil. Carbonic acid also acts upon the mica mineral derived from granite, breaking it into potassium and magnesium carbonates along with oxides of

iron and aluminum. The third mineral derived from granite, quartz, however, is relatively inert, being composed of silicon dioxide (silica), better known in the form of sand.

Granite thus weathers by physical and chemical processes, liberating potassium, magnesium, iron, and aluminum compounds together with a clay mineral and silica. Although several of these elements are needed for plant growth, aluminum can be toxic, and many of the other important elements plants need are absent, such as phosphorus and calcium. The absence of calcium, in particular, is important because the acids that accumulate in the soil are not neutralized, and the soil becomes acidic in reaction. In other words, it has a low pH (see sidebar on page 54).

The degree of acidity of a soil is important because it affects the growth of plants. Not only does acidity have a direct effect on plant roots, it also influences the availability of many other elements in the soil. For most plants a pH of less than 5 begins to cause problems. Many elements that are essential for plant growth, such as potassium, phosphorus, calcium, and magnesium become increasingly difficult for a plant to obtain at such high levels of acidity. Iron and aluminum, on the other hand, are more soluble in very acid solutions, but both can be toxic to plants, so this is not a great advantage. As will be described later (see "Vascular Plants in the Tundra," pages 149–152), some plants are able to cope with extreme acidity, but they are relatively few. The

---

The pH scale of acidity and alkalinity. Neutrality is marked by pH 7. The scale is logarithmic and negative, so pH 4 is 10 times as acidic as pH 5. Various familiar substances are recorded along the scale.

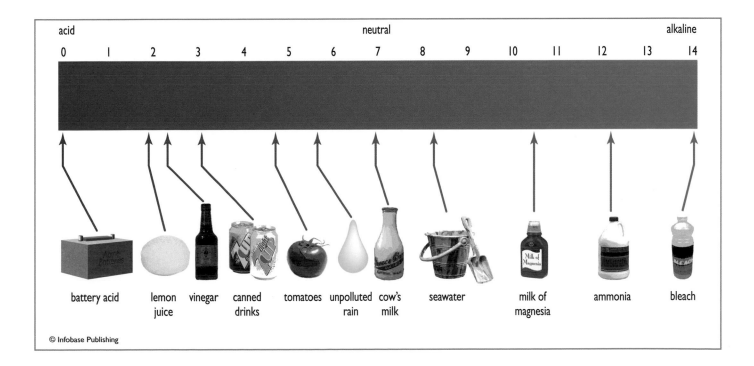

acid							neutral							alkaline
0	1	2	3	4	5	6	7	8	9	10	11	12	13	14

battery acid lemon juice vinegar canned drinks tomatoes unpolluted rain cow's milk seawater milk of magnesia ammonia bleach

© Infobase Publishing

Acidity and pH

An acid can be defined as a source of hydrogen ions, or protons, H^+. The more hydrogen ions present in a solution, the greater the acidity. Chemists therefore measure the acidity of a solution in terms of its hydrogen ion concentration. But this varies greatly, and the numbers involved would be extremely great, so a scale, called the pH scale, has been devised in order to express this concentration in terms that can be more easily handled. As in the case of any very long scale of numbers, the simplest way of reducing it is to express the numbers as logarithms. Using a base 10 logarithm, the difference between one unit and the next is a factor of 10 times. In the case of pH, the scale is also expressed in a negative fashion, so that high numbers represent low hydrogen ion concentrations and low numbers indicate high concentrations, in other words, greater acidity.

The pH scale is shown in the diagram on page 53. The scale runs from 0 to 14, and the figure of pH 7 is taken as an indication of neutrality, neither acidic nor basic. At this pH, hydrogen ions have a concentration of 10^{-7} molar, or 0.00000013 ounces per gallon (0.0000001 g/l). Above the figure of 7 the pH indicates basic conditions, and below 7 the pH denotes acidity. The diagram shows the pH of certain common solutions to demonstrate how the scale works out in practice. It is important to remember that the scale is logarithmic, so pH 4, for example, indicates 10 times the acidity of pH 5 and 100 times the acidity of pH 6.

acidity of soils in regions of the tundra underlain by acid rocks therefore limits the types of vegetation that can develop. In some areas, however, less acidic rocks, including calcareous limestone, form the tundra bedrock, and here very different soils develop with neutral or slightly alkaline pH. In general, more of the essential elements for plant growth are easily available at neutral pH, but very high pH can cause problems. Above pH 8.5, phosphorus, manganese, and iron can be difficult for a plant to obtain.

Detritus left by a retreating ice cap or glacier, however, has the advantage of being fresh and young. Soils are said to "mature," and this often means that some elements are progressively lost from them, leaving them poorer. The young soils of recently deglaciated terrain are freshly ground into fine particles, exposing large surface areas for weathering activities and releasing many chemical elements into the

water that permeates them. Even rocks relatively poor in such elements are able to supply the requirements of many plants at this stage in their weathering, so freshly exposed tundra areas are rarely deficient in the necessary materials for plant growth.

Living organisms rapidly colonize any site where they can make a living. Only the most severe environments on the planet, such as the vent of an active volcano, are totally devoid of life. The rocks of the tundra exposed by retreating ice are quickly invaded by algae, lichens, bacteria, various protists, invertebrates, mosses, and higher plants. All of these have their impact on the physical and chemical environment and in particular on the rock particles that form the skeleton of the developing soil. Photosynthetic organisms, such as some bacteria, lichens, and green plants, add organic matter to the soil in the form of dead litter, and other bacteria, together with the animal life in the soil, use this as an energy source, converting dead organic compounds to carbon dioxide. As has been described, carbon dioxide dissolves in water to produce carbonic acid, and this acid attacks the rocks, dissolving some of its component parts.

Some lichens (see sidebar "Lichens," page 143) grow flat upon the surface of rocks, and they secrete organic acids that attack their substrates, releasing elements for their continued growth. In the process they wear away the rocks they inhabit. Higher plants produce roots, some fine and some robust, all of which can penetrate into the fine crevices of rock particles and expand to split them apart. The weathering process thus accelerates when plants invade a habitat as a result of these *biological weathering* activities.

The combination of physical, chemical, and biological weathering gradually wears the rock particles into smaller parts and releases increasing quantities of chemical elements into the local environment. In the European Alps geologists have tried to measure the overall rate of rock weathering, and they have come to the conclusion that on average 0.004 inches (0.01 cm) of rock are being worn away each year. As the rocks are degraded, a soil is gradually being born.

■ SOIL FORMATION AND MATURATION

Very soon after the glacial detritus has been abandoned by the retreating ice, living organisms invade and a true soil begins to develop. Apart from the physical particles, the permeating pockets of atmosphere, and the increasing quantities of life and organic material in the soil, there is one additional soil component that plays a vital role in soil formation, namely water.

Water is usually, though not invariably, a common feature of most polar and alpine tundra soils. Its abundance is due mainly to the low evaporation rate in low temperature conditions rather than to any excess precipitation, although this may be high in some alpine sites. Water moves through the soil, displacing air from its abundant small pockets and hollows and clinging tightly to small soil particles, held by the forces of *capillarity*. Some water becomes chemically bound to crystals in the soil's mineral particles and is very difficult to displace. A very simple but very expressive classification of soil water consists of three components, gravitational water, capillary water, and chemically bound water. Gravitational water does not remain in the soil for very long. Following a soaking by snowmelt or rainfall, the soil is saturated with water, but within 24 hours of the water supply ceasing much of the water drains away under the influence of gravity. From the point of view of plant life, this component of the soil water is of little value because it is present for such a short period of time. Gravitational water, however, is mobile, and it does influence the nature of the soil by taking with it some of the smaller particles and the dissolved chemicals in the soil solution. The removal of elements from the soil by the water moving through it is called *leaching*.

Capillary water is held longer in the soil because of the forces of surface tension. Thin layers of water on small soil particle and small columns of water penetrating into the complex architecture of the rock fragments are held in place by these surface forces, and the pull of gravity is not sufficient to remove them. Plant roots exert greater pressures than gravity. Water is constantly being lost from the leaves of plants as vapor moves out of the leaves and into the atmosphere, and this creates a strong pull on the columns of water that extend unbroken from the leaf down the stem and into the root. As water is withdrawn up the stem, so the root tissues are placed under a negative water pressure, and this has the effect of drawing in water from the soil outside the cells. It is the capillary water in soils that forms the main resource for the plant's water requirements because this water is available over long periods of time and yet is not held so strongly in the soil that it cannot be accessed by the suction of the plant root. The third kind of water in the soil, the chemically bound water, is too strongly held for plant roots to withdraw it, so it is effectively useless to them.

A podzol soil profile from polar tundra in northern Finland. The surface layer consists of organic litter derived from the vegetation, below which is a pale layer that has been leached of its iron (A horizon). Lower in the profile is a deposition layer (B horizon) containing iron, aluminum, and organic matter leached from above. *(Peter D. Moore)*

The Soil Profile of a Podzol

Soil scientists, or *pedologists,* who seek to understand the development of a soil over time usually dig a pit so they can observe a cross section, or profile, of the soil. A soil profile reveals layers that have resulted from the operation of soil processes as the soil has matured, and one of the most important processes in temperate and tundra soils is leaching. Leaching occurs where water moves downward through the soil profile, dissolving some of the components in the upper soil layers and sometimes depositing these lower down. Downward leaching of this sort can occur only if the input of water to the site from the surface is greater that the evaporative forces that pull water back into the atmosphere, so it is a process limited to cooler or wetter parts of the world. Leaching is assisted by acids dissolved in the descending water because these help dissolve elements held in the soil. Some of the acidity is derived from carbon dioxide in the atmosphere and in the soil (produced by respiring bacteria, plant roots, and small animals). Other acids come from atmospheric pollution, including sulfuric and nitric acids. The soil organic matter also adds to the acidity of the soil water because dead and decomposing material liberates many acidic compounds, such as polyuronic acids and polyphenols. All of these contribute to the acidic nature of the water as it leaches the surface layers of soil.

In the case of the podzol or spodosol profile, which is found mainly in the Low Arctic and in adjacent boreal regions of coniferous forest, the outcome of the leaching is a very distinctly layered soil profile, which is shown in the diagram. The top layer of the soil profile consists of a layer of organic matter produced by the plants that occupy the site and is usually called the O horizon (denoting organic). Often this can be subdivided into L, F, and H layers, standing for litter layer (relatively intact plant fragments), fragmentation layer (broken and comminuted parts of plants), and humic layer (dark, compacted material in which few fragments are recognizable). There is very little mixing between these layers, and at their base is a distinct line of demarcation, separating the O horizon from the pale gray zone below in which organic matter is scarce. The lack of mixing indicates that few soil invertebrates, such as earthworms, are active in the soil profile. Some detritus-eating insects, such as springtails and flies, are present in the O horizon, but they do not take the organic matter farther down the profile. A clear layer of organic matter that decays slowly and is not incorporated into the deeper soil is called *mor humus.*

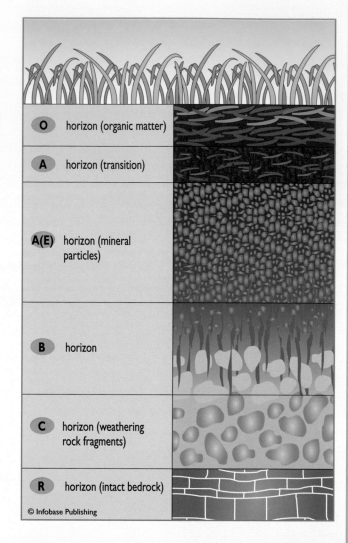

Podzol soil profile. A cross section of a soil can be divided into a series of layers, or horizons. The uppermost O horizon consists of decomposing organic matter derived mainly from the litter of plants on the surface. The A horizon may have some mixing between the organic matter and the mineral soil, but much of this layer (sometimes called the E horizon) has been stripped of organic matter, together with the iron and aluminum from degraded clay minerals. This process is called eluviation (hence E horizon) and leaves the soil in this layer bleached and pale. The B horizon is where much of the leached material is deposited, including an organic layer and a red-stained layer of iron oxides. The C horizon consists of rock fragments in the course of weathering and breakdown, and the R horizon is the intact bedrock.

Organic acids are produced in the O horizon, and these contribute to the low pH of the descending water, which dissolves many soluble materials out of the A horizon below. This horizon becomes bleached and relatively colorless as elements such as iron are leached from it. Many elements, including calcium, potassium, and phosphorus, are also leached, as is aluminum. The acid waters also carry down any organic compounds present in the horizon, and they even cause the clay minerals in the soil to migrate down the profile, leaving the A horizon devoid of all but bleached silica particles. It is said to be an *eluviated horizon*. Some soil scientists prefer to label it the E horizon. In the B horizon, immediately below the A horizon and usually separated by a sharp transition, many of these leached materials are redeposited. Organic compounds may emerge from solution and create a thin dark line. Iron and aluminum are then deposited, forming a reddish orange layer that can become hard and concreted, forming an *iron pan*. This iron-rich layer may be so tough that it prevents water from moving downward and causes a block to drainage. Below this B horizon is the subsoil, consisting of fragmented and weathering bedrock in the C horizon. The nature of the C horizon depends on the original geology of the area.

The capillary water present varies with the architecture of the mineral skeleton of the soil. If a soil consists of large particles, such as sand or gravel (see sidebar "Particle Size," page 50), then there are relatively few capillaries in which water can be held. Most of the water present is of the gravitational type and is quickly lost. Soils with large particles are thus free draining, and plants living upon them may experience water shortages under dry conditions. Soils with an abundance of fine particles, such as silt and clay, however, are rich in fine capillaries and have a high capacity for water retention. They are less prone to water shortage, and plants generally find them more suitable for growth. They may, however, suffer from the opposite problem, namely long-term water saturation or waterlogging. Excess water over long periods creates problems for plant roots because the water occupies all the spaces where air would normally penetrate the soil, so that roots can effectively be drowned (see "Waterlogging and Oxygen Shortage," pages 121–123). The ways in which soil particles are sorted, especially by water action (see "Sorting by Water and Wind," pages 49–52), following their deposition results in some tundra sites having ridges with high quantities of large particle sizes and other sites having an abundance of fine fragments. So drought-prone soils and waterlogged soils can occur in close proximity.

The movement of water through a soil has an important impact on its maturation. The removal of chemical elements from a soil by leaching leaves the soil progressively more deficient in some compounds. Materials may be removed completely from the site and washed out into streams, rivers, lakes, and ultimately the oceans, or they may simply be redeposited deeper in the soil. If this happens, then the soil takes on a layered form that can best be seen in cross section by digging a soil pit. Such a cross section of a soil is called a *soil profile*. A soil type that illustrates the development of a soil profile most clearly is the *podzol* or *spodosol* (see sidebar on page 56 and the picture of a soil pit exhibiting a tundra podzol on page 55).

Podzols or spodosols occur in tundra environments where soil drainage is good and where precipitation is relatively high. In situations where drainage is poor and the soil is waterlogged, the downward leaching process does not take place. Here the soil profile takes on a relatively uniform gray appearance, with only the surface accumulation of organic matter being apparent. The grayness is due to the lack of oxygen in the waterlogged conditions. As a consequence, any iron present in the soil is reduced from its Fe^{+++} form (ferric iron, which is a red-orange color) to its Fe^{++} form (ferrous iron, a blue-gray color). Soils of this type are called *gleysols,* and they are frequent in tundra habitats where water accumulates or moves very slowly through the soil profile. A close examination of the profile of a gleysol may reveal thin orange tubes leading through the uniform gray matrix. These are old root channels in which the original plant root tissue has died and decayed, leaving an open tube connecting to the air above. Air permeates these tubes and oxidizes the ferrous iron present, turning it from blue to the orange color of ferric iron.

Permanent waterlogging and the maintenance of a high water table in the soil can lead to the formation of increasingly thick layers of undecayed organic matter on the surface of a soil. If the organic O horizon is constantly saturated with water, the lack of oxygen affects the process of decomposition. Oxygen diffuses about 10,000 times more slowly when dissolved in water than it does when moving through air, so organisms that require oxygen for their respiration find it much more difficult to obtain adequate oxygen resources for their needs. Fungi and aerobic bacteria that feed upon organic matter in the soil and assist in its decomposition are suppressed by the lack of available oxygen, and organic materials decay much more slowly. As a consequence the deposition of plant litter and detritus on the soil surface

The extent of continuous and discontinuous permafrost in the Northern Hemisphere

may occur more rapidly than the decomposer organisms can consume them, and peat begins to accumulate. Once the peat layer is 16 inches (40 cm) deep, the landscape can be termed peatland and the soil type is called a *histosol*.

Because some soils are constantly being disturbed by the advance and retreat of glaciers and the redistribution of materials by wind and water action or by avalanches and rock slips, many soil profiles in the tundra never have the chance to mature. Often they are represented by thin layers of churned up rock particles lying on top of the parent rock. Shallow, immature soils of this type are given the name *leptosols*. These are common in the High Arctic and in the alpine zones of high mountains.

There is one additional factor that must be considered in any survey of tundra soils, namely the impact of frost. All tundra soils experience freezing at some time during the year, and some are present in such permanently cold conditions that their lower horizons are permanently frozen. Such soils are called *cryosols,* and their permanently frozen horizons are called *permafrost*. The extent of permafrost in the

Arctic region is shown on the map below. If the mean annual temperature is below freezing, then ice will persist in the soil from one winter to the next, creating the conditions needed for the existence of permafrost. The surface layers of such soil, however, thaw in the warmth of the Sun each summer, but they usually remain wet and often waterlogged because their deeper layers consist of an impermeable layer of ice, so drainage is impeded.

Permafrost can be divided into continuous permafrost, where the entire landscape has permanently frozen deep soils, and discontinuous permafrost, where frozen soil is patchy and interspersed with areas that seasonally defrost. As can be seen from the map, the discontinuous permafrost lies along the more southerly parts of the tundra and is sometimes occupied by boreal forests (taiga) of birch and coniferous trees. Discontinuous permafrost is also present in some of the mountain regions of the world, especially in the high plateau of Tibet, which is shown on the map. Continuous permafrost is present in conditions of extreme cold in the High Arctic and in those areas where conditions are continental and relatively unaffected by warmer oceanic currents. Thus, continuous permafrost is most widespread and extends farthest south in eastern Canada, Greenland, and Siberia. In Siberia, Alaska, and the Northwest Territories of Canada the permafrost is even found offshore, resulting in the formation of permanent ice around these edges of the Arctic Ocean. Northern Europe, on the other hand, is clear of continuous permafrost despite lying at the same latitude as northern Alaska. This is due to the influence of the warm Gulf Stream current in the North Atlantic (see "The Oceans and Climate," pages 11–14). Some estimates suggest that continuous and discontinuous permafrost areas occupy up to 26 percent of the total land surface of the world.

Permafrost can extend deep into the Earth's crust. On the North Slope of Alaska, the permafrost can be more than

Tundra polygons become clearly marked in winter, when snow accumulates in the hollows between the elevated vegetation of the polygon centers. *(David Marchant, National Science Foundation.)*

a third of a mile (0.6 km) deep, while in Canada it can extend down to two-thirds of a mile (1.1 km). Permafrost is even deeper in Siberia, where depths of 0.9 miles (1.4 km) have been recorded. Wherever the surface layers of soil absorb enough of the Sun's energy during summer, however, they are able to melt and produce what is know as an *active layer*, in which living organisms can recover from winter dormancy and begin to feed and breed once more. The depth of the active layer depends upon the climate and microclimate of the site. In flat meadow soils, only the upper 12 to 17 inches (30 to 40 cm) may thaw in summer, while on drier sites the thaw may extend to a depth of about three feet (1 m). Most of the roots of tundra plants occupy the upper four to 10 inches (10 to 25 cm) of the soil, so plant growth in the summer is able to occur while rooted within the active layer, unaffected by the permafrost down below.

Frost activity, particularly in the active layer above the permafrost, creates conditions that can lead to the development of some unusual features of the landscape. Just as the geomorphology of glaciated regions is profoundly affected by moving ice above the ground, so the geomorphology of the *periglacial* regions (those parts of the world that experience extreme cold but have no ice developed above their ground surface) are affected by the ceaseless activity of freezing and thawing.

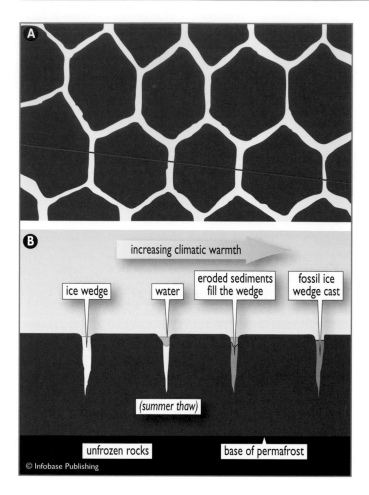

■ PERIGLACIAL FEATURES

When soil layers are subjected to intermittent freezing and thawing, as occurs in the active layer of permafrost regions, the structure and the layering (*stratigraphy*) of the soil are affected. An observer can see these effects in the form of patterns on the soil surface, often taking the form of honeycomb patterns of polygons. Patterning of this type is familiar in all kinds of situations, not just in conditions of extreme cold. When an area of wet mud dries in the summer heat of a temperate or tropical area, the soil often cracks up into a series of polygons. As water is lost, the soil contracts, and it fractures along planes of weakness that create the honeycomb effect. This happens in shallow tundra soils, where summer dryness can have precisely the same effect. Water penetrates the weak fractures and expands as it freezes in the winter, thereby exaggerating the polygonal patterns and making them permanent. Once they have developed, these polygons become further accentuated by the patterns of vegetation that develop over their surfaces (see "Polar Desert," pages 65–66). The scale of polygon development is very variable and is partly related to the depth of the soil. In the shallow soils of the High Arctic, polygons are only a few inches across, whereas the deeper peat soils of the tundra wetlands (see "Tundra Wetlands," pages 74–78) develop polygonal forms that can measure 100 feet (30 m) in diameter.

(above) Arctic polygons. (A) Soil surfaces in the Arctic are often broken into polygonal forms as they are regularly subjected to freezing and thawing. Water accumulates in the crevices between the polygons, and as it freezes it expands, forcing ice wedges deep into the soil. These are shown in the cross section of the soil (B). The scale of the pattern is very variable, ranging from a few feet to several hundred feet.

(below) Stones in tundra soils may be forced to the surface by ice action. As the soil cools in winter, ice crystals form in the upper layers. Stones lose heat faster than the surrounding soil, so ice accumulates below them; as the ice expands it forces the stone upward. Frost-heaved stones often form patterns, stripes or polygons, on the soil surface.

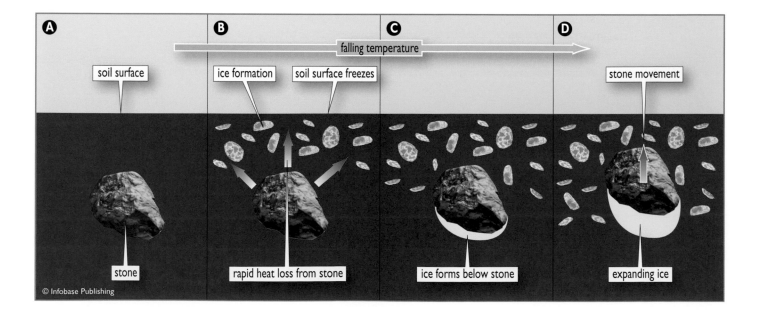

Each winter the cracks around polygons freeze, and each summer they thaw, in the case of the larger structures forming small moats around the polygon centers (see photograph on page 59). In cross-section these frozen cracks take the form of ice wedges driving down into the soil, as shown in the diagram on page 60. At this stage some of the soil material around the banks of the moats is eroded into the crevices between polygons and forms an infill. These infilled ice wedges have survived in many parts of the world that were once subjected to periglacial conditions, such as at the end of the last glaciation. They are fossil evidence recorded in recent geological sediments indicating very cold conditions in the past, and they have proved valuable to geologist when reconstructing the former extent of periglacial environments and climatic history. Ice wedges form only in very cold conditions, where the mean annual temperature is 21°F (–6°C) or below, so their remains provide a precise indication of former climate.

Freezing and thawing of the active layer also affects the arrangement of stones in the soil profile. A stone embedded in the active layer becomes surrounded by ice during the winter. When the soil begins to thaw, the layer beneath the stone is protected to some degree from the warmth penetrating from above, so the ice below the stone survives for longer and may even last through the summer. The next freezing cycle then begins with this surviving ice block, and the expansion of the water as it freezes beneath the stone forces it upward through the soil. The outcome is that stones are gradually pushed upward through the soil until they arrive at the surface, where they then remain, as shown in the diagram. This process can take place even in deep peat deposits, heavy stones being forced upward through less dense peat.

In patterned ground where polygons form the freezing and thawing is most pronounced in the crevices where ice penetrates, so the stone-heaving activity is greatest in these parts of the soil. As a result, stones are pushed to the surface along the lines of the greatest frost activity, forming patterns of stones accumulated in lines between the polygon centers. Stone polygons are most common over flat areas, but if the ground is sloping, then the polygons take on an increasingly linear arrangement depending on the angle of slope.

On steep slopes lines of stones develop along the contours forming parallel patterns rather than polygons. Stone stripes are often found in regions that were once marginal to glacial activity or in regions left behind by glacial retreat, even when the retreat was long ago. Stone stripes have been found in regions as far from the present Arctic as New England and southern Britain. They are found in alpine as well as polar

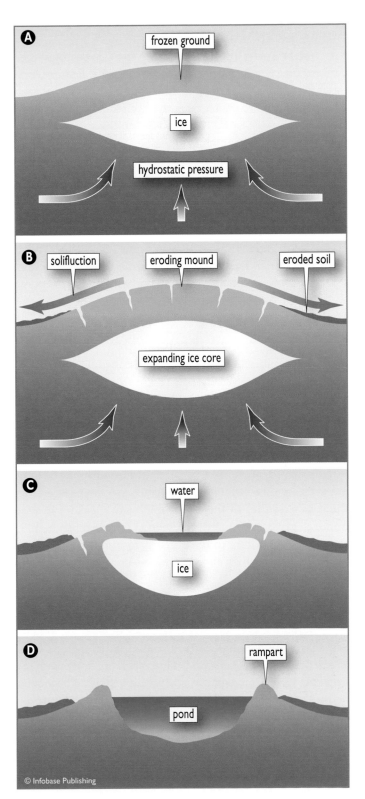

The formation and collapse of a pingo. Pingos are mounds with ice cores that may form along the spring line of a valley or in an old lake basin in the tundra. (A) Water under pressure freezes as it approaches the surface of the ground, expanding and creating an ice lens below the surface. (B) This grows and can create a mound hundreds of feet in diameter. (C) The soil on the ice surface eventually erodes, exposing the ice core and resulting in meltdown, especially in a warming climate. (D) The collapsed pingo takes the form of a circular pool surrounded by a bank or rampart.

tundra and occur in the alpine zone of the equatorial mountains of East Africa.

Slopes also create soil instability, especially when a saturated active layer has been formed by summer thawing and overlies ice permafrost below. Soil begins to slump downhill, leaving upper slopes denuded and causing an accumulation of soil slurry at the bottom of the slope. This is called *solifluction,* and the mixed mass of transported soil is given the term *head.* This is a characteristic soil process in periglacial environments.

Water moving through the soils of the tundra in summer can produce some complex geomorphological features, one of which is the *pingo,* which is shown in the diagram on page 61. The word *pingo* comes from the Inuit language, meaning a large hummock or hill. Water beneath the soil, and especially if it accumulates beneath the permafrost, can become pressurized and burst upward to the surface. This can occur in valleys along spring lines or in flat alluvial river valleys or old lake beds. As the water approaches the ground surface it pushes up a mound of ice covered with soil and rock debris. In winter the subterranean water freezes, creating an even larger ice mound and resulting in a massive blister on the surface of the landscape. Pingos are dynamic structures, and their anatomy is constantly changing with the pressure of water and ice from below. The soils and sediments on the crest of the pingo are unstable sitting on top of the mound of ice, so they begin to erode down the sides of the mound. The stretching of frozen soils and ice as the mound grows from below creates strains over its surface, and deep crevasses develop over the pingo. As summer warmth penetrates to the ice core of the mound, so a meltdown begins and the pingo collapses to form a rounded pool surrounded by a circular rampart. In the course of time this pool becomes colonized by plants, and silts accumulate to fill the hollow. Pingos vary greatly in size. Small frost mounds may be only a few feet across, but the word *pingo* is usually reserved for the larger features of this sort, which can measure several hundred yards in diameter and several hundred feet high.

A similar feature to the pingo is found in some Arctic wetlands, where the ice mound forms in a peat matrix and is called a *palsa.* The rise and fall of the palsa mound will be described in detail in a later chapter (see "Tundra Wetlands," pages 74–78).

As in the case of other periglacial features, pingos can survive in a "fossil" form. Landscapes in regions once affected by periglacial conditions, such as the south of England, have geomorphological features that can be identified with past pingos. Rounded ponds with high ramparts or a series of crescentic ramparts on a valley side can indicate that the area once held these massive frost mounds in a former colder time.

The activity of frost upon the soils of the tundra region thus creates a great wealth of varied geomorphological features that contribute to the diversity of the tundra landscape and the range of habitats available for vegetation and wildlife.

■ CONCLUSIONS

The tundra regions of the world consist of a wide range of rock forms, including some of the oldest rocks on the surface of the Earth. These rocks lie in a series of crustal plates that are constantly on the move, creating geologically unstable and volcanic conditions around their edges. The Antarctic occupies a single plate, while three plates are in contact in the Arctic region. The mountains of the world have been formed as a result of tectonic forces, including the collision and buckling of the crust and the upthrusting of rocks during volcanic activity in the vicinity of subduction zones. Sedimentary, metamorphic, and igneous rocks are thus all found both in polar and alpine tundra habitats. The chemistry of these rocks is very variable and has considerable influence on local ecosystems.

The cold climate of all tundra habitats ensures that water is often present in the form of ice, and precipitation often occurs as snow. The accumulation of snow over long periods of time leads to the development of glaciers, ice caps, and ice sheets. There are currently two ice sheets present in the world, over Greenland and Antarctica, but ice caps and glaciers are more frequent and widely distributed. All of these ice structures grow as a result of snowfall and decay by ablation. Their growth leads to ice movement out from a central area of accumulation or from an elevated location and their spread downhill toward the lowlands or the sea. If they reach sea level, they may extend away from the land as an ice shelf or they may become fragmented as icebergs. In winter sea ice may add to the extent of ice around the edge of polar continents and islands.

Glaciers develop in valleys and fuse to form ice caps in many mountainous areas of the world. As glaciers move downhill, they carve out rounded valleys with steep sides, which are apparent when the glacier has retreated, as are hanging valleys previously formed by tributary valleys. The load of rock detritus carried by a glacier can build up beneath the ice as lodgement till, or it may be deposited at the terminus of a glacier, forming a terminal moraine. Lateral moraines and drumlins are also geomorphological features associated with the deposition of glacial detritus.

Tills and other debris left behind by glacial retreat or accumulating beneath a glacier are sorted by the action of running water. Meltwater below and at the snout of a glacier sorts particles of detritus according to size. Large fragments do not travel far, while fine particles may be carried many miles by water and deposited in lakes or alluvium far down a valley. Esker ridges of gravel and sand develop below the glacier, while silts and clays accumulate in the banded, varved sediments of glacial lakes as glaciofluvial processes sort the detritus.

Surface rocks and detritus are exposed to the action of frost, causing physical breakdown or weathering. The chemical activity of rainwater and its dissolved acids, together with the biological activity of microbes (bacteria and fungi) and plants, further contribute to the decay of rocks. The outcome is the formation of tundra soils. Many soils in the tundra are fresh and immature, often disturbed, and very shallow. Some, however, have been stable long enough to mature, which involves the development of a soil profile in which different horizons are apparent. Upper horizons may be leached of some of their chemical contents, while lower horizons form zones of deposition. Some tundra soils are waterlogged and may even develop organic peat deposits on their surface.

Tundra regions that are not under permanent ice are said to be periglacial. The constant activity of frost creates various geomorphological features typical of the very cold climate. Soils are split into patterned, polygonal blocks separated by deep cracks that freeze to form ice wedges. Frost action in the soil heaves stones to the surface, arranging them in polygons and stripes according to the degree of slope, and deep in the soil the ice may never thaw, forming a layer of permafrost. Water under pressure in the permafrost soils can lead to the formation of ice mounds, or pingos, along valley sides and bottoms, creating a hummocky landscape.

In all these ways, through varied geology, geomorphology, and soils, the tundra regions of the world are characterized by distinctive and diverse scenery and landscapes. They are habitats in which geological forces are intensely operative and mobile. These forces present challenges and opportunities for those organisms that live in the tundra, including humans.

3

Types of Tundra

Tundra occurs in a very wide variety of forms. As has been described (see "Patterns of Tundra Distribution," pages 1–6), it is a biome that contains two major types, polar tundra and alpine tundra. But even within these two forms, a range of habitats and vegetation types exists that affect their animal life and their history of human occupation and exploitation. All tundra habitats have a low annual temperature, but they vary greatly in their daily and seasonal fluctuations (see "Arctic Climate," pages 14–16, and "Alpine Climate," pages 21–26). The tundra climate also covers a wide range of precipitation patterns, from the extreme drought of the very high latitudes and high altitudes to the heavier precipitation of the boreal forest transition and the tree line region of mountain slopes.

■ TUNDRA FORMS

The different climates and microclimates of the tundra affect the general form of vegetation that develops. In general, it is the climate that determines vegetation, particularly the extremes of climate during unfavorable periods. The extreme cold coupled with ice-blasting from the frozen particles in high winds have an extremely destructive effect on plant growth and hence on the survival of animal life in the tundra. Low water availability and long periods of darkness add to the stress of polar tundra living, and the nature of vegetation is closely determined by these extremes of climate. Climatic stress limits the distribution of different species of plant, but it also has a more general effect on the *vegetation structure* (see sidebar below),

---

Vegetation Structure

Structure is the term used by ecologists and vegetation scientists to describe the architecture of vegetation. Just like a building, the plants that cover the ground are constructed in a distinctive manner. Plants compete with one another to gain access to the vital necessity for their growth, sunlight. Growing tall, therefore, places an individual plant at a distinct advantage over its neighbors, but in the tundra extending above the general canopy brings its own hazards, particularly the prospect of damaging winds and low temperature. In tropical forests there is no such limitation, and tall growth is generally the most competitive option. In the tundra there are advantages to being close to the ground. There is a trade-off between growing up into a sunny location and playing safe by maintaining a low stature. But even low-lying vegetation has its own architecture. Viewed from the perspective of a small mammal, such as a lemming,

the vegetation consists of a series of columns supporting a roof of foliage. Beneath that canopy conditions may be darker, more humid, and cooler in summer, but it is also protected from snow penetration, wind, and excessively low temperature during the winter. Depending on the density of the canopy and the amount of light that penetrates, there may be a cover of small plants such as mosses and lichens on the ground. These form the basal layer in the relatively simple architecture of the tundra.

In some types of tundra the structure may be even simpler, lacking the canopy of foliage and consisting simply of a low carpet of mosses and lichens. This is trodden underfoot by the lemming, but to a smaller animal, such as a springtail or a beetle, even a moss canopy provides some cover from the climatic severity of the air above. The term *structure*, therefore, is a relative one depending upon the scale at which it is studied.

---

which is the term ecologists use for the architecture of vegetation. Vegetation structure has a profound influence on the microclimate close to the ground and thus influences the conditions experienced by small animals. It is the main characteristic used in the classification of different tundra habitats.

■ POLAR DESERT

The very high latitudes around the North and South Poles are characterized by high pressure systems that generate very low levels of precipitation (see "Global Climate Patterns," pages 6–9). In the Northern Hemisphere these regions are called the High Arctic and include the most northern parts of Canada and the high latitude islands of the Arctic Ocean (see map on page 3). Even the most favorable months are cold; the mean July temperature usually falls between 36 and 41°F (2 to 5°C), and frost is possible at any time. The combination of low temperature, low precipitation, and long, dark winters can result in the total absence of vegetation where these conditions are at their most extreme, but living organisms have a remarkable capacity to cope even under the most rigorous stress, so scattered plants are found wherever there is some shelter or where the cover of ice and snow is occasionally lifted.

Snow cover is one of the most important factors in determining the presence and the extent of vegetation in the polar desert. Being a desert, the snowfall is very little, but its presence has both beneficial and harmful effects for plants under the extreme conditions of the polar desert. A cover of snow in the winter has an insulating effect, protecting the soil surface and its plant life from the most extremes of low temperature. But a snow cover can also interfere with the commencement of growth in the early summer, and the persistence of snow shortens the growing season. For any given location there is an optimum snow cover for life in the polar desert.

Russian scientists on the island of Zemlya Aleksandra in the High Arctic found that vegetation was most lush in situations where the winter snow cover was one foot (0.25 m) in depth and disappeared by mid-June. Under those conditions, mosses and lichens survived through the cold dark winter but experienced a summer growing season that was sufficiently long for sustained growth. Thicker snow cover led to less prolific vegetation growth; once it attained a depth of three feet (1 m), mainly in the protected hollows, no vegetation could survive beneath it. Less snow cover led to the development of some vegetation, but the coverage was poorer than in areas with a thin winter blanket.

Snow cover is not the only factor that can tip the balance of plant development in the polar desert. The presence of stones on the soil surface is also influential. Stones project-

ing through the snow in the early summer absorb the Sun's heat and cause snowmelt in their immediate vicinity. Plants growing near stones may thus have a longer growing season than those in habitats that are less stony. A surrounding of stones can create a kind of stone hothouse that encourages plant growth and development. This effect may also occur beneath a snow or ice cover. When light passes through the frozen snow in the summer, it is absorbed by the dark surface of the soil or the plants beneath and is then radiated as heat. This creates a minute greenhouse and leads to the ice and snow melting from below, forming small caves of ice just an inch or so deep where the vegetation begins its growth. Under these extreme conditions, where plant survival is on a knife-edge, the small effects of microclimate may be critical in determining whether life is sustainable.

Vegetation in the polar desert is nevertheless extremely sparse. There is a complete plant cover only in the most favorable locations, and the structure of the vegetation is very low and simple, rarely extending more than four inches (10 cm) above the surface of the soil. The total vegetation cover, however, is often less than 50 percent of the soil surface. By far the most abundant plants are mosses, liverworts, and lichens, all of which lie very close to the ground. These are very primitive plants in the sense that they appeared early in the course of plant evolution, but they are evidently well suited to the extreme environmental conditions of the High Arctic. Mosses and liverworts have a very simple structure and give the impression of being delicate and vulnerable plants, but they have a remarkable capacity for survival under cold, dry, and dark conditions over long periods of time, so this may be the secret of their success (see "Tundra Mosses and Liverworts," pages 145–147). Lichens are also small and delicate in appearance, but they, too, are very drought resistant and manage to cope where few other plants can survive (see "Tundra Lichens," pages 143–145). Stone fields are usually covered by lichens and mosses.

There are few flowering plants in the polar desert, and most of those present are small grasses and sedges. The High Arctic grasses are mostly tussock forming rather than spreading. In patterned ground where the soil surface is broken into small polygons, the grasses often occupy the cracks between the polygons, sending their roots into the crevices to obtain scarce water in summer. Other herbaceous plants are quite scarce and include saxifrages, mouse-ear chickweeds, and members of the cress family. There are also some poppies (*Papaver* species). In hollows where water accumulates, the cotton grasses, such as *Eriophorum scheuchzeri*, predominate, forming small marshes, but this occurs only in the less drought-prone areas, such as the eastern Queen Elizabeth Islands in Canada.

Many of the flowering plants are dependent on the mosses for their initial colonization. Seeds lie dormant within the protection of the moss cushions and eventually germinate in the slightly warmer and damper conditions found there. Root development takes place within the organic detritus of the moss layer, and then the roots extend down into the soil, especially into the moist cracks between the polygons. Deep root penetration is impossible, however, because ice wedges and frost lenses persist in the soil through the summer, so roots are restricted to the upper soil layers, especially the organic layers immediately beneath the mosses.

Trees and tall shrubs are totally absent from the polar desert, but there are some woody plants that are able to survive by clinging close to the ground surface. The Arctic willow (*Salix arctica*), for example, forms small, dense clumps, and several species of the mountain avens (*Dryas* species) scramble over bare soil and rock surfaces. But no plant raises its growing points more than a few inches above the ground.

The soils of the polar desert are extremely unstable, especially where they occur on slopes. Each summer the surface layers of soil melt while deeper layers remain frozen, so there is a constant slippage of soil surfaces downhill in the process of solifluction. As the soil moves, so does the vegetation cover, and plants are thus carried around the habitat and may colonize new areas by this physical dispersal of entire plants. On steep slopes the slipping of soil and vegetation creates a series of steps or terraces, giving the impression of parallel paths along the contours. Although these are often used by grazing mammals, they are created by solifluction. The frost-heaving of stones in the soil brings them to the surface (see "Periglacial Features," pages 60–62), and this plows the soil, causing lower layers to be heaved up to the surface. The vegetation of the polar desert must therefore be capable of withstanding constant physical disruption and disturbance.

The landscape of the polar desert is thus open and treeless but by no means bare of vegetation. The domination of mosses and lichens in this vegetation is illustrated by the data from a Russian survey of one Arctic island. Ecologists listed one grass, nine herbaceous flowering plants, 12 mosses, nine liverworts, and 45 lichens. Many of the plants of the High Arctic are found throughout the entire region, encircling the North Pole. An example is the widespread grass *Phippsia algida*. These organisms are said to be *circumpolar* in their distribution. This may seem surprising, but, in fact, the dispersal distances involved are not exceptionally great. The standard Mercator projection of the globe, as used in most atlases, gives the illusion of uniform length for the lines of latitude, but this is clearly not the true case. The Arctic Circle is very much shorter than the equator because it is near the top of the global sphere rather than around its middle. When the Arctic is viewed from a polar perspective, as in maps such as the one shown on page 3, it becomes apparent the dispersal of seeds around the region is not a major problem. The ocean and the pack ice areas are inhospitable, but transport around the edges is quite possible.

Some plants, such as the purple saxifrage, *Saxifraga oppositifolia*, occur around the Arctic in a circumpolar distribution but are also present on many mountain ranges farther south. Plants and animals that are found both in the polar and alpine tundra habitats are said to have an Arctic-alpine distribution.

Animal life in the polar desert is low in diversity, but some types have relatively high population densities at favorable times. The moss and lichen patches are rich in *detritivores,* animals that scavenge for dead plant materials and organic matter. Worms are frequent, including segmented earthworms and nonsegmented nematode worms. Larvae of craneflies (Tipulidae) also abound. Birds are present in summer and take advantage of the period of growth and productivity. The ptarmigan (*Lagopus mutus*), the sanderling (*Calidris alba*), and the red knot (*C. canutus*) are particularly frequent. All three of these birds have a circumpolar distribution, breeding in northern Canada, Greenland, and Siberia. The curlew sandpiper (*C. testacea*), however, is confined to Eurasia, while the white-rumped sandpiper (*C. fuscicollis*) breeds only in the polar tundra of North America. Of the large mammals, the musk ox (*Ovis moschatus*) is by far the best adapted to a permanent life in the polar deserts. This animal has had a circumpolar distribution in the past, but hunting eliminated the musk ox from Siberia and Alaska. Reintroduction programs may allow it to recover its former range.

At its southern edge the polar desert merges into semidesert, where the plant cover is greater and the diversity of species increases. The polar semidesert regions of Canada, for example, contain more than 20 species of flowering plant compared with 10 in the true polar desert areas. The cover of grasses and dwarf rushes, including the hairy woodrushes (*Luzula* species) gives a dry grassland appearance to the semidesert vegetation, and some biogeographers have named this vegetation type *polar steppe*. Normally, however, the term *steppe* is reserved for temperate grassland, the Asiatic equivalent to the North American prairie. The semidesert is also used by grazing mammals to a greater extent; collared lemming (*Dicrostonyx groenlandicus*) and caribou (*Rangifer tarandus*) are found here. But dwarf shrubs are absent, or only represented by ground-hugging plants, so the structure of the vegetation is very simple, and bare soil remains an evident feature of the landscape.

■ DWARF SHRUB TUNDRA

Moving south out of the High Arctic and into the Low Arctic, vegetation forms a virtually complete cover over the tundra soils. Cushions of herbs are joined by creeping and mat-forming species so that the landscape is completely green in summer. Saxifrages, mosses, and woody plants form low mounds, often up to three feet (1 m) in diameter, and the general richness of the flora is higher. Meadows of grasses and sedges are present, especially where snowmelt creates flat, waterlogged habitats. Standing water is a frequent feature of the landscape, therefore, with extensive areas dominated by cotton grasses (*Eriophorum* species) and the water sedge (*Carex aquatilis*). These are taller plants than those found in the polar desert regions, so they

create a different microclimate beneath their canopy. The sedge wetlands are important feeding and breeding sites for many wetland birds. They are also the main breeding site for many aquatic insects, including blood-feeding blackflies and mosquitoes.

Polygon structures remain frequent in the Low Arctic, and the larger ones of these are often associated with wetland patches. The diagram below shows cross sections of two types of polygons, one with a concave center and the other with a convex one. The ditches that surround the polygons are water-filled in summer, so they provide an ideal habitat for tall sedges. The concave polygon centers also retain water from snowmelt in the summer, and wetland patches develop in these. Permafrost still occupies the lower soil levels, however, restricting the penetration of roots, so only shallow-rooted wetland plants are able to grow. The

Profiles of Arctic polygons and their vegetation. (A) Low-centered polygons are depressed in their center, so water accumulates in the middle, creating a shallow marshy wetland. (B) High-centered polygons are elevated in the middle, and their centers are relatively dry. They are subject to erosion, however, and the eroded sediments accumulate in the surrounding ditches.

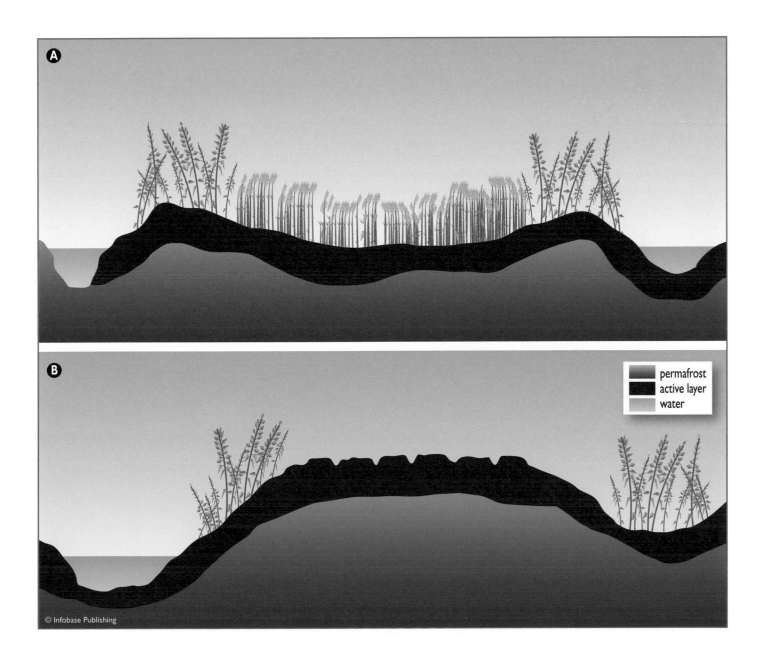

presence of permafrost also prevents the drainage of water through the soil, leading to water retention and the development of anaerobic gley soil profiles (see "Soil Formation and Maturation," pages 54–59).

Polygons with convex centers are better drained and do not develop into wetland habitats. The elevated part of the polygon may suffer soil erosion during snowmelt, in which case soil is carried into the surrounding ditches, and the height of the center is gradually lowered. Better drainage in summer means that the polygon center does not become waterlogged, and the soil profile has a more efficient supply of air that encourages the growth of plant roots and the activity of detritivores and microbes. Eroded soil surfaces are dark in color and are therefore efficient absorbers of sunlight, leading to warm soils, and this also encourages plant and animal growth. It is possible that such locations become so well drained that they suffer drought in the summer, in which case only plants that are drought adapted can survive there. Among these are the dwarf shrubs.

Dwarf shrubs are among the most dominant plants of the Low Arctic tundra. They are woody in structure but limited in stature, as shown in the photograph below. The height of the canopy of dwarf shrubs is usually between 16 and 24 inches (40 to 60 cm). It often forms a patchy cover, alternating with open patches of mosses, lichens, and herbs. The presence of these dwarf shrubs creates a very different set of microclimatic conditions from those of the polar desert. Beneath the shelter of the canopy the air is relatively undisturbed by strong winds, so humidity rises, which is favorable

for most invertebrate animals. Although shaded from direct sunlight, the subcanopy layer is kept fairly warm because the dark colors of the canopy absorb the light and become heated. At night, when temperatures above the canopy fall and frost may form on the upper leaves, beneath the canopy the heat is retained, and the insects survive unharmed. So the presence of a dwarf shrub component in the vegetation affects the entire ecosystem.

Some of the dwarf shrubs are evergreen, especially members of the heather family Ericaceae, and the closely related crowberry family Empetraceae (see sidebar "Why Be an Evergreen?" pages 116–117). Among these are relatives of the blueberry, including the mountain cranberry (*Vaccinium vitis-idaea*) and the tundra bilberry (*V. uliginosum*), together with the alpine bearberry (*Arctostaphylos alpina*) and the black crowberry (*Empetrum nigrum*). Other dwarf shrubs are deciduous, producing leaves in the summer and losing them in the fall. These include the red bearberry (*Arctostaphylos rubra*) and the dwarf birches (*Betula* species), together with various alpine willows. At the end of summer the dying leaves of these deciduous shrubs change from green into a range of yellows, reds, and browns and lend a surprising degree of color to the tundra landscape. The fall leaves are accompanied by a range of fruit, especially

Dwarf shrub tundra with dwarf birch (*Betula nana*) in northern Norway   (*Peter D. Moore*)

red berries that attract the attentions of feeding ptarmigan and willow grouse (*Lagopus lagopus*).

Two species of dwarf birch are found in the Low Arctic, *Betula glandulosa* and *B. nana*. The latter is very variable, and some botanists regard one of its races as a separate species, *B. exilis*. In the dwarf shrub tundra zone they grow no higher than about two feet (0.6 m), but they can grow a little higher at the southern edge of their range. All of them are evidently palatable, as their leaves and shoots are eaten by grazers such as ptarmigan. The dwarf birches are often most abundant in areas where glaciers have recently retreated, so they may represent a stage in the colonization and vegetation succession in these sites.

Dwarf shrubs are not confined to the dry regions of the Low Arctic tundra, they also colonize wet sites, especially where there is a cover of *Sphagnum* moss, usually indicating very acid conditions. The vegetation of these tundra bogs is very similar to that of the more southerly mires, the *muskeg* of boreal regions. Where *Sphagnum* grows, decay processes are usually inhibited, partly because of the wetness that this spongelike moss maintains in its vicinity and because *Sphagnum* produces a range of chemicals that affect microbial activity and the movement of nutrients in the soil. The consequence is that peat develops beneath the moss cover as the undecomposed remains of plants, including those of the dwarf shrubs, accumulate below (see "Tundra Wetlands," pages 74–78). When areas of peat become eroded by frost action and snow melt, the bare peat surface is often colonized by crowberry and by a plant related to the raspberry and dewberry, *Rubus chamaemorus*. This plant has a range of English names, including salmonberry (after its large pink fruits), baked apple berry (in Newfoundland), and cloudberry (in Europe). The Inuit name for this plant, when translated, means "man with no clothes on," again presumably referring to the pink fruit. Its range extends south into the temperate zone.

The dominance of dwarf shrubs instead of the grass and moss combination of the polar desert is partly related to summer temperature. In Greenland, for example, dwarf shrubs achieve importance only in the more protected and warmer parts of the coastal inlets. It is also related to snow cover. The polar desert generally has only a thin covering of winter snow because of the low precipitation of the region, but south of the polar desert winter snow is more extensive and deeper. The taller branches of the dwarf shrubs are undoubtedly protected from winter blasting in high winds by the snow cover. Conditions below the snow, however, are still cold, and winter temperatures at ground level even under deep snow are usually freezing. This is because heat is lost to the extremely cold permafrosted soil beneath, so the snow cover is unable to protect from this heat loss. But the temperature is manageable for frost-adapted plants, and warm-blooded mammals are even able to maintain activity under these conditions. In Siberia, for example, Middendorff's vole (*Microtus middendorffii*) moves around and breeds beneath the winter snow drifts, as do several other species of vole and lemming. They usually breed late in the winter, between April and June, so that the productive summer lies ahead and they will have plenty of fresh plant growth on which to feed. Stoats and weasels that prey upon these small mammals also continue hunting below the snow through the winter months. They use the tunnels of the voles and lemmings as a means of travel beneath the snow, preferring the soft snow of drifts, hollows, and ravines to the thin compacted snow of the more exposed areas.

In areas of extreme exposure where wind strips the ground of its winter snow blanket, dwarf shrubs are more stunted, rarely exceeding a height of eight inches (20 cm). Lichens and mosses are more evident on these stony ridges, and some of the shrubs assume a prostrate form, lying close to the surface of the ground. The alpine azalea (*Loiseleuria procumbens*) and cassiope (*Cassiope tetragona*) are characteristic dwarf shrubs of these habitats. Arctic poppies (*Papaver* species) are also well adapted to this bleak environment. Stony ridges and mountain slopes in the tundra are often given the name *fell fields*. This expression is derived from the Norse word *fjell*, which means "stony mountain." These habitats are also prone to drought in the summer, so lichens thrive on open stones and dry soil, while mosses become dry and shriveled through the summer but recover well once their moisture supply is renewed. Flowers abound on fell fields, and pollinating insects, including bees, also thrive in the short summer season.

Snow cover affects not only the plants but also the larger grazers, including caribou and musk ox (*Ovis moschatus*) in the dwarf shrub tundra. In summer this habitat is ideal for big herbivores and contains an abundance of fresh greenery and new shoots, but snow cover buries the food so these animals find it hard to excavate large volumes. Caribou solve the problem either by migration or by remaining in regions of polar desert and semidesert, where the snow cover is less. The musk ox is uncomfortable with snow more than two feet (0.6 m) in depth, so it remains confined to the polar desert and the most northern parts of the dwarf shrub tundra.

Snow depth is variable depending upon the topography of the terrain. Windswept ridges have little snow cover, while sheltered hollows and valleys retain greater depths, especially if they are north facing, shaded from the impact of the low Sun (see sidebar "Aspect and Microclimate," page 19). Deeper snow melts more slowly in the early summer and produces large quantities of meltwater, and this water supply encourages the development of wetlands in adjacent low-lying areas (see "Tundra Wetlands," pages 74–78).

The coastal plain of Alaska and Yukon Territory bears a distinctive type of dwarf shrub tundra that is sometimes

Tussock-heath tundra at the edge of the Barents Sea in northern Norway   (Peter D. Moore)

called cottongrass tundra or tussock-heath tundra, as shown in the photograph of Norway above. The undulating slopes between the mountains and the Arctic Ocean are covered by a hummocky vegetation dominated by the cottongrass *Eriophorum vaginatum*. This plant is actually more closely related to sedges than to grasses and has densely packed basal shoots with fibrous sheaths, creating compact tussocks from which the young leaves and flower heads sprout. The "cotton" part of its name comes from the fruit, which is covered by a dense mass of long, cottonlike white hairs that assist in fruit dispersal by wind. The cottongrass tussocks are interspersed with dwarf evergreen shrubs and dwarf birch, and the hummocky structure of the vegetation is an ideal environment in which small mammals can construct tunnels and for nesting sites for such birds as ptarmigan. The dense mass of shoots provides very effective temperature insulation so that the tussock is an ideal location for invertebrate residence through the rigors of winter. In summer the cottongrass tundra is especially rich in insects, particularly mosquitoes, midges, and blackflies.

The development of dense, fibrous tussocks also protects the cottongrass against the effects of fire. Just as the fibers prevent the tussock center from becoming cold, they also create a steep temperature gradient in times of burning, and the tussock center does not become overheated. The growing buds of the plant are thus protected, and so are the invertebrates that make the tussock their home. The presence of charcoal among the cottongrass fibers indicates that this habitat has been exposed to regular fire in the course of time, possibly spreading from the boreal forest and forest tundra to the south. The cottongrass tundra does not extend into the colder parts of the Arctic semidesert region farther north on the Arctic Ocean islands, but even in the Low Arctic the effects of low temperature are evident on the ground. Part of the surface irregularity of the cottongrass tundra is caused by the occurrence of *frost boils*. These are mounds of soil that are forced up by the expansion of ice forming beneath the surface. Water may accumulate in certain places and lead to these wet patches developing into frost boils, or the tussocks of vegetation may insulate some parts of the soil and leave the open areas more susceptible to freezing. The outcome is that an irregular soil surface develops with small mounds several inches in height interspersed with hollows, adding to the hummocky nature of the terrain.

The dwarf shrub tundra is thus a varied environment including both evergreen heath and cottongrass tussock habitats, bare stony ridges, and wet peaty regions. The common feature of all these habitats, however, is the lack of tall vegetation, especially trees.

■ TALL SHRUB TUNDRA

To the south of the dwarf shrub tundra zone lies a region of denser, taller vegetation. Less severe winters, greater snowfall, and deeper snow cover allow shrubs to project a little higher above the ground surface without being stunted by fierce winter winds and ice storms. The structure of the vegetation is thus more complex, with a higher canopy reaching three to seven feet (1 to 2 m) in height. Often the arrangement of shrubs is patchy, with tall clumps of dwarf trees intermingled by patches of dwarf shrubs and open ground.

Snow cover in winter is usually between six and 15 feet (2 to 5 m), so there is a much more effective insulation of the vegetation layers than that found farther north in the dwarf shrub and polar desert zones. Summers are longer and warmer, so the snow melts completely, apart from isolated and protected snow patches, and the vegetation canopy and the soil is heated by the Sun. The surface soil layers thaw to a depth of about four feet (1.5 m), but the permafrost remains intact below this active layer. As a result, the activity and growth of plant and animal life is greatly enhanced during the summer months, and the biological diversity of the system is higher.

Some of the shrubs in the tall shrub tundra are also found in the dwarf shrub zone but grow more luxuriantly here. The dwarf birches, for example, grow taller, up to seven feet (2.5 m) in these southerly locations. But there are also new species of willow and alder that join them in creating the more diverse environment, and some of these are not confined to the tundra regions. The mountain alder, or speckled alder (*Alnus crispa*), for example, is a frequent member of this community but is also found in wet habitats in the temperate regions of North America. There are many species of willow within this tall shrub community, among them the Alaska willow (*Salix alaxensis*). This species has its main distribution east of Hudson Bay, running through Alaska and into Siberia on the other side of the Bering Strait. It is absent from the western Arctic regions of North America and from Greenland, northern Europe, and western Russia. A species that straddles the Bering Strait, but is absent from the North Atlantic regions of the Arctic is said to have an *amphi-Beringian* distribution pattern. A species that is distributed on both sides of the Atlantic but does not extend into eastern Siberia or western North America has an *amphi-Atlantic* distribution. An example of an Arctic plant with this pattern is the pink lousewort (*Pedicularis hirsuta*).

Some of the taller shrubs of this zone have a remarkable shape. A dense mass of shoots and foliage quite close to the ground form a base to the shrub, then tall, bare stems rise from this mass several feet before once again bushing out into branching leafy stems. The Russians have an expression for these that describes their odd appearance: They call them "trees in skirts." Their structure reflects the snow cover in winter, which buries the lower bush but leaves the upper stems emerging from the snow. Wind and suspended snow and ice particles are most abrasive just above the surface of the lying snow, so the emergent stems are stripped of leaves and growing buds. A little higher the abrasion is not as intense, so lateral twigs and branches survive.

Growing in the protection of the taller shrubs, some robust herbs are able to survive in the tall shrub zone of the tundra, one of which is the fireweed, or rosebay willowherb (*Epilobium angustifolium*). This is an upright herb that can grow up to four feet (1.2 m) in height and has large pink flowers. As its English name implies, it is particularly associated with sites that have been burned because it likes open, disturbed habitats, especially if they are enriched with nutrients, which is usually the case after fire. It is an arctic-alpine plant that is found in mountainous regions of the Old and New Worlds as well as in the Arctic but is one of the very few plants from this type of habitat that has proved invasive and a successful weed in urban and agricultural regions. It extended its range and its populations considerably in Europe during World War II, when it took advantage of bomb-damaged sites within cities. In the tundra it grows in patches that add summer color to the landscape and provide an attraction to pollinating insects.

The width of the tall shrub tundra is variable and often rapidly grades into a zone where true trees, even if stunted in their growth, are able to survive. Such vegetation is more correctly termed forest tundra or the forest tundra *ecotone*. An ecotone is a boundary zone between one distinctive community of plants and animals and another. Sometimes the forest tundra zone abuts directly onto the dwarf shrub tundra or tussocky cottongrass tundra. The regular occurrence of fire may encourage such sharp transitions in the tundra vegetation.

■ FOREST TUNDRA

The limit of forest growth and the beginning of the tundra is a very blurred boundary in most areas. The trees of the boreal forests become less dense and shorter in stature as one proceeds north, and it is often difficult to determine precisely where the timberline lies. For this reason biogeographers often prefer to regard this as an ecotone, a gradual merging of one biome with another that can conveniently be termed forest tundra.

The map of this boundary on page 14 shows that it does not simply follow a line of latitude but meanders around the Arctic, sweeping far from the North Pole in the Hudson Bay region and around the Bering Sea but coming farther north in central Siberia and eastern Canada. The fact that the tree line does not coincide with a precise latitude suggests that the factor controlling its position is not the pattern of day length. The number of hours of sunlight received at any location is determined by its latitude. As discussed earlier (see "Arctic Climate," pages 14–16), temperature varies in a more complicated way than day length, and this is an important factor for plant growth, including that of trees. The mean July temperature provides a measure of summer warmth, and the boundary between the forest and the open tundra corresponds quite closely to the 50°F (10°C) July isotherm. (An isotherm is a line that joins all geographical locations having a particular mean temperature.) The conclusion to be drawn from this correlation is that tree growth requires warm summers, and that once the summer warmth falls below a critical level trees cannot attain their full stature. The blurred boundary, or ecotone, between forest and tundra is caused by the fact that trees may be able to survive in a stunted form even beyond the zone in which they grow most effectively. This is the zone that is given the German name *krummholz* (see sidebar "Timberlines," page 22).

Where the forest tundra zone is found in a northerly location, as in Canada and central Siberia, it is usually quite narrow, forming a belt about six miles (10 km) wide. Where the boundary zone lies farther south, however, as in western Alaska, it is much more diffuse and wide, up to 30 miles (50 km) in width.

The forest tundra zone thus contains most of the tundra species from the tall shrub zone plus scattered trees that are best suited to the boreal forests to the south, as shown in the photograph on page 73. Although the traditional image of the northern boreal forest is one of dark, evergreen conifers (gymnosperms), there are also deciduous conifers and broadleaved deciduous flowering (angiosperm) trees that are important in its constitution. Precisely which tree species mark the limit of tree growth in the forest tundra zone varies from one geographical region to another.

In North America various species of spruce are frequently the final representatives of the boreal forest in the far north. On the well drained ridges and mountain slopes it is usually white spruce (*Picea glauca*) that forms the forest limit, while in wet, low-lying regions, especially on peatland, black spruce (*Picea mariana*) is often the most northern tree. This is an unusual conifer because its lower branches may become buried and take root in the soil, leading to the development of rings of spruce around their parent tree. In the forest tundra ecotone spruces are often shorter and have a more narrow, columnar shape than the one they achieve in their preferred climatic conditions farther south. They grow slowly, which adds to their stunted and gnarled appearance, and usually fail to produce ripe seed. It is not seed supply that limits the invasion of spruce into this zone, however. Spruce seed is winged and dispersed by wind, so the abundance of seed produced farther south is adequate to ensure

colonization. Germination, establishment, and survival of the tree are evidently controlled by the climate, especially the summer warmth.

On peaty soils the larch, or tamarack (*Larix laricina*), a deciduous conifer, is often associated with black spruce. This tree has needle leaves like the spruces, though lighter green in color, but they turn yellow-brown in fall and are then lost, leaving the tree bare and dead looking through the winter months. Pines and firs are rarely important forest tundra trees in North America, but there are many deciduous trees that are frequently found in this zone, including birches, willows, alders, and cottonwoods.

The dwarf birches of the northern shrub tundra zones are joined in the forest tundra ecotone by taller and more robust species of deciduous trees. The paper birch (*Betula papyrifera*) is one of the most common trees of the ecotone in North America, its range extending across from Alaska to Labrador. Clearly, evergreen leaves are not an essential prerequisite for tree life in the far north (see sidebar "Why Be an Evergreen?" pages 116–117). Paper birch is often accompanied by black cottonwood, or balsam poplar (*Populus balsamifera*), which extends from the limit of tree growth in the Arctic southward into the temperate forests. This species and the related quaking aspen (*P. tremuloides*) are often found well north of the limit of evergreen conifers. In Alaska poplars can be found 50 miles (70 km) or more beyond the conifer range. The balsam poplar is a tall tree, often the tallest within the forest tundra zone, achieving heights of 100 feet (30 m) or more in sheltered sites, and its patches are often very evenly sized, suggesting that small populations establish at a site at the same time. It spreads vegetatively by suckering from the roots, so this also enables the species to establish pure patches. The quaking aspen is the same species that forms aspen parkland on the northern edges of the midwestern prairies, and in the forest tundra zone it is particularly associated with areas that have been affected by fire.

Mountain alder (*Alnus crispa*) is mainly a western species in the North American Arctic but is also found in the forest tundra transition of the western mountains. It is a particularly influential tree in habitat development because its roots are equipped with nodules containing bacteria able to fix atmospheric nitrogen into a form that can be used by the plant. When the plant sheds its leaves in fall, these nutrients are passed on to the remainder of the ecosystem (see "Element Cycling in the Tundra," pages 93–98), thus enriching the entire community. Alders generally favor moist locations, such as along riverbanks, where they produce dense thickets of vegetation. They are often accompanied by a range of willow species, such as the beak willow (*Salix bebbiana*), which also forms a dense growth of red-colored twigs along riverbanks. Riverside species are said to be *riparian* in their distribution patterns.

The forest-tundra transition in northern Ontario, Canada. Black spruce trees (*Picea mariana*) are interspersed with wetland muskeg. *(Peter D. Moore)*

The forest tundra of the Eurasian region bears a strong superficial resemblance to that of North America, but there are, in fact, virtually no tree species found in both areas. Many of the genera are the same, such as birch (*Betula*), spruce (*Picea*), and larch (*Larix*), but the species are different. The forest tundra of Eurasia is much longer than that of North America, and there is a greater variety of tree species involved in its composition. In the far east of Siberia it is a pine species, *Pinus pumila,* that forms a shrubby limit to tree growth, and in the west in Finland, it is another pine, the Scots pine (*P. sylvestris*), that usually manages to extend farther north into the tundra ecotone than its accompanying conifers, such as Norway spruce (*Picea excelsa*). It is surprising and at present inexplicable why pines should be so important in the Eurasian forest tundra and yet relatively insignificant in the equivalent habitat in North America. In central Siberia, between the two pines, the forest limit is usually achieved by a species of larch, the Russian equivalent of the tamarack. The Siberian larch (*Larix sibirica*) and the dahurian larch (*L. dahurica*) are two of the most widespread of these trees. Like the tama-

rack, these species are able to extend their roots laterally in the soil and thus avoid problems with penetrating the hard permafrost below.

The trees that contribute to the vegetation of the forest tundra zone thus vary in different parts of the Arctic, and both evergreen and deciduous forms are found in the ecotone. The timberline is rarely as distinct and clear as that found in the equivalent zone of alpine tundra sites. Dense forest gradually becomes more open and sparse toward the north, and tree growth becomes slower and their forms more stunted, forming the krummholz community. The factors that determine the position of the ecotone vary from one geographical location to another. Trees in the forest tundra zone may not produce fruit, as this is one plant function that is lost when it is under stress. This can result in a poor supply of seeds but this rarely limits colonization.

Even if seed is present at a site, climatic conditions may be unsuitable for germination to take place. All seeds have particular temperature requirements for germination, and the correlation between the timberline and July mean temperature may reflect the warmth requirement of germinating seeds. Alternatively, seeds may germinate, but survival and growth of the seedling may be possible only if the summer proves sufficiently warm, so this could determine the northern limit of tree growth. Winter conditions are also influential in tree survival. Extremely low temperature, or very high winds, or a combination of the two, are harmful to the growing points of stems and branches (see sidebar "Plant Life-forms," page 112). The buds on the tips of branches are insulated by scales, but they are easily damaged. The outcome is stunted growth, as found in the dwarf trees of the krummholz zone.

Microclimatic conditions can strongly modify all these factors (see "Microclimate," pages 18–21). South-facing slopes and deep sheltered valleys north of the main timberline may contain isolated fragments of forest, surviving in the favorable local conditions. This is less evident in flat, uniform regions, but in more mountainous areas with rugged and varied terrain the forest tundra ecotone becomes extremely ragged.

The study of timberlines and the factors that control them is of great importance in understanding the ecology of the tundra and will prove even more so in the future. As the world temperatures continue to rise, the effects in the Arctic are likely to be detected first in movements in the timberline and changes in the vegetation of the forest tundra ecotone (see "Biological Responses to Climate Change," pages 224–226).

◼ COASTAL TUNDRA

The tundra of North America and Eurasia has an extensive coastline, greatly indented and supplied with an abundance of islands. One might expect, therefore, that the tundra would be rich in maritime habitats, such as sand dunes and salt marshes, but this is not the case. There is very little tidal movement in the Arctic Ocean, so the intertidal zone is very narrow. This limits the opportunity for salt marsh development. Some salt marshes have been described in Alaska, the Northwest Territories, and the Hudson Bay area, but these consist mainly of muddy habitats with very little plant cover. Often the herbaceous plants are less than two inches (5 cm) in height. The short turf of these marshes, however, provides an important feeding habitat for snow geese (*Chen caerulescens*). Low density grazing has a distinctly positive impact on the marshes because the droppings of the geese are rich in nitrogen, and they stimulate productivity. Under a high intensity of grazing, however, the structure of the marsh is damaged.

Sand dunes require wind-blown sand in order to grow, and in temperate areas with a high tidal movement the source of sand is usually the intertidal zone. Expanses of sand are left bare to dry and then become mobile as wind blows across them. In the Arctic, however, the low tidal movement means that such expanses of bare sand are rare, and therefore sand dunes are also very limited. Coastal sandy ridges are often colonized by succulent plants, such as the sea purslane (*Honckenya peploides*).

In some coastal regions of the tundra the presence of pack ice on the sea limits the development of coastal veg-etation. As shown on the map on page 3, permanent ice is present along the north coast of Greenland, Ellesmere Island, and the more northern of the Queen Elizabeth Islands of Canada. Summer pack ice is still present beyond this limit, affecting many of the other islands of the Queen Elizabeth group together with parts of Baffin Island, Banks Island, Victoria Island, northern Svalbard, Franz Joseph Land, and various polar desert islands north of Siberia. Winter sea ice is much more extensive, so the opportunity for developing low-lying coastal salt marshes is limited to the milder regions of the Low Arctic, such as the southern edge of Hudson Bay. In most tundra regions the characteristic vegetation extends right to the edge of the sea.

Sea cliffs are distinctive in two respects. First, they receive salt spray directly from the ocean, so this creates an additional chemical stress for plants and animals inhabiting them. Second, cliffs are often used by seabirds for roosting and nesting, and this has further chemical implications resulting from the guano (dung) they deposit at their roost or nest sites. To this is added the decaying remains of regurgitated food, such as fish, and the dead chicks and broken eggs that invariably accompany colonial nesting sites. These various types of detritus are the source of significant additional inputs of nutrients to the ecosystem (see "Element Cycling in the Tundra," pages 93–98). In moderate amounts the nitrates and phosphates derived from these sources stimulate plant growth, and some lush green vegetation may result from local bird activity. But heavy depositions of uric acid from the excretion of birds and deep guano deposits may kill the vegetation and leave the cliff sites relatively bare. Some plant species and some lichens are said to be nitrophilous, meaning that they are particularly favored by high doses of nitrates or ammonia. One of these is the bright yellow lichen *Xanthoria parietina*, which can be very conspicuous on bird cliffs.

◼ TUNDRA WETLANDS

Wetlands are abundant in the Arctic despite the fact that precipitation is generally low, often less than 16 inches (40 cm) per year. The low overall temperature of the air reduces evaporation rate, so the water present in the ground often remains there, saturating the soil. The soil itself may be shallow, and where it lies over the hard rocks of the Precambrian Shield, as in eastern Canada, water does not drain into the impervious rock. Even in deeper soils, the presence of permafrost creates an impenetrable layer that ensures water is trapped near the surface. Dry, gravelly ridges remain free of waterlogging, but most low-lying sites outside the polar desert zone are liable to saturation and wetland development.

Open pools of water are frequent, especially in peaty regions. These eventually become colonized by sedges and cottongrass species that invade the pool from the edges. How quickly the infilling process takes place depends upon the supply of eroded silt that enters the pool from its surroundings. The streams that flow into pools carry a suspended load of soil fragments. These sediment in the pools because the water movement slows, and the capacity of slow-moving water to carry sediment is poorer. The presence of plants in a pool, rooted in the basal mud and extending their shoots up into the air (emergent plants), has the effect of further slowing the water movement, so the plants encourage sedimentation and accelerate the process of infilling. These plants also add their own dead remains to the accumulating deposits.

Some tundra pools lie in the middle of flat peat lands and receive little in the way of eroded sediments from elsewhere, and these may remain as pools for long periods of time. Most of the input of the material deposited in these pools comes from dead organic matter, so the basal sediments are dark brown and muddy. The summer sunshine penetrates the clear water and warms these muds, stimulating the growth of filamentous green algae at the bottom of the pool. As the algae photosynthesize they produce oxygen, which dissolves in the water and creates a well-oxygenated environment in which microbes become active. The microbes, fungi and bacteria, feed upon the dead organic matter at the base of the pool and thus intensify the decomposition process. The result is that organic debris in such a pool can be broken down rapidly, so the pool is very slow to infill. The sediments are lost by decomposition almost as fast as they arrive.

The ditches around the edges of polygons in the Arctic are interconnecting, narrow, linear wetlands, interspersed with the sedge and cottongrass marshes of the concave polygon centers, as shown in the diagram on page 67. In some areas of the dwarf shrub tundra zone extensive areas are dominated by sedge and cottongrass meadows, where flat or gently sloping land is constantly saturated by water from melting ice and snow percolating through the soil. Wetlands are thus created and maintained in the tundra despite the low precipitation.

Sedge meadows, polygon mires, and open pools are all wetlands that are fed by flowing water through the habitat. Ecologists use the word *rheotrophic* to describe this type of wetland, where the supply of nutrient elements to plants comes from both rainfall and, more importantly, moving water. *Rheo-* means "flow," while *-trophic* indicates the nutrition of the ecosystem, so the word literally means "fed by the flow." The majority of wetlands in the tundra are rheotrophic in nature. The alternative system of nutrient supply is found in the *ombrotrophic* mire, which receives its elements solely from rainfall (*ombro-* means "rain").

One important species of the tundra wetlands is the bogmoss *Sphagnum*. There are many different species of this moss, and several species are often found growing together, each occupying slightly different microhabitats. Some species prefer very wet conditions and live underwater; others live just at the water level, forming wet carpets; and others form dense tussocks and become elevated several inches above their immediate surroundings. All the species are remarkable in two major respects: They hold water like sponges, and they change the chemistry of their environment.

Sphagnum has an unusual cell structure. The largest of its cells, both in the leaves and the stems, are dead and empty and have holes in their walls through which they become filled with water. Their remaining cells are living and green and form a network around the dead cells. It is the dead cells that act as storage vessels for water and give the moss its spongelike character. The walls of the cells are chemically complex and contain molecules called *polyuronic acids*. These acidic compounds have the capacity to attract and hold positively charged particles (cations) that they extract from the surrounding water, exchanging them for hydrogen ions. Elements such as calcium, potassium, and magnesium are bonded by the polyuronic acids, and the hydrogen ions that replace them in the water render the local conditions more acidic. So the overall outcome of the presence of *Sphagnum* in a habitat is that water is retained, creating an increasingly wet set of soil conditions, nutrient elements required by other plants are taken up and retained by the moss, and the water becomes more acidic. The bogmoss thus creates its own type of habitat, and relatively few other plants are able to live under the conditions it generates.

The nutrient poor, acidic, wet conditions that the bogmosses create thus make life difficult for many of the more robust and potentially competitive plants with which it shares its habitat. Other organisms that find life with the bogmoss uncomfortable are the soil microbes. The fungi and bacteria responsible for the decay of organic matter in the soil are inhibited by the wetness, the acidity, and the nutrient poverty of the *Sphagnum*-dominated habitat. Decomposition slows down, and some of the organic detritus, including the dead remains of the *Sphagnum*, are left as an undecomposed residue, accumulating as peat.

At the southern edge of the tundra, peat-forming habitats dominated by bogmosses and dwarf shrubs become increasingly abundant. Some regions of forest tundra and even the boreal forest regions themselves have extensive cover of these wetlands, known as *muskeg*. In Canada these habitats are characteristically occupied by scattered black spruce or tamarack together with evergreen dwarf shrubs, including Labrador tea (*Ledum groenlandicum*) and leatherleaf (*Chamaedaphne calyculata*). In oceanic areas, where precipitation is much higher than is typical for tundra conditions, this type of habitat grades into *blanket bog*. Examples are found in Labrador and Newfoundland, and

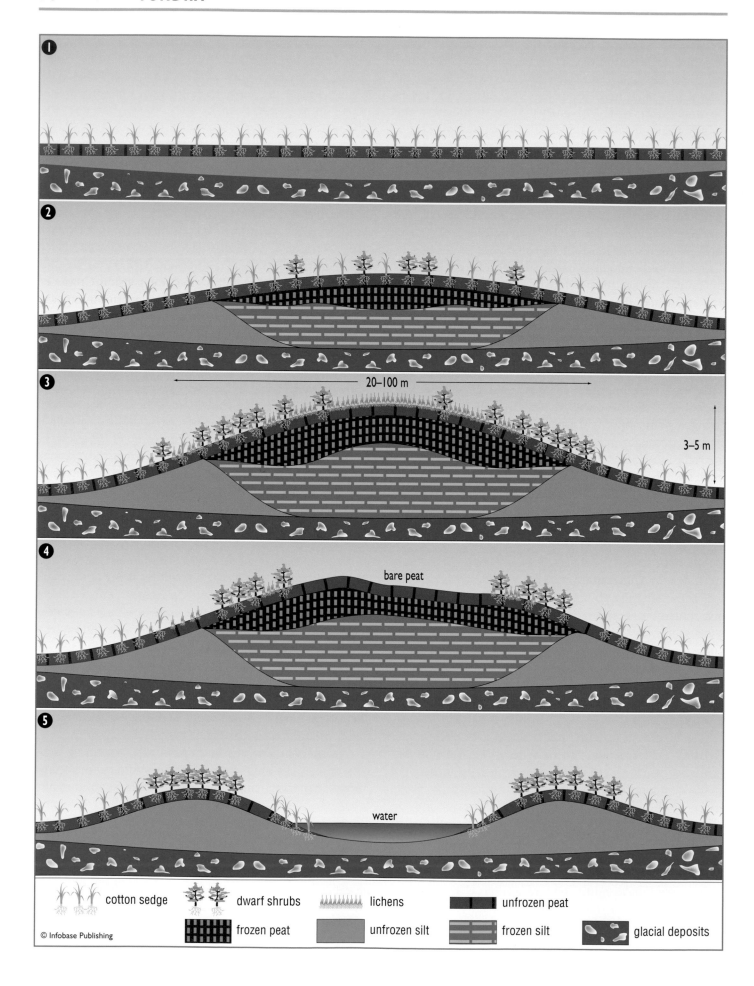

1

2

3

20–100 m

3–5 m

4

bare peat

5

water

	cotton sedge		dwarf shrubs		lichens		unfrozen peat

	frozen peat		unfrozen silt		frozen silt		glacial deposits

similar habitats are found in western Norway in Europe. Ecologists usually reserve the word *bog* for peat lands that are ombrotrophic. The accumulation of peat beneath their surfaces is sufficient to elevate the entire ecosystem above the influence of underlying soils and rock. Water arrives at the bog surface entirely by rainfall, which generally means that the nutrient supply to the wetland is very poor. Blanket bogs are a particular kind of ombrotrophic wetland in which the peat cover is almost independent of the topography, developing over hilltops, valleys, and even moderate slopes, forming a blanket over the entire landscape.

One other very distinctive type of peat-forming wetland occurs in the Arctic, namely the *palsa mire*. They are generally found in the southern parts of the tundra zone, where the annual average temperature is around freezing (32°F; 0°C) and where the summer growing season for plants is 120 days. They are extremely well developed in northern Norway, Finland, and Russia. A palsa mire looks like a low-lying area of pools and sedge meadows with occasional elevated masses of peat in a variety of conditions, some covered in dwarf shrubs, some in lichens, and others an eroding mass of peat. These protrusions can be six to 10 feet (2 to 3 m) in height and up to 150 feet (45 m) across. When viewed in cross section, they are found to have a frozen core of ice, covered by a relatively thin layer of peat.

Detailed studies of palsas in Finland have revealed that they pass through a series of developmental stages as shown in the diagram on page 76. The palsa hummock forms from a relatively level wetland of sedges or cottongrass underlain by permafrost. This habitat is rarely entirely level but is slightly undulating as a result of the uneven growth of the vegetation. In winter snow lies on the surface of the wetland and drifts in the wind, being blown off the elevated parts of the surface and accumulating in hollows. Snow forms an insulating blanket, preventing some frost penetration, and where it is absent, on the slightly raised areas of the wetland surface, frost penetration is deeper. The result is a buildup of ice beneath the surface of the mire during winter, and this may not melt down completely in summer. The expansion of

(opposite page) The rise and fall of a palsa mound. (1) The Arctic wetland surface is relatively flat. (2) Any slight irregularities, such as slightly higher ground, result in less snow cover and poorer insulation on the raised areas, so ground ice persists through the summer and swells as more ice forms. (3) The ice core continues to grow and raises the mire surface above the surroundings, as a result of which it becomes drier and clothed with dwarf shrubs and light-reflecting white lichens. (4) Eventually the top of the palsa mound begins to erode as a result of water runoff, and bare black peat is consequently exposed, which rapidly absorbs summer sunlight and instigates an ice core meltdown. (5) The palsa mound collapses, leaving a pool surrounded by a circular rampart.

water as it freezes means that the formation and persistence of ice creates a higher mound, which results in less snow cover and further freezing. So there is a positive feedback in this system whereby the ice core generates yet more ice.

As the ice core grows and persists from year to year, the mound of the palsa rises above the surface of the wetland, and the elevated peat surface becomes better drained than its surroundings. Over the years the sedge vegetation on the palsa hummock is replaced by evergreen dwarf shrubs and by white and gray lichens, forming a pale summit region. The pale color reflects sunlight, leading to cool soil conditions through the summer, and this further protects the ice mass from summer meltdown. But the continued rise of the palsa eventually leads to instability and splitting of the thin peat cover, stretched laterally by the ice expansion. Crevasses appear in the peat layer, and wind and water erosion exposes more of the underlying dark peat, absorbing summer sunlight and becoming heated. The final collapse of the palsa occurs when the ice core becomes overheated in summer, and there is a rapid meltdown, leading to the formation of a pool surrounded by a ring of peaty ramparts.

In a palsa mire the various stages in development are often found alongside one another, rising young palsas emerging from sedge lawns and the remnants of pools and ramparts. Some ecologists have come to the conclusion, however, that palsas start life at certain times in history, such as when the climate is particularly cold. This would account for the fact that groups of even-aged palsas are often found together. Some palsas have been recorded in Finland that date back more than 5,000 years, so the full palsa cycle must be a very long-lasting process. Even the life span of a palsa may vary, however, depending upon the local conditions, the microclimate, and even the direction in which a palsa mound faces. The conditions needed for the initiation of a palsa mound have been put to the test, however. By sweeping an area of sedge meadow in a palsa region of Finland through a number of winters, peat land scientists have created the first stage of an artificial mound. An ice core has started to form under the swept area, where snow was not allowed to lie and where frost penetration has taken place. The first human-generated palsa has begun its growth.

Palsa mires are quite complex in their water flow systems, their hydrology. The sedge meadow stage is clearly rheotrophic, being fed by water flowing through the low-lying marsh. But as the palsa becomes elevated above its surroundings by the ice core, water is received at its summit only in the form of rainfall (or snowfall), so the palsa mound is ombrotrophic.

The Arctic wetlands as a whole thus contain a mixture of different types of ecosystems, ranging from rheotrophic to ombrotrophic. On the whole the flow-fed systems predominate, especially in the drier regions to the north. In the tundra forest ecotone peat lands are both more abundant and are more likely to develop into an ombrotrophic form. It is perhaps ironic that the northern dry lands support a high proportion of the world's wetlands.

ANTARCTIC TUNDRA

The Antarctic continent lies in a region of high atmospheric pressure and is climatically a polar desert. The fact that the continent is also occupied by the world's largest ice sheet means that vegetation is absent over the entire area with the exception of some coastal fringes, including the Antarctic Peninsula and its associated islands. Over the ice sheet there is no vegetation at all. There are some algae that exist embedded in the surface layers of snow and ice, often turning the snow a pink or red color, but no true plants can exist here. Some fungi also grow in the snow, feeding upon the deposition of dust from the atmosphere that lands on the surface of the ice sheet. Only along the narrow coastal strip, where some rocks lie exposed by retreating ice, is there a sparse growth of lichens and mosses.

On the western edge of the Antarctic Peninsula there is a greater abundance of mosses and liverworts. Along with the associated islands of South Sandwich, South Orkney, and South Shetland, the Antarctic Peninsula is sometimes called the Maritime Antarctic. Parts of these lie outside the limits of summer pack ice, and their slightly milder climate allows some vegetation development. But even here there are only two species of flowering plants along with about 100 bryophytes (mosses and liverworts) and 150 lichens. Diversity is thus extremely low. Its poverty is all the more remarkable considering the flora of the southern tip of South America, Tierra del Fuego, less than 1,000 miles (1,600 km) to the north. Here there are 417 species of vascular plants alone.

The two flowering plants of the Maritime Antarctic, both found on the western side of the Antarctic Peninsula, are tussock-forming species. One is a grass, *Deschampsia antarctica,* and the other is a pearlwort, *Colobanthus quitensis,* both occupying soil-filled crevices among the Antarctic rocks. The limit of flowering plant growth in the Southern Hemisphere is thus 68°S. At 68°N lies the island of Iceland, and this has 329 species of flowering plant. Even Sptizbergen, Svalbard, at 78°N, has 110 species. This illustrates the benign impact of the Gulf Stream in the North Atlantic, taking warm water from the Caribbean north into the waters of the Arctic Ocean.

The lichens are by far the most colorful element in the flora, painting the rocks with yellow, orange, red, and black pigments. The function of these pigments is likely to be protective rather than photosynthetic. There is a very high level of ultraviolet radiation in the Antarctic, and this can be very damaging to unprotected cells, especially the genetic material. The intense pigmentation filters out most of the harmful rays and protects the delicate cells within (see sidebar "Lichens," page 143).

Beyond the Maritime Antarctic lie the subantarctic islands, and on these there is some development of tundra vegetation. These islands lie closer to the southern polar front, so they receive more precipitation than the Antarctic mainland or the Maritime Antarctic. The subantarctic islands are consequently much richer in plant species, with 56 flowering plants, 16 ferns, more than 400 mosses, and about 300 lichen species. Where conditions are harsh, the ground cover of vegetation may be less than 50 percent, consisting of cushion plants and tussocks of grasses and sedges, similar to the Arctic polar semidesert vegetation. But more northerly islands and sheltered locations have taller vegetation, with tall grasses and herbs up to six feet (2 m) high. Macquarie Island, south of New Zealand, has a large, cabbagelike plant called *Stilbocarpa polaris*. This has large rounded leaves with stalks up to three feet (1 m) in length, but its size decreases with altitude, rarely achieving eight inches (20 cm) in height at 1,000 feet (300 m) altitude. Wind is the main limiting factor on these subantarctic islands. The intensity of the wind is such that most exposed areas have the vegetation trimmed down to a low level, producing a ground-hugging cover that is sometimes referred to as feldmark. In structure it is equivalent to the Arctic dwarf shrub tundra.

Macquarie Island lies at a latitude of 55°S and carries a vegetation of tall herbaceous vegetation but no trees. At the same latitude in the Northern Hemisphere lie the boreal coniferous forests of Alberta, Quebec, Scotland, Sweden, and central Siberia. It is not until one reaches the Auckland Islands at 50°S that the first trees are found. On some of these islands there are forests of rata trees (*Metrosideros* species), members of the myrtle family. So the "timberline" in the Southern Hemisphere lies at an equivalent latitude to Vancouver, Winnipeg, and the southern tip of England.

The difference in vegetation with respect to latitude in the two hemispheres is only partly explained by the redistribution of energy by the ocean currents in the Northern Hemisphere. There is also the major difference in the arrangement of continental landmasses to be considered. The landmass in the south lies right over the South Pole, hence it lacks any oceanic influence, is extremely cold, and bears its immense ice sheet. Beyond this continent islands are the only land present, and these are blasted by some of the strongest winds on the Earth's surface. Not only are these islands exposed to extreme climatic conditions, but their isolation makes them difficult for plants to invade. Unlike the Northern Hemisphere, where these latitudes are occupied by continents that heat up in summer and have extensive land surfaces over which trees can move and equilibrate with climate, the south is fragmented, permanently cool, and buffeted by strong winds.

Vegetation may be sparse in the Antarctic, but other forms of life are not. By far the richest habitat of these southern regions is the Southern Ocean itself, and many of the creatures that harvest the products of that ocean, especially the seals and the birds, require land on which to breed. Although the islands of the Maritime Antarctic and the subantarctic are poorly vegetated, they supply adequate breeding habitats for a very wide range of birds, from penguins to albatrosses. Their lack of predatory mammals means that ground nesting is secure; only predatory birds are a threat to unguarded eggs and chicks. Tussock grassland is a perfect breeding ground for many albatrosses and penguins, though some members of the latter group are even capable of breeding in the total absence of any tundra vegetation. Similarly, seals are able to breed along coastlines above the zone hunted by orcas and leopard seals, so the islands provide a safe haven for these also. Only the arrival of human hunters placed such colonies at risk.

The Antarctic, then, has relatively little tundra habitat compared with the Arctic. Most of the tundra that is present consists of fragments along the continental coasts and on isolated islands.

■ ALPINE TUNDRA

The tundra habitats of the high mountains of the world have some features in common with the Antarctic tundra because they are also patchy and fragmented. Isolation means that travel between the tundra locations is difficult for mountain plants and animals. Hence, there may be little genetic exchange between sites, and each mountain may develop its flora and fauna separately. It is not surprising, therefore, to find that mountains show a great deal of variation in their vegetation and animal life, but there are some features that they all have in common.

Very high mountains, such as those of the Himalayas, are also dry because extremely cold air cannot retain much water vapor (see "Alpine Climate," pages 21–26), and most water has been lost as air has ascended from the lowlands. The lee side (downwind) of mountain summits is also drier than the side exposed to prevailing wind. But most mountains have higher precipitation than the polar desert ecosystems of the high latitudes. Low temperature and high winds, however, are common features. Day length varies with latitude, but growing seasons, as in the polar tundra, are limited, in this case because of low temperature rather than lack of sunlight.

The result of these general features of alpine climates is that vegetation at high altitude is subjected to a very similar range of stress to that of the polar tundra. The permanent snow and ice of the glaciers, as in the polar regions, is virtually devoid of life. Only a few unicellular organisms of the

group known by biologists as *protists* survive here. Some of these are photosynthetic and motile, such as the unicellular green organisms *Chlamydomonas* and *Chloromonas*. These have tiny whiplike structures by which they are able to swim in a film of water, and they replicate rapidly when the summer brings additional warmth. They have the green pigment chlorophyll, but they also can become red in color, in which case they may cause the surface snow to become pink. Microscopic predatory protists feed upon the photosynthetic organisms in the snow so that a complicated but unseen food web is constructed in the surface layers.

Where permanent snow and ice give way to exposed rocks, low-growing plants, including mosses, lichens, and cushion-formed flowering plants are able to establish themselves. Frost action and wind trimming of any growing points extending above the ground surface limit the height of this alpine vegetation. Some of the species of plants and animals in this alpine tundra are the same as those of the Arctic tundra, especially where the mountain ranges run uninterrupted from north to south, as in the case of the Rocky Mountains of western North America. Even in regions where this is not the case and mountains form isolated east-west running chains (as in Southern Europe), many plants and animals have wide distribution patterns extending into the Arctic. The likely explanation in this case is that these organisms had a much less fragmented distribution pattern at the end of the last ice age, when large areas of tundra habitat existed and linked the mountains together. As climate became warmer, the arctic-alpine plants and animals withdrew into the high latitudes and the high altitudes. The fragments of alpine vegetation in the Appalachian Mountains, such as Mount Washington, can also be regarded as relict patches of a formerly widespread postglacial tundra.

The landscape of the high but ice-free areas of mountains therefore closely resembles that of the Arctic tundra, and the term *fellfield* is often used to describe it. Lichens color the rock surfaces, small mounds of moss occupy crevices, and rounded tussocks of saxifrages are scattered over the ground and send up large flowers on stalks to attract pollinating insects in the spring. Pioneer plants of the high mountains include the aptly named sky pilot (*Polemonium eximium*), which holds the altitudinal record, growing close to the summit of Mount Whitney, California, and the alpine gold (*Hulsea algida*), a member of the sunflower family. These are opportunistic plants that establish themselves in rock crevices, where their one major advantage is that they face no competition from other species. Dwarf shrubs are also present, some hardly worthy of the term shrub because their woody parts lie prostrate on the ground or even beneath the surface. The least willow (*Salix herbacea*) and the net-leaved willow (*S. reticulata*) are examples of these low-lying shrubs in the European Alps. Ground cover near the snow line is very incomplete, with more than 50 percent or more of the soil or rock surface uncovered.

Few mammals occupy the fellfield habitat, but the American pika (*Ochotona princeps*) and various marmots (*Marmota* species) do manage to survive. The pika remains active through the winter, feeding upon its larder of stored hay, while the marmot fattens in the fall and sleeps through the winter period. These small mammals are predated upon by large raptors, such as the golden eagle (*Aquila chrysaetos*). In summer jackrabbits (*Lepus californicus*), snowshoe hares (*L. americanus*), and bighorn sheep (*Ovis canadensis*) move into the region and exploit the season of higher productivity. There are some resident insects, such as the ice crawlers (*Grylloblatta* species), and a number of springtails (Collembola) that consume dead organic material, and spiders use their webs to catch any flying insects blown up to the fellfields from the lower altitudes. Some butterflies migrate into the flower-rich alpine zone during summer to take advantage of the nectar supply.

Soils in the spring are often wet because of the melting snow and ice. The high angle of the Sun, especially in low-latitude mountains, causes rapid rise of temperature during the day, and this results in fast decay of ice in the spring. Wet marshy areas, sometimes called "flushes," are common in alpine habitats and are rich in flora and invertebrate fauna. These are the sites where many of the most spectacular of the alpine flowers are found, including the gentians. Showy flowers are a frequent adaptation for alpine plants because flying insect pollinators are relatively scarce, and the plants must compete for their attention (see sidebar "Pollination," page 126).

The topography of most mountain tops is generally more varied than polar tundra. Extensive flat areas are scarce, while steep slopes, cliff faces, rock scree, and deep gullies are more common. This means that microclimate is diverse, and the vegetation cover varies accordingly. Aspect is an extremely influential factor, and considerable differences in vegetation occur depending on the extent to which a slope or cliff face is exposed to sunlight. Deeply shaded locations (north-facing in the Northern Hemisphere) may retain snow patches until late in the year or even through the entire year. Each spring they contract as the temperature warms, and each fall they grow again. As a consequence, a pattern of plant zonation develops around snow patches depending upon the tolerance of different plants to varying periods of snow burial. In the zones farthest from the summer snow limit are the species that require longer periods for growth and fruit production, such as the alpine snowbell (*Soldanella alpina*). The inner zone, which may be clear of snow for only a few weeks in the year, is usually dominated by mosses and liverworts. One species of moss that is virtually restricted to this habitat is *Polytrichum sexangulare*. As in the case of high altitude pioneer species, its highly spe-

cialized requirements ensure that very few other plants are able to compete with it for space or resources.

Ridges between these snow hollows are exposed to high intensity sunlight in the alpine summer, and the main environmental problem is drought. Some of the plants found in this type of location, such as pussy paws (*Calyptridium umbellatum*) and members of the buckwheat family (Polygonaceae), have desert relatives. These dry alpine habitats are the mountain equivalent of the polar desert.

Dwarf shrub vegetation, equivalent to the polar tundra zone, is commonly found in the alpine tundra, especially when the underlying rocks are acidic. Alpine laurel (*Kalmia polifolia*), blueberry (*Vaccinium occidentale*), and several species of manzanitas, azaleas, and rhododendrons occupy this zone in the mountains of the western United States. In Europe this zone is particularly extensive and is dominated by members of the heather family (Ericaceae). In the Himalaya Mountains, various species of rhododendron and juniper form a low scrub, while on the tropical mountains of East Africa one heath species, the giant heath (*Erica arborea*), is the dominant species. The presence of a zone of scrubby heathland of relatively low, evergreen shrubs, however, is a feature of many of the alpine habitats of the world. Why evergreen leaves should prove advantageous to a plant under these conditions is still widely disputed among botanists (see sidebar "Why Be an Evergreen?" pages 116–117).

Meadows—dominated by grasses but containing an abundance of other flowering plants—are also a common feature of many mountain regions, as shown on page 81 in the photograph from Yellowstone National Park in the United States. The Rocky Mountains, the Cascades, and the Sierra Nevada all have a range of meadow habitats, including both wet and dry examples. Although grasses, such as *Calamagrostis*, *Festuca*, and *Poa* species, together with sedges dominate these meadow areas, they are made colorful in early summer by an abundance of flowers. Lupines (*Lupinus* species), paintbrushes (*Castilleja* species), and phloxes (*Phlox* species) abound, and these tall, brightly colored flowering herbs make the alpine meadows one of the most impressive of mountain habitats. The meadows often occupy valleys and gentle slopes and are an important habitat for mountain insects, especially butterflies and moths. They are grazed by deer, especially the mule deer in the United States and the chamois in Europe, and it is possible that this grazing is an important factor in maintaining the habitat free of invasive shrubs and trees. Small mammals, especially voles and gophers, burrow in meadow soils and also consume the vegetation. Farmers have long used these habitats to support herds of grazing animals, especially sheep in summer, and this has led to extensive damage to their diversity. In Europe alpine meadows are often harvested for

Alpine meadows, such as this one in Yellowstone National Park, are renowned for their abundance of colorful flowers. In the brief summer, when pollinating insects are still relatively scarce, plants compete to attract insect visitors. *(Bryan Harry, Yellowstone Digital Slide File, Yellowstone National Park)*

hay during the summer. The harvested grasses and herbs are dried in the sun and are stored for use as fodder for cattle kept in stalls through the winter. This mowing management is far less damaging to plant diversity than grazing. The extensive human exploitation of alpine meadows throughout the world has confused the general picture of vegetation zonation with altitude. Sometimes the timberline has been lowered by felling and grazing, producing artificial meadows at lower altitudes.

Alpine meadows, especially wet alpine meadows, bear a considerable resemblance to the polar tundra grasslands, as shown in the photograph on page 82. Some plant species, such as the cottongrasses, are found in both polar and alpine habitats, but the polar regions tend to have a lower overall diversity of species. In the Rocky Mountains about 40 percent of the plant species of the alpine wet meadows are also found in the polar tundra, but in the Sierra Nevada only 20 percent are found in both regions. Stone patterns and even peaty polygons are sometimes found in the wet alpine meadows, but there is no permafrost, so the action of frost heaving and patterning is limited. In the northern Appalachian Mountains, the vegetation is closely related to that of the Canadian Arctic. In part this is due to the climate, which is often cloudy and foggy, quite different from that of the Rockies.

At the junction between the alpine tundra and the forest of lower altitude, a zone of stunted trees, or krummholz, often occurs, depending upon the factors that control the timberline at any given location. If grazing, human management, or fire are the main determinants of the forest limit, then the boundary with the forest may be sharp, but in situations where climatic factors, particularly temperature and wind action, are important a transition zone of krummholz is often present. As in the case of the boundary between polar tundra and boreal forest, adequate summer warmth together with the strength and abrasive quality of the wind often control the establishment of trees in the transition zone. Even when established, trees in this zone rarely achieve full growth and usually fail to set seed.

What species of tree forms the timberline varies between mountains and may even vary with aspect on a single

Wetlands are frequent in alpine tundra regions, especially in hollows and valleys where water collects. This Austrian wetland is dominated by cotton grasses (*Eriophorum scheuchzeri*).   (*Peter D. Moore*)

their history, and such variation is nowhere more evident than in the tropical mountains.

■ TROPICAL ALPINE HABITATS

Mountains in the low latitudes lie in regions of warm climate. Therefore, they need to be exceptionally high if they are to bear tundra vegetation. There are such mountains, but they are widely scattered, and they are often volcanic in origin. This means that they have passed through periods of extreme geological instability, and they may be relatively young compared with other mountain systems. It is not surprising, therefore, that their alpine plants and animals are unusual.

Situated at 19°N in central Mexico is the mountain Citlatepetl (Pico de Orizaba). Its summit lies at an altitude of 18,700 feet (5,700 m), and the timberline is around 12,500 feet (3,800 m). Like many tropical mountains, it bears very few species that can truly be regarded as alpine, and those flowering plants that have invaded this habitat fail to grow at very high altitude. The paintbrush species *Castilleja tolucensis* takes the record as the highest of this mountain's flowering plants, but even this species fails to grow above 15,000 feet (4,570 m). Flowering plants grow at considerably greater altitudes in the mountains of higher latitudes, such as the Himalaya and the southern Andes of South America, sometimes reaching around 20,000 feet (6,000 m). Why is this mountain together with the other mountains of Mexico so poor in alpine species that are capable of existence at very high altitudes? The answer is probably the geological youth of these mountains coupled with their isolation and low latitude position. Because they are young and volcanic, the flora of the Mexican mountains must have accumulated and evolved relatively recently. Their isolation from other mountainous regions means that they have not received immigrant species by normal dispersal processes. And their low latitude means that the mountains of Mexico were not surrounded by the extensive lowland tundra habitat that extended over much of temperate North America at the end of the last ice age, so they were not in the position of such mountains as Mount Washington, which received arctic-alpine communities retreating from the advancing forests of the postglacial period.

Mosses and lichens spread by microscopic, dustlike spores that are carried in the air, so they have no problem in dispersing even to the most isolated of mountains. The zone immediately below the snow line (about 16,400 feet; 5,000 m) on Mexican mountains is dominated by mosses and lichens and bears considerable resemblance to the alpine fellfields of temperate mountains except for the lack of flowering plants. Once within the range of flowering plants, however, grasses dominate the vegetation and produce alpine tussock grass-

mountain. In the central Rocky Mountains south-facing slopes usually have a timberline of pines, while north-facing slopes have spruce and fir, particularly subalpine fir (*Abies lasiocarpa*) and Engelmann spruce (*Picea engelmannii*). In the Sierra Nevada Mountains whitebark pine (*Pinus albicaulis*) is one of the commonest trees in the subalpine zone and is often found contorted in the krummholz zone. In the European Alps a deciduous conifer, the European larch (*Larix europea*), is usually the last of the forest trees before the tundra zone is reached, sometimes accompanied by the arolla pine (*Pinus cembra*). In the Himalaya Mountains Himalayan birch (*Betula utilis*) and Himalayan fir (*Abies spectabilis*) form the timberline.

Alpine tundra habitats, therefore, have many similarities with polar tundra but also some differences. They are much more variable from place to place, largely because of

land. As in many such communities, grazing and fire are common in this zone. The timberline is usually composed of pine forest, mainly of the species *Pinus hartwegii.*

Farther south, in Venezuela, the mountains of the northern Andes are very similar to those of Mexico in their habitats. The alpine zone of the Andes Mountains is called the *páramo,* and its main characteristic, especially in the north, is that it is both cold and wet. Warm trade winds from the Caribbean bring moisture to Venezuela, and this results in high humidity in the mountains these winds encounter. Unlike most high mountains, drought is rarely a problem here. One result of this is a great deal of temperature variation from one day to the next, depending upon cloud and fog density. As is often the case in tropical mountains, daytime temperatures can be very high, while frost can occur at night, so plants and animals need to be able to cope with constantly alternating extremes. Ground-hugging plants dominate the vegetation, but there are also some giant herbs belonging to the sunflower family (Asteraceae). These plants, members of the genus *Espeletia,* have thick, tall stems covered with dead leaf bases and a crown of upright hairy leaves at the top. Plants with relatively short, fat stems and densely spaced leaves are called *pachycaul,* and this type of growth is found on many of the high tropical mountains. The covering of dead leaf bases serves an important function, insulating the inner living parts of the stem from night frosts. The flowers of some species are pollinated by humming birds.

There are some true trees that are able to survive in the alpine zone of the Andes, such as *Polylepis sericea,* a member of the rose family (Rosaceae). This is an exceptional tree that manages to grow regularly at 13,800 feet (4,200 m) and has been recorded as high as 16,400 feet (5,000 m), making it the highest tree in the world in the altitudinal sense. It frequently grows well above the main timberline in the Andes, around 10,500 feet (3,200 m), and it seems to depend upon sheltered microsites that act as traps for solar heat, usually among boulders. The rocks also act as heat absorbers, releasing at night the heat they gain during the day, thus elevating the local nighttime temperature. In these condition the local temperature can average up to 16°F (9°C) warmer than

The alpine zone of high mountains in equatorial regions is home to some remarkable plants. Giant senecios, such as the species *Dendrosenecio elgonensis* shown here on Mount Elgon, Uganda, East Africa, are treelike forms of a plant group that is normally herbaceous. *(Peter D. Moore)*

surrounding areas, often warmer than conditions 3,000 feet (1,000 m) lower at the treeline.

The timberline itself in the northern Andes is unusual, consisting of very moist and stunted *cloud forest,* sometimes called *elfin forest.* The trees are low in stature, gnarled in appearance, and lie in a layer of almost constant cloud and mist. The cool and moist conditions encourage the growth of mosses, lichens, and ferns, covering the branches of the trees and hanging in festoons of dripping vegetation. The relative humidity rarely falls below 100 percent. This cloud forest is the high-altitude equivalent of the rain forest that lies at lower altitudes. The cloud forest often passes directly into the alpine zone, but sometimes there is a narrow zone of shrubs present. Biogeographers are not entirely sure about what factors determine the timberline in the tropical mountains. Humidity decreases above the timberline but remains high. The likely factor is frost. The timberline often lies at an altitude where regular nighttime frosts first set in.

In Southeast Asia the tropical rain forests are occasionally interrupted by high mountains, such as the high mountain ranges of New Guinea and Mount Kinabalu in northern Borneo. This mountain rises to an altitude of almost 13,500 feet (4,100 m), and although this is not high enough for snow cover, it has geological evidence of past glaciation. The summit is now bare rock, plates of granite split by frost action. Scattered plants consist largely of grass tussocks and occasional dwarf shrubs, evergreen species of *Rhododendron* and *Leptospermum.* These form tight cushions on the rock, sending woody roots deep into the crevices. Rainfall is abundant even at high altitude, and the lower slopes around 10,000 feet (3,000 m) are clothed with open scrub vegetation descending into moss forest. The moss forest of Southeast Asian mountains is similar to the cloud forest of the Andes, humid and rich in mosses and lichens hanging from the low tree canopy. Orchids are also abundant at the timberline here, and one of the important trees is a conifer, the celery pine (*Phyllocladus* species).

The high mountains of East Africa have been intensively studied by ecologists. Unlike the Andes Mountains, which form a long chain running down the western side of South America, the tropical African mountains are scattered and isolated from one another. Mount Kenya and Mount Kilimanjaro lie to the east of Lake Victoria, and Mount Elgon, the Ruwenzori Mountains, and the Virunga Volcanoes lie to the north and west of the lake. Mounts Kenya and Kilimanjaro together with the Ruwenzori Mountains are high enough to have glaciers despite the fact they all lie close to the equator. The alpine zones of the mountains are clothed with tussocky grasses and small shrubs, such as the white, hairy-leaved species of the genus *Helichrysum.* The white color of this shrub reflects much of the sunlight in daytime, preventing overheating, but conserves heat at night by means of its dense hairs, illustrating the main problem of life in the tropical mountain tundra, namely, rapidly changing temperature.

The pachycaul plants of the Andes have their equivalent species on the East African mountains, but here they are close relations of the temperate ragworts (genus *Senecio*), called *Dendrosenecio* (*Dendro-* means "tree"). A view of these trees near the summit of Mount Elgon in Uganda is shown in the photograph on page 83. These tree groundsels bear a superficial resemblance to Joshua trees in their rugged, branched structure, but their lower trunks are densely covered with dead leaf bases, similar to those of *Espeletia* in the Andes and serving a similar insulating function. The night temperature often drops to 25°F (−4°C), but within the leaf bases the temperature is maintained above 36°F (2°C), preventing frost damage to the stems. The tree groundsels are not alone among the pachycaul alpines of East Africa; giant lobelias (*Lobelia* species) are also present in the high-altitude alpine tundra. These are similar to the *Dendrosenecio* plants but usually form stout, unbranched trunks with flowers at the top.

The isolation of the East African mountains, far from any other mountain group and surrounded by tropical forests and savannas, has led to the development of a distinctive vegetation, quite different from that of any other mountain range in the world. Most tundra plants from other regions have not been able to reach these remote mountains, so a new range of alpines has evolved out of the tropical forest and savanna species that were available. Hence, local ragwort and lobelia species have taken on the role of tundra species and have developed the adaptations that plants of these stressful environments require. Close analysis of the individual mountains reveals that each of the main mountains has species that are found there and nowhere else. These are said to be endemics. *Dendrosenecio, Lobelia,* and *Helichrysum* all have endemic species that are restricted to each of the high mountains. *Dendrosenecio elgonensis,* for example, occurs in the alpine tundra of Mount Elgon and in no other locality of Earth. The East African mountains, therefore, provide a laboratory of evolution similar to that of the Galapagos Islands that inspired Charles Darwin to devise the theory of evolution by natural selection. In many respects they are islands, isolated from one another and from the rest of the world by uncrossable oceans of forests.

The lower alpine zone of the African mountains is occupied by evergreen shrubs, dominated by the tree heather (*Erica arborea*). This zone is equivalent to the evergreen dwarf shrub zone of the polar tundra, but the shrubs here are not dwarf. The tree heather grows to a height of 16 feet (5 m) and can form dense and impenetrable forest on the mountain slopes. It merges with bamboo forest at its lower levels, which is the equivalent to a timberline on the East African mountains at an altitude of around 10,000 feet (3,000 m).

The tundra habitats and vegetation of alpine tundra are thus much more varied than that of polar tundra. Climatic conditions vary with latitude and with local conditions, such as wind directions and the proximity to an ocean, which brings additional humidity. The plants that compose the alpine vegetation have a more diverse set of origins, often being derived from the local flora, especially in the case of the tropical mountains. But there are certain features that the alpine sites have in common. The overall cold conditions have led to the plants and animals becoming adapted in similar ways, whatever their origins.

■ CONCLUSIONS

Tundra is not a uniform biome but contains a range of different types. Polar and alpine tundra are similar in their low annual average temperatures but differ in their daily and seasonal variations. But the vegetation in all tundra habitat has a common structure. The term *structure,* when applied to vegetation, means its architecture, and all tundra vegetation is low in stature, either forming a carpet close to the ground or developing a series of cushion forms constructed of evergreen dwarf shrubs or tussock grasses.

The most extreme form of tundra is the polar desert, which develops at very high latitudes of the High Arctic. Not only is the growing season short and the prevailing temperature low, there is very little precipitation. Vegetation cover is usually incomplete, consisting largely of lichens and mosses with scattered flowering plants forming mats or cushions. Farther south, in the Low Arctic, a complete vegetation cover develops, consisting mainly of dwarf shrubs or extensive areas of tussock grassland. Cottongrasses and sedges may dominate some habitats. This vegetation gives way to tall shrub tundra in which the structure of the habitat is more complex, and dwarf species of trees, such as dwarf birches and willows, become more abundant.

At its junction with the northern boreal forest, the Arctic tundra forms a transition zone, the forest tundra. Here there is a greater predominance of true trees, including birch, alder, aspen, and cottonwood, and the overall growth of vegetation is taller. Trees that dominate the forests of the boreal regions may extend into this zone, but often in a stunted and contorted form. The gnarled, dwarf forests of this transition zone are called krummholz. Spruce, larch, tamarack, and pine all contribute to this krummholz vegetation in different regions of the Arctic.

Coastal regions in the Arctic have few truly maritime communities, such as salt marshes and sand dunes, because there is very little tidal movement in the Arctic Ocean and hence only a narrow band of intertidal zone. Arctic vegetation often extends right down to the shore. The most important coastal habitats are the cliff communities of nesting and roosting birds and shore platforms where seals are able to breed.

Despite low precipitation in the Arctic, wetlands are frequent because of the low evaporation rate of surface water. Sedge and cottongrass marshes are important habitats for many migrant wildfowl and waders. The bogmoss, *Sphagnum,* plays an important part in the development of many peat lands, creating deep deposits of acidic peat where conditions permit. In these peatlands palsa mounds may develop, rise, and collapse, generated by the development of ice mounds within the peat and destroyed when erosion and solar heating of exposed dark peat leads to a meltdown.

Polar tundra in the Antarctic regions is far less well developed because of the extent of the Antarctic ice sheet and the general lack of land in the appropriate latitudes of the Southern Hemisphere. Antarctic islands and the coastal region of the Antarctic Peninsula are the main locations of polar tundra in the south. Diversity of species is low, but some groups of organisms, such as albatrosses and penguins, have their main centers of diversity in the Antarctic.

Alpine tundra is much more fragmented and scattered than polar tundra because of its location on high mountaintops. The basic vegetation structure of fellfields is similar to that of Arctic tundra, with mat-forming and cushion-forming evergreen plants predominating. Alpine meadows, dominated by grasses, sedges, and flowering herbs, are present in many mountain areas but sometimes owe their origin and maintenance to grazing pressures by domestic animals. Dwarf shrub vegetation forms a zone below the open alpine tundra, just as it does in the Arctic, and there is often a tall shrub or krummholz zone at the transition with forest. The timberline itself is generally determined by temperature, but wind is also important together with fire and grazing pressure, especially if human influences are strong.

Tropical mountains are often well supplied with rainfall, and they frequently develop a zone of cloud forest below the alpine zone. The tundra is also usually wet and misty and experiences large fluctuations of temperature between day and night. This exposes plants to considerable stress, and some unusual adaptations are found, such as the pachycaul habit. The isolation of many tropical mountains, such as those of East Africa, has also led to unexpected evolutionary developments of alpine forms from tropical groups. Tropical alpine tundra, therefore, is an ideal habitat in which to study the evolution of new species. Many of the structural characteristics of tropical alpine tundra habitats are similar to those of the polar tundra, but the composition of the vegetation and its history is very different.

4

The Tundra Ecosystem

Every individual organism on the Earth interacts with its surroundings, and the organisms living in tundra habitats are no exception. A terrestrial organism living on the surface of the land is in contact with the atmosphere and the soil, but it will also interact with other members of its own species. An individual organism, especially if it is a relatively advanced animal, will have social interactions with other members of its species, and a collection of individuals of the same species interacting in this way is called a *population.* A collection of penguins living together in a breeding colony is an example of a population. An organism also interacts with other living creatures belonging to different species. Some of these may be its prey, while others may be its predators; some will be harmful parasites, while others compete with it for the same food resources. A collection of different species living together and interacting in all these different ways is called a *community.* Finally, a collection of living organisms coexisting in a community interacts with the nonliving world that forms a setting for all life. Plants are rooted in the soil, from which they absorb the minerals they need to grow, and they take up carbon dioxide gas from the atmosphere, which they convert into sugars in the process of photosynthesis. Animals receive most of the energy and the chemical elements they need in their food, which may consist of plant materials or prey animals. Animals may also drink water or absorb it through their surfaces and supplement their mineral intake in this way; some may even eat soil if they run short of certain elements. A community of different species of animals and plants living in the physical and chemical setting of the nonliving world is called an *ecosystem.*

■ WHAT IS AN ECOSYSTEM?

The concept of an ecosystem is an extremely useful one to ecologists and conservationists. It provides an approach to the study of the natural world that can be applied at a range of different scales. A single tussock of sedge in the tundra can be regarded as an ecosystem. Using this approach one can study the ways in which the energy and mineral elements contained in the growing plant are consumed by grazing animals or decomposed by fungi and bacteria. The microbes living in the soil beneath the sedge are then eaten by invertebrate animals, which in turn may be fed upon by carnivores, such as beetles, and these larger creatures attract visiting insectivorous birds, which consume the beetle and thus harvest the energy that it contains. Each organism is obtaining energy from its food and uses some of this energy in such processes as growth and movement (see sidebar on facing page). As the bird flies away with its prey, it is removing energy from the sedge tussock and transporting it elsewhere. In this example the ecosystem concept is being applied on a very small scale. It is possible, on the other hand, to regard the entire tundra as an ecosystem, in which case the single clump of sedge is simply a part of a greater whole in which the photosynthesis of the vegetation is trapping the energy of sunlight, storing it in leaf and root tissues, and eventually providing an energy source for the grazing herbivores and the assemblage of microbes and animals inhabiting the decomposing materials of the tundra soils. The insectivorous bird is now part of the same ecosystem, performing its own part in the organization of the whole.

Although it is possible to use the ecosystem idea at many different scales, all ecosystems have certain features in common. All ecosystems have a flow of energy through them. The source of energy for most ecosystems is sunlight, which is made available to living organisms by the photosynthesis of green plants. There are some bacteria that can photosynthesize, and there are some that can obtain energy from nonsolar sources, such as chemical reactions with inorganic materials, including the oxidation of iron. But these are generally of little significance in most ecosystems compared with the contribution of green plant photosynthesis. Some ecosystems, such as a mudflat in the estuary of a river, import energy from other ecosystems. Dead plant materials, rich in the energy derived ultimately from sunlight, are brought into this type of ecosystem, and these imported

Energy

Energy is the capacity to do work. All animals, plants, and microbes need energy for their daily life because they need to grow, or move, or reproduce. When an animal forages for food, it is using some of its energy in the pursuit of yet more energy. When a plant absorbs elements from water in the soil, it expends energy. Many of the biochemical processes that take place in individual cells also demand the expenditure of energy. Animals need energy to move, hunt, and reproduce. So all living creatures constantly demand energy.

Energy can be divided into two major types, kinetic and potential. Kinetic energy is the outcome of the motion and the mass of an object. Water running along a stream, a rock falling down a mountain slope, and the wind blowing in the trees all exhibit kinetic energy. Potential energy is stored in an object, ready to be released. A rock perched on the top of a cliff has the potential to release kinetic energy if it falls over the edge. Many chemical compounds contain energy in the bonds that hold the molecules together. A sucrose molecule, for example, can be burned to release its component carbon atoms as carbon dioxide, and it releases kinetic energy in the form of heat as it does so.

Living organisms can capture the energy they need in either kinetic or potential form. A green plant, for example, absorbs electromagnetic radiation, which is a form of kinetic energy, when it intercepts sunlight. The green pigment chlorophyll is able to capture this energy and transfer it to chemical potential energy for use in the trapping of carbon dioxide molecules and reducing them to sugar. When a herbivore eats part of a plant, it removes energy-rich chemicals, including sugars and starch, which can then be transformed into animal tissues or be respired to provide the kinetic energy needed for movement and other activities.

Energy occurs in many different forms and can be converted from one form into another. A burning piece of wood produces light and heat; falling water can turn a turbine and generate electricity. The movement of energy between its many forms obeys two fundamental laws, the laws of thermodynamics. The first law of thermodynamics is concerned with energy conservation. It states that energy cannot be created or destroyed, so each energy transfer can be described by an energy budget equation that must balance. The second law of thermodynamics states that no transfer of energy from one form to another can be 100 percent efficient. A turbine can never capture all the kinetic energy released by falling water, and the electricity generator will never be able to convert all the energy released by the rotation of the turbine into electricity. All energy transfers are accompanied by energy losses, often in the form of waste heat. When a person eats a potato, the energy contained in the stored starch is not completely transformed into human flesh. Wastage and losses may account for as much as 90 percent of energy intake, and only a small proportion of the total energy is captured by the consumer. The second law of thermodynamics explains why the transfers of energy from Sun to plant, from plant to herbivore, and from herbivore to carnivore all involve losses and wastage.

sources of energy and chemical elements supply the needs of the animals and microbes that feed upon them. As energy flows through an ecosystem, one can distinguish certain groups of organisms that play different roles. The plants are *primary producers,* fixing solar energy into organic matter; they are said to be *autotrophic* in their nutrition, which literally means that they can feed themselves. Some of the energy they trap from the Sun is used in the energy-consuming activities of the plant, such as the uptake of chemical elements against a concentration gradient from the soil. Energy needed for such purposes is released from storage and is liberated by the process of *respiration.* The remaining energy is used in building new plant materials: leaves, stems, roots, flowers, and seeds. Herbivores are *primary consumers;* they are dependent on plants for energy, so they are said to be *heterotrophic,* meaning that they need to be fed by others. Predatory animals are also heterotrophic. They, too, depend ultimately on plants, but indirectly because they feed upon the herbivores or upon other animals that eat herbivores. These are *secondary* and *tertiary consumers,* respectively. They occupy different positions in a hierarchy of feeding, sometimes referred to as a *food web* in the ecosystem. All these organisms release energy according to their immediate needs from stored chemicals, using the process of respiration in their cells.

The waste materials produced by egestion (defecation) and excretion by living organisms, together with the dead parts or dead bodies of those individuals that escape consumption by predators and survive long enough to die a natural death, are used as an energy source by the decomposer organisms. Animals that eat dead plant and animal materials and derive energy from them are called *detriti-*

vores. But ultimately it is the microbial *decomposers* in the ecosystem, the bacteria and fungi, that are responsible for the degradation of all the residual, energy-rich materials in the process of decomposition. Nothing is wasted, nothing is lost. All the energy entering the ecosystem is finally used up and is dissipated as heat, released by the respiration of all the various plants and animals involved in the food web.

While energy flows through the ecosystem and is eventually lost, chemical elements are recycled around the ecosystem. Carbon atoms, for example, are taken up by plants in a gaseous form as carbon dioxide, and these atoms become incorporated into carbohydrates. They may be stored in this form or as fats or may be converted to proteins by the addition of the element nitrogen derived from the soil. Carbon compounds may also be converted into different kinds of molecules by the addition of other elements, such as phosphorus to make phospholipids, or sulfur to construct some types of amino acids, the building blocks of proteins. The materials built into plant bodies are consumed by animals. A proportion passes through the body of the consumer to be voided as waste, while some becomes incorporated into the body of the animal. Respiration results in the release of carbon back into the atmosphere as carbon dioxide gas, while the nitrogen and phosphorus together with other elements are lost to the organism by excretion or by death. These elements enter the soil and are gradually released to the environment in the process of decomposition. Chemical elements are thus recycled and can be used over and over again. Energy passes once through the ecosystem and is finally dissipated, but chemical elements may cycle round and round the ecosystem indefinitely. The constantly turning wheel of element motion is called a *nutrient cycle.*

In addition to the cycle of nutrients, however, there is also usually an import and export of elements to an ecosystem. A stream entering an ecosystem will bring dissolved minerals from the ecosystems in the catchment area. Rainfall will also bring a supply of elements, its richness depending on how close the site is to the ocean. Animals may migrate into an ecosystem and bring elements from outside, such as the salmon that migrate up rivers to breed and die, bringing elements they have collected in their ocean feeding grounds. Plant materials, including twigs and leaves, may be washed or blown into a pond, supplying a source of elements from another ecosystem. But just as these processes bring elements into an ecosystem, they can also take them out. Streams can leave an ecosystem, animals can move out, and plant material may be blown away. So the study of an ecosystem involves a consideration of its energy and nutrient imports and exports in order to understand the balance of supply and loss in both energy and mineral elements. Before an ecosystem can be fully understood, it is necessary to calculate its energy and nutrient budgets.

■ PRIMARY PRODUCTIVITY IN THE TUNDRA

The food webs of the tundra are based upon plants and the energy they fix from the Sun in the process of photosynthesis. Energy capture depends upon the pigment chlorophyll, and the tissues containing the pigment must be exposed to light in order to intercept the incoming radiant energy. The most important organ used by most plants for light capture is the leaf, which is often broad, flat, and thin in order to expose a maximum surface area to the sunlight. Stems and branches are used to support the leaves and hold them in an exposed position, but they can also be used for photosynthetic activity. Many stems, especially in herbaceous plants and small shrubs, are green and assist in the vital process of energy gathering. Even parts of the reproductive organs may be green and photosynthetic, such as the sepals on the outer part of the flower and the wall of the carpel in the developing fruit. A plant cannot afford to waste any opportunity of using an organ exposed to light in order to boost its food production. There are some organs, however, such as the petals (colored to attract pollinating insects) and bark (dead outer cells protecting the water-conducting elements in the wood), where other functions are even more important than energy trapping. Roots are important for stability and for taking up water and nutrients from the soil, but their position below the ground obviously precludes any photosynthetic activity. As a plant grows, therefore, it is faced with a strategic problem in the various ways it can allocate its growth energy. Should it allocate materials to the production of more leaves? Or should it place more emphasis on structural support? Plants have evolved a number of different strategies in their patterns of *resource allocation* (see sidebar on page 89).

However a plant allocates its resources, the total amount of energy that accumulates in an area during the course of a year provides the energy resource for the animals that live in that area. Ecologists need to measure this total primary productivity so they can gauge the efficiency of an ecosystem's vegetation as a support system for its animal and microbial life. Measuring the primary productivity, however, is difficult. The most commonly used method for terrestrial vegetation studies is to harvest sample areas of the plant community at different times during the growing season. The researcher selects an area of ground, often about a square yard in area, and completely crops all its vegetation down to soil level. The next step is to sort the harvested material into the different species contained in the vegetation and then to dry the samples in an oven. The drying process is designed to remove all water so that the results are not affected by the degree of wetness of the samples. Care must be taken, however, not to let the temperature rise too high, because this might cause the evaporation of other components of the plants, such as oils, which may constitute important reservoirs of energy. For detailed studies of energy flow it is useful to conduct one additional analysis to obtain the energy content, or calorific value, of the cropped material. This is achieved by reducing the dried samples to ash in a bomb calorimeter. A bomb calorimeter is a piece of equipment in which a known weight of material can be ignited to release the energy it contains. The rise in temperature of the calorimeter is directly related to the energy released, so the energy content of the sample can be calculated.

As a result of this type of study scientists can calculate how much new plant material is added to a given area of the ecosystem each year, and this can be expressed either as dry weight per unit area per year, or as energy (calories) per unit area per year. There are several problems with this type of study, however, that need to be considered. The first problem is that of sampling. Almost all types of vegetation are extremely varied and nonhomogeneous. Selecting relatively small samples as representatives of the entire ecosystem can therefore be misleading. The only way to overcome this problem is to take many samples, but the analytical process is time consuming, and time is limited in all research projects, especially in one where successive samples are needed through the course of a growing season. The second problem is that it is almost impossible to sample the plant growth that is taking place below the surface of the soil. Roots are very difficult to separate from the soil in which they grow, and very fine roots have a short life span, often only a few days, so it is well nigh impossible to estimate all the productivity that goes on beneath the ground.

A third problem is loss of energy from the ecosystem even while the productivity is taking place. Some of the energy fixed by plant photosynthesis is used by the plant in its own activities and metabolism, so what accumulates at the site over time is actually the excess of productivity over and above the needs of the plant. This is called the *net primary productivity* as opposed to the *gross primary productivity*, which is the total amount of energy fixed in a given time. In a natural ecosystem under field conditions, however, there are additional drains on the fixed energy. Grazing animals, both large and small, will be taking their crop of

Resource Allocation

As a plant gains in weight, the newly acquired material must be moved to appropriate parts of the plant body. In the early stages of plant growth the new products of photosynthesis may be directed toward the construction of new leaves, in which case all the investment is used for the development of productive organs in which more photosynthesis can occur. But root development must keep pace with the growth of the shoot so that the plant remains stable in the soil and a supply of water and mineral nutrients is maintained for continued transpiration and growth. As the shoot continues to grow it needs extra support, so instead of making photosynthetic organs, such as leaves, the plant must expend an increasing amount of its energy on support structures, including stems and branches. In the case of shrubs and trees, support tissues include wood, which holds the plant relatively rigid and assists in the development of a complex canopy, enabling the leaves to reach the light and continue their productive function. This allocation of new materials to different parts of the plant is called *resource allocation*.

As plants grow, therefore, they must devote an increasing amount of their energy to nonproductive support tissues, especially roots and rigid stems. In its early stages of growth a plant invests energy almost at a compound rate of interest. This means that every portion of new material is reinvested in production, so the growth is potentially logarithmic. But the older the plant becomes, the lower the proportion of new production (the "interest" gained) that can be reinvested in this way. As more support tissues are needed, the growth rate falls from a logarithmic one and becomes slower, eventually reaching a plateau as the plant attains its maximum size.

Eventually the plant begins to reproduce, and the investment of new material is directed away from both productive and support tissues into the development of flowers and then fruits. In many annual plants the proportion of energy invested in reproduction is very high because such species rely heavily on seed production for survival into the next growing season. Long-lived perennials devote proportionally less energy to seed production (see sidebar "r- and K-Selection," page 125).

Grazing animals, which rely on plant products for their food, thus have a number of options for feeding. They may attack leaves directly or may feed on roots or timber. Alternatively, they may gain energy from nectar, or pollen, or the seeds that are subsequently produced by the plant. Resource allocation by plants thus presents consumer organisms in the ecosystem with a range of feeding strategies.

energy from the plants even while the experiment is proceeding. It is not difficult to fence an area to exclude large grazers, such as caribou, but keeping out all invertebrate grazers, from caterpillars to grasshoppers, is very difficult. A full study of productivity therefore requires further work on the losses due to these small grazers. It is also possible that some plant material will die during the experiment. Old leaves may wither and fall, providing a source of energy for the detritivores and decomposers. One way of checking this source of loss is to place trays within the vegetation that collect any falling plant tissues.

As a result of all these complications to the study of plant productivity, the results available from studies around the world are quite variable. The table below shows a range of productivity values from a number of different types of tundra habitat and also data from some other ecosystems for comparison. These data are expressed as pounds dry weight per square foot per year (or kg m^{-2} y^{-1}), and it is important to remember that productivity is a *rate,* not a quantity. Harvesting a sample of vegetation at a given point in time provides information about the quantity of vegetation present per unit area, called the *biomass.* The productivity can be obtained only by measuring changes in the biomass over the course of time.

Although the values for productivity are very variable both within and between the different tundra habitats, there is an evident pattern in the data shown in the table. The productivity values for a polar desert are extremely small, which means that the rate of energy input into the polar desert ecosystem is low and the amount of heterotrophic life (animals and microbes) that it can support is

therefore correspondingly low. Moss and lichen tundra is similarly poor in productivity. Arctic sedge marshes and alpine sedge meadows are more productive. Indeed, these habitats match shrub tundra in their productivity and even overlap with that of forest tundra. It is important to remember that the growing season in these tundra habitats is extremely short, often less that two months, so the relatively high values for productivity in the sedge marshes are all the more remarkable. It explains why so many migratory animals, including caribou, geese, and wading birds, find it worthwhile to travel to these regions for the summer season.

Comparing the tundra productivity figures with those from other global ecosystems, they are generally low, but again the short growing season must be taken into account. Temperate marshes have several extra months in which to grow and, tropical marshes can continue their growth throughout the year, so it is not surprising that their overall productivity figures are higher than those of the tundra. At the bottom end of the scale, the polar desert has a similar range of productivity to that of the hot deserts of tropical and subtropical regions.

Biomass is the total amount of above-ground material (mainly vegetation) present in an ecosystem, and there is often a general relationship between biomass and productivity. But this is not always the case. Polar tundra has very little biomass, only about 0.03 to 0.06 pounds per square foot (0.1–0.2 kg m^{-2}), while tundra sedge marshes have a biomass of about 0.6 to 0.9 pounds per square foot (2.0–3.0 kg m^{-2}). This difference is reflected in their respective productivity figures. Shrub tundra generally has a higher

Primary Productivity Values for a Range of Tundra and Other Ecosystems
(in Pounds per Square Foot per Year [Figures in Brackets Are in Kg m^{-2}y^{-1}] in Dry Weight)

TUNDRA ECOSYSTEMS	NET PRIMARY PRODUCTIVITY
Polar desert	0.003–0.009 (0.01–0.03)
Moss tundra	0.018–0.022 (0.06–0.07)
Arctic sedge marshes	0.03–0.09 (0.1–0.3)
Shrub tundra	0.03–0.09 (0.1–0.3)
Forest tundra	0.06–0.44 (0.2–1.5)
Alpine sedge meadows	0.03–0.12 (0.1–0.4)
Alpine fellfield	0.06 (0.2)
OTHER ECOSYSTEMS	
Temperate sedge marshes	0.13–0.30 (0.5–1.0)
Temperate reed beds	0.30–0.73 (1.0–2.5)
Tropical papyrus marshes	0.60–2.93 (2.0–10.0)
Tropical rain forests	0.29–1.17 (1.0–4.0)
Hot desert	0.003–0.058 (0.01–0.20)

biomass than sedge marshes, ranging between 1.2 and 1.8 pounds per square foot (4.0–6.0 kg m^{-2}), but its productivity is similar to that of the marshes, so the relationship between biomass and productivity does not hold in this case. Herbaceous vegetation is evidently more efficient at making a rapid start to growth when the daylight and warmth of spring arrives, while the bursting of buds and production of new leaves by shrubs is evidently slower. The woody habitat has other advantages, however, especially the ability to form a dense cushion form that aids survival in drier locations through the ice blasts of winter. In such situations evergreen leaves may also prove advantageous (see sidebar "Why Be an Evergreen?" pages 116–117).

The low angle of the Sun during the growing season in the polar tundra may also be helpful to the sedges. Holding their leaves in a vertical position may enable them to capture energy that comes to them at an angle only a little above the horizontal. The low solar angle is compensated for in part by the very long summer days, so leaves are able to continue their photosynthetic activities for a full 24 hours. Alpine tundra habitats do not have this advantage, but the high angle of the Sun does increase the productive potential of the fellfield type of tundra, allowing this vegetation type to achieve greater productivity values than the polar equivalent, the moss tundra.

■ FOOD WEBS IN THE TUNDRA

The primary productivity of tundra habitats forms the energy base of the entire ecosystem. The energy fixed in photosynthesis and stored in plant tissues provides a source of food for all the heterotrophic organisms found within the ecosystem. There are two major pathways along which energy can travel. Energy-rich plant tissues may be consumed by herbivorous organisms and subsequently predators, or the plant organ may survive until its eventual death, when it will enter the detritivore and decomposer system. But the track of energy movement is actually more complex than this because energy may move between these major pathways as it proceeds through the ecosystem. Fungi are important decomposers, but they may be eaten by tiny soil invertebrates, such as springtails, which may then form the prey of beetles, themselves a source of food for insectivorous birds, and so on. Energy has thus moved from the decomposer system back into the herbivore-predator system. Similarly, the insectivorous bird may die and fall to the floor, where it is invaded by bacteria, and its energy enters the decomposer system.

Simplified food web for a tundra ecosystem

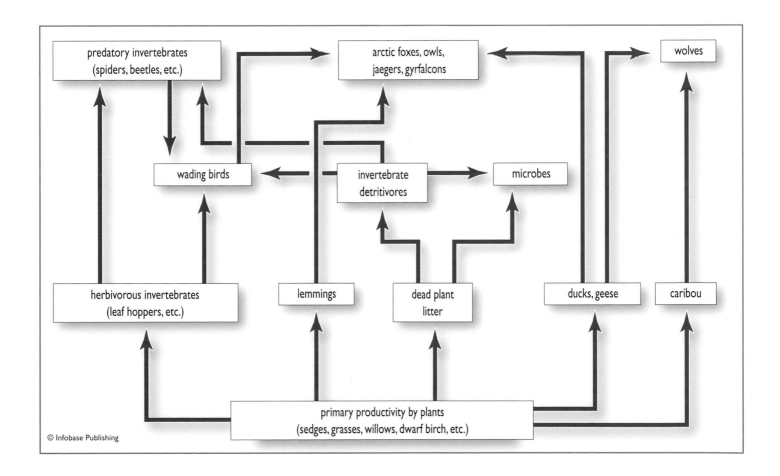

It is possible, therefore, for energy to move on complex tracks through the ecosystem, but it cannot circulate indefinitely. Eventually it is used by an organism to produce work and is then lost as heat in the process of respiration. If it were possible to follow the course of one unit of energy through its course of movement subsequent to its being trapped by a plant in photosynthesis, it would move through a series of stages as it became transferred from one organism to another. A sedge plant in the tundra might use some of its fixed energy to produce seeds, which could be consumed by seed-eating birds, such as a Lapland longspur or a willow ptarmigan. These may fall prey to one of the top predators of the Arctic landscape, such as a peregrine falcon. Energy has thus moved along a *food chain,* eventually being released by the respiration of the predator as it conducts its energetic hunting, or perhaps then entering the decomposer system when the predator dies. But the complex possibilities for energy movement in an ecosystem mean that the range of possible food chains is extremely large, and ecologists prefer to think in terms of a *food web.* A highly simplified example of a tundra food web is shown in the diagram on page 91.

When the feeding relationships of an ecosystem are simplified in this way, it is possible to detect a series of feeding levels, called *trophic levels,* in the food web. The first trophic level (at the base of the diagram) is occupied by the primary producers, the plants of the tundra. The primary consumers, whether invertebrate, bird, or mammalian grazers, form the second trophic level, and this is followed by one or two layers of predators, belonging to the third and fourth trophic levels, respectively. For example, an invertebrate grazer, such as a cranefly larva, could be consumed by a predatory wading bird, such as a golden plover (the third trophic level), only to be eaten in turn by a gyrfalcon (the fourth trophic level). As energy moves from one trophic level to the next, however, some is inevitably lost, and the efficiency of transfer is often only about 10 percent.

One consequence of the inefficiency of energy transfer in ecosystems is that the amount of energy available gradually decreases as it moves up from one trophic level to the next. There is a larger quantity of biomass of vegetation in a tundra sedge marsh, for example, than there is of herbivores. Similarly, if one could gather together all the herbivores in the marsh, from caterpillars to caribou, their combined mass would be several times greater than that of their predators. The loss of about 90 percent of the energy at each stage of transfer means that the biomass present within each trophic level declines with each step up the trophic levels of the food

A wolf pack chases a group of elk. Wolves are the major predators of elk, and by picking off the weaker young and elderly animals, they control elk populations and select for genetic strength. *(Jim Peaco, Yellowstone Digital Slide File, Yellowstone National Park)*

web. By the time the energy resources reach the top predator level (including gyrfalcons and arctic foxes), there is only a very small quantity of living material present per unit area of the tundra landscape. This arrangement is known as a *pyramid of biomass*. The energy pyramid is broad at the base, supported by the mass of vegetation, and becomes smaller at each trophic level until the top predators form the very small amount of biomass supported at the summit. This energy structure is common to most terrestrial ecosystems, including the tundra.

The potential length of food chains and the complexity of food webs are therefore limited by how much energy is available at the base. In other words, the rate of primary production and the quantity of plant biomass sustained in an ecosystem have a profound effect upon the range of different organisms supported and the complexity of their interactions (see "Patterns of Biodiversity," pages 137–139). In the tundra most food chains have a maximum of about four links within them. Very little energy remains after this series of energy transfers, so higher trophic levels cannot be supported. A simple example of a short food chain is one in which elk feed upon vegetation and are preyed upon by wolves (see photograph on page 92). There are just three links in this chain.

Each trophic level may also affect the level upon which it feeds. Caribou grazing, for example, especially when directed toward small shrubs, can strongly influence the biomass and productivity of the ecosystem. In one study of caribou grazing on willow, the biomass of the willow was reduced by 50 percent, and the reproductive success of the willow was also curtailed. Such a reduced biomass means that the entire architecture of the ecosystem is changed, and this alters the environment for other species, such as leaf beetles (genus *Gonioctena*) and sawflies (Hymenoptera). Caribou also have a marked effect on tundra landscapes that are dominated by mosses and lichens. Grazing and trampling reduces the depth of the moss cover, and this results in a better penetration of light and warmth to the soil below, raising soil temperatures.

Goose grazing is an important local ecological factor in some tundra habitats, such as the salt marshes around Hudson Bay, where large flocks of snow geese (*Chen caerulescens*) graze during the summer. Low-intensity grazing can stimulate the growth of salt marsh grasses by enhancing the rate of turnover of certain elements, especially nitrogen, but heavy grazing strips off the entire cover of grasses and herbs, leaving bare mud surfaces and reducing productivity.

The short growing season in the tundra also means that a relatively complex food web, such as the one shown in the diagram on page 91, can be maintained only for a few weeks or at best months. Just as primary productivity is limited to a short season, so is the activity of most other trophic levels. As productivity falls at the close of the growing season, the animals in the food web are faced with stark alternatives: hibernate, migrate, or die.

ELEMENT CYCLING IN THE TUNDRA

Just as energy is transferred from one trophic level to another, so are the elements that make up the bodies of the food. Most of the nutrient elements needed by caribou, for example, are obtained from the vegetation they feed upon, though they may supplement this with their drinking water, and, in conditions of nutrient scarcity, may even lick rocks, eat soil, or consume the shed antlers and the bones of other individuals. The caribou loses elements when it urinates, defecates, sheds antlers and hair, and dies. The elements thus lost enter the soil, where they are released by microbial decomposition and are available for uptake by the vegetation once more. The nutrients in the tundra thus circulate in a nutrient cycle, as shown in the diagram on page 94.

The nutrient cycle of the tundra, like the food web, is relatively complex. Additional trophic levels ensure that some elements do not pass straight from the caribou to the soil but provide additional routes along which they may pass. Wolves prey upon caribou, so an atom of nitrogen, for example, obtained by a caribou grazing on lichens can pass on to a wolf and pass back into the soil via its urine. Unlike energy, however, there is no loss of nutrients to the ecosystem as a consequence of transfer between trophic levels. Nutrients can move out of ecosystems, but they are never entirely lost in the way that energy is lost. This is why ecologists speak of nutrient cycling as opposed to energy flow.

As in the case of energy, the movement of nutrient elements around the ecosystem is determined to a degree by the activities of the organisms present. The faster the rate of primary productivity, the higher the rate of nutrient uptake from the soil, and the more intensive the grazing and predation, the faster nutrients are circulated between trophic levels. The speed of decomposition of dead materials also determines how fast elements are liberated and made available for plants once more (see "Decomposition in the Tundra," pages 98–99). Low temperature and short growing season limit the rate of nutrient cycling in the tundra and confine the biological part of the cycle to the summer period.

One of the problems resulting from the relatively short period of nutrient-cycling activity during the summer is that many mobile animals leave during the winter shut down and seek more comfortable and productive winter quarters (see "Behavioral Adaptations: Hibernation or Migration?" pages 130–133). Caribou herds, for example, whose adults and calves have fattened on the growing vegetation of the sum-

The nutrient cycle in a tundra ecosystem. Decomposer organisms within the soil liberate elements from the organic materials upon which they feed. Elements are imported from the oceans by seabirds that feed on marine life, which then deposit feces at their roosting and nesting sites.

mer tundra, may move to more southern regions where they can shelter in the boreal forest margins. A caribou herd can be considered a nutrient reservoir on the hoof. The migrating animals transport quantities of nutrients, including nitrogen, phosphorus, potassium, calcium, and so on from the northern tundra to their wintering ground.

There are also nutrient losses as the streams and rivers that cross the tundra during the summer snowmelt erode soils and leach nutrients into the sea. Even the tundra lakes and wetlands can become a drain upon the nutrient resources because elements in their sediment are locked away from the living organisms of the ecosystem and form a kind of fossil reservoir of elements that enter the slow-moving geological cycles of the rocks (see "The Rock Cycle," pages 28–29). Since elements are lost in so many ways, is the nutrient status of the tundra constantly diminishing? The answer is that this is not the case. Just as elements are being lost from the tundra, they are also being gained.

There are several different ways in which elements enter the tundra ecosystem, which can be classified into geological, meteorological, and biological mechanisms. Geological sources of elements are based upon the breakdown of rocks, the weathering processes whereby ancient elements trapped

for millions of years are released back into circulation (see "Geology and Rock Weathering," pages 52–54). The grinding activity of glacial ice, the freezing and thawing of rock particles, and the chemical consequences of microbial activity in the soil all assist in the degradation of rocks and the liberation of elements into the environment.

The atmosphere is also a source of nutrient imports to the tundra, both directly in supplying the gases carbon dioxide and nitrogen, and indirectly through the meteorological processes within the atmosphere. Precipitation in the form of rain or snow is rarely pure. As water droplets form in the atmosphere and as they fall toward the ground, they collect dust fragments and dissolve chemical atoms and molecules that they encounter on their journey. Some of these atmospheric components result from human pollution, often many thousands of miles away, and other molecules are entirely natural, derived from wind-eroded soils or from the spray of the oceans.

The influence of the oceans on the chemistry of precipitation can be seen in the table on page 96. The sites considered here are not specifically from the tundra, but they illustrate the effect of proximity to the sea upon the quantity of certain elements deposited in rainfall. As might be expected, the amounts of sodium (derived from sea salt, sodium chloride) are much higher in regions close to the sea. Magnesium is another element that is relatively abundant in seawater, so regions near the oceans receive a large input of the element in the rainfall. Some other elements, such as potassium and calcium, may also be found in greater quantities in the rainfall of oceanic areas, but this depends upon the chemistry of other sources, such as dust from soil erosion. Sodium and magnesium, therefore, are clearly two elements that are more abundant around the fringes of the oceans, and this may have an influence on animals and plants. Sodium is not required by plants for their growth, but it is needed by animals for their nerve function. Animals living far from the sea may suffer from a shortage of sodium, but this is an unlikely problem in oceanic sites. Some aquatic plants accumulate sodium, which is why certain grazing animals, such as moose, often spend much of their time in summer grazing in shallow bodies of water (see photograph below). Magnesium is required by plants for the construction of chlorophyll, so an additional input to an ecosystem from oceanic precipitation can be beneficial.

Neither sodium nor magnesium is usually limiting to the productivity of an ecosystem, however. The elements that often limit plant productivity and therefore ecosystem development are nitrogen and phosphorus. Nitrogen is an essential component of all amino acids, which are the building blocks of proteins, so no plant or animal can exist without a supply of this element. Phosphorus is essential as a component of phospholipids, the basis of cell membrane construction, and is the element used in the energy relations

of the cell in the form of adenosine triphosphate (ATP). Both of these elements are also brought into tundra ecosystems from pollutant materials derived from human activities beyond the tundra region. Ecologists have conducted experiments in which tundra habitats have been artificially supplied with additional quantities of nitrogen and phosphorus in the form of fertilizer, and these treatments have greatly enhanced plant productivity. They have also changed the composition of the vegetation, so the nutrient status of tundra ecosystems is clearly an important factor in determining their biodiversity and species content.

Nitrogen is an abundant element in the atmosphere, where it accounts for approximately 79 percent of the gas volume. But the gaseous form of nitrogen, the molecule N_2, is relatively inert, and neither plants nor animals are able to obtain it directly. Nitrogen-containing pollutants occur mainly in the form of oxides of nitrogen, derived from the

A young moose grazing in wetland. The moose is a boreal forest animal that sometimes wanders into the tundra wetlands, where the vegetation is rich in sodium, an element that can be scarce in the forest. *(Carolyn McKendry)*

Quantities of Certain Elements Delivered to an Ecosystem by Precipitation Each Year
(Contrast Coastal with Inland Sites; Data Expressed as Pounds per Acre [Kg/ha])

	SODIUM (NA)	POTASSIUM (K)	CALCIUM (CA)	MAGNESIUM (MG)
United States				
Long Island (coastal)	126 (142)	6 (7)	9 (10)	17 (19)
Hubbard Brook (inland)	2 (2)	1 (1)	3 (3)	1 (1)
United Kingdom				
Cornwall (coastal)	176 (198)	4 (4)	8 (9)	18 (20)
Kent (inland)	18 (20)	3 (3)	9 (11)	4 (4)

combustion of fossil fuels, especially coal, oil, and natural gas. Oxides of nitrogen dissolve in water droplets in the atmosphere to produce nitrous and nitric acids, thus generating acidic rain. Although these acids can be damaging to plant tissues when they fall directly upon them, in the soil they provide a source of plant nutrition, generating nitrate ions (NO_3^-), which can be absorbed by plant roots and converted into amino acids. Atmospheric pollution in this situation acts as a general fertilizer for the tundra ecosystem, and one might therefore suppose that it is to be welcomed. The problem is that some plants respond to the fertiliza-

The polar bear is the top predator of the Arctic tundra. By feeding mainly upon seals polar bears provide a link between the food web of the ocean and that of the terrestrial tundra. *(Jurgen Holfort, International Arctic Research Center, University of Alaska)*

tion more rapidly than others, and their strong growth may suppress and even eliminate the smaller and slower-growing species. Atmospheric fertilization must therefore be regarded as a conservation threat to the tundra environment (see "Pollution of the Tundra," pages 226–229). The rate of atmospheric fertilization varies considerably in arctic and alpine habitats, depending on their location in relation to industrial sources and wind movements. In northern Alaska deposition of nitrogen reaches levels of 0.4 ounces per square foot (1 g m^{-2}) per year.

Phosphorus is also present in the atmosphere as a pollutant derived from industrial sources and the use of agricultural fertilizers in temperate regions. Its natural source in the tundra ecosystem is the weathering of rocks, and phosphorus is relatively scarce in this form, so it can become limiting to plant growth and ecosystem development. Arctic soils are usually very poor in phosphorus, so the aerial input, which amounts to only about 0.0004 ounces per square foot (0.001 g m^{-2}) per year, is nevertheless significant.

Biological sources of nutrients in the tundra include the possibilities of import with the arrival of migrating animals. Just as caribou take elements out of the tundra when they leave on their migrations, they bring elements back with them on their return. It is reasonable to suppose, however, that the overall budget from these movements is likely to balance out. In the case of wading birds, however, it is possible that they arrive in the tundra well stocked with nutrients following their winter in warmer climes. Some ecologists have tested this hypothesis in northern Greenland and Canada by examining the isotopes of carbon in their eggs. Using carbon isotopes (different forms of the element) it is possible to distinguish between sources of the carbon, whether derived from marine invertebrates obtained during their winter feeding or from terrestrial sources used since their arrival in the tundra. The results showed that the carbon in the eggs of waders was derived from the tundra environment, so the birds had not relied upon imported sources of nutrients to provide for their

Nitrogen Fixation

Atmospheric nitrogen occurs in the form of the dinitrogen (N_2) molecule, which has a very strong set of bonds between the two nitrogen atoms and is consequently a relatively inert material, but certain microbes are able to break this bond and to reduce nitrogen gas into the ammonium ion NH_4^+ and subsequently to an organic form. Among these are some free-living organisms, such as the photosynthetic blue-green bacteria (Cyanobacteria) and nonphotosynthetic microbes, such as *Azotobacter* and *Clostridium*. Sometimes the cyanobacteria are found in a symbiotic relationship with plants, such as the water fern *Azolla,* and the cycads, a tropical group of gymnosperms. Other nitrogen-fixing microbes are found only in association with higher plants, including *Rhizobium* and *Bradyrhizobium,* a group of bacteria commonly known as *rhizobia,* found in the root nodules of members of the pea plant family (Fabaceae), and others associated with such plants as alder (*Alnus* species), bog myrtle (*Myrica gale*), sweet fern (*Comptonia* species), and mountain lilacs (*Ceanothus* species). The process of fixation consumes a large amount of energy, provided by the molecule ATP, which is why so many nitrogen-fixing microbes are found in association with photosynthetic plants. The plant produces the energy, and the bacteria do the fixing. So when a plant develops a symbiotic relationship with a nitrogen-fixing bacterium, it takes on a very expensive partner in energetic terms. But the payoff is worthwhile, because the plant is able to enhance its protein supplies.

Vegetation containing plants of the pea family enhances the nitrogen levels of the soil, thus acting as a fertilizer. The Ancient Greeks knew about this means of soil improvement and used beans to improve their agricultural soils. Clover (*Trifolium* species) is often used in the same way by modern farmers, who find that up to 270 pounds per acre (300 kg ha^{-1}) of nitrogen can be added to the soil each year by this means. Some calculations suggest that 150 million tons of nitrogen are fixed from the atmosphere each year by natural biological fixation over the surface of the Earth.

breeding activities. The summer tundra is evidently sufficiently rich in its nutrient supply to support these immigrant breeding birds.

Some birds act as major sources of nutrients to the tundra on a local level. Where the tundra meets the ocean many marine birds and some mammals, such as seals, use the land as a platform on which to roost or to breed. Bird colonies, such as those of the common murre (*Uria aalge*) and razorbill (*Alca torda*) occupy the cliffs along rocky shores, and the ground becomes covered in their droppings, known as *guano.* This material is derived from the oceans in the form of fish, and the droppings are very rich in nitrogen and phosphorus, so the ground is heavily fertilized by these nutrient inputs from the oceanic ecosystem. This source of nutrient addition to some tundra habitats is included in the diagram on page 94.

Polar bears (see photo on page 96) also act as a link between the oceanic and terrestrial ecosystems. Their main source of food is the seal, which derives its nutrient elements from the oceans in the form of fish and crustaceans. The polar bear, on the other hand, spends most of its time on land, bringing some marine elements into the terrestrial soils whenever it defecates.

Certain microbes possess the ability to reduce atmospheric nitrogen to ammonia and thus bring it into the nutrient cycle of ecosystems (see sidebar above), and these constitute an important biological source of nitrogen to an ecosystem.

Biological nitrogen fixation is relatively slow in tundra ecosystems. The fixation rate for Arctic ecosystems rarely exceeds 0.16 ounces per square foot (0.4 g m^{-2}) per year. This is only about one 10th the fixation rate in most temperate ecosystems. In tundra ecosystems, both in polar and alpine sites, lichens are important sources of nitrogen fixation. Lichens are compound organisms consisting of a fungus and an alga, or blue-green bacterium, living together in symbiosis (see sidebar "Lichens," page 143). If the photosynthetic partner is a blue-green bacterium, the lichen is able to fix nitrogen, but the rate of fixation is low in comparison to that achieved in temperate environments because of low temperature and short growing season. As already explained, the input of nitrogen from atmospheric pollution is about four times this value, which underlines the potential impact of this type of pollution on the nutrient cycle of the tundra ecosystem.

The general picture of the tundra ecosystem as far as element budgets and cycling is concerned is one of nutrient poverty. Tundra rocks are often nutrient poor, and weathering supplies a limited input of plant requirements. Biological fixation is slow, as is the rate of nutrient turnover. Perhaps this is why tundra plants often develop a very extensive root system. Compared with plants from other biomes, tundra plants allocate more biomass to roots, thus assisting in gathering the essential nutrients that are in such short supply (see sidebar "Resource Allocation," page 89).

■ DECOMPOSITION IN THE TUNDRA

Tundra soils do not lack microbes. Many tundra regions are remote, either occurring on islands in extensive oceans or on mountaintops surrounded by very different habitats, but this has not led to poverty in their microbial populations. Bacteria and fungi disperse by means of spores. These are extremely small, dustlike structures that become suspended in the air and can be transported by currents high into the atmosphere. They are washed out by rain and can arrive in locations many thousands of miles from their point of origin. Many land in the ocean or on frozen ice sheets, where they die or remain in a dormant state for long periods of time. But others fall upon soils where they can germinate and establish themselves, many feeding upon the dead remains of plants and animals in the ecosystem where they find themselves. Some fungi and bacteria may take up close relationships with other organisms in the ecosystem. They may combine together to form a lichen (see sidebar "Lichens," page 143), fungi may become associated with plant roots to form mycorrhizae, some bacteria may invade the roots of certain plants and develop nitrogen-fixing nodules, but some remain independent and consume the organic detritus they find in the soil.

The respiration of soil microorganisms results in decomposition or decay, and the rate at which this takes place is limited by two main factors in tundra ecosystems, low temperature and low nutrient availability. In general, chemical reactions take place more slowly at low temperature, and this applies to decomposition. Microbial decomposition is very similar to the digestive processes of an animal, except that the enzymes involved in breaking down the food resources are secreted into the general environment instead of being contained within a gut. A microbe may, for example, secrete an amylase enzyme that breaks down starch in a dead plant cell. It then absorbs glucose through its cell wall and respires it in a conventional way. The microbes are employing an external digestive system, and the chemical reactions taking place in their environment are strongly influenced by temperature. Decomposition, therefore, becomes slower as tundra temperatures fall and increases in the summer as the soil becomes warmer. Microclimate is important in this respect because radiant energy from the Sun can raise the soil temperature to 60°F (15°C) even when the air tem-

Nitrification

Nitrogen is present in living plants, animals, and microbes in a number of different forms but most frequently as a constituent of amino acids. Amino acids are the components of proteins, and proteins are present in the cell as enzymes, chemical catalysts that are responsible for regulating all the metabolic processes within the cell. When an organism dies, microbes use proteins, among other substrates, as a source of energy, and they release the energy by breaking the bonds in complex organic molecules. The decay of proteins first involves breaking the large molecule into amino acids and then a process called *ammonification,* or nitrogen *mineralization.* In this reaction the nitrogen component of amino acids is removed from the carbon skeleton and released in the form of ammonium ions, NH_4^+. A wide variety of microbes are capable of this part of the decay process.

Ammonium ions are retained by soils because of their positive charge. Clay particles in the soil are negatively charged, so they adsorb these cations and hold them in a loose bond. Most plants, however, are unable to absorb nitrogen in this form, so the ammonium nitrogen is unavailable to much of the vegetation. There are other microbes in the soil, however, that are capable of oxidizing the ammonium ions to nitrate ions.

$$2NH_4^+ + 3O_2 \rightarrow 2NO_2^- + 4H^+ + 2H_2O$$
$$2NO_2^- + O_2 \rightarrow 2NO_3^-$$

This conversion is called *nitrification.* It is carried out in two stages, as shown here, and involves two different groups of bacteria. The fist stage is conducted by bacteria of the genus *Nitrosomonas,* and the second by those of the genus *Nitrobacter.* These microbes are *chemosythetic autotrophs,* meaning that, like photosynthetic autotrophs, they can feed themselves, but they obtain their energy not from light but from inorganic chemical reactions. The oxidation of ammonium ions and nitrate ions releases energy, so the microbes that conduct this process are making a living by regulating this chemical reaction. The outcome of nitrification is the production of an oxidized form of nitrogen, nitrate ions (NO_3^-), and most plants use this form of nitrogen as their major source of the element.

perature remains at the freezing point (see "Microclimate," pages 18–21).

Some decomposition occurs even at the freezing point in the tundra because certain strains of bacteria and fungi (*psychrotrophic* strains) are capable of maintaining some activity even at such low temperature. These microbes have a range of physiological adaptations that allow them to operate under such extreme conditions (see "Cold Stress," pages 109–110). But despite the presence of these specialized organisms, decomposition rates in the tundra soil are strongly temperature dependent.

Microbes, like all other organisms, need various elements to make up their bodies, and the elements required are similar to those of other living things. If nitrogen is in short supply, microbes can be limited in their rates of protein production, and a shortage of phosphorus limits their ability to develop membranes and ATP. The general nutrient poverty of the tundra thus has its effect upon microbial populations and therefore decomposition rates. This limitation is easily demonstrated experimentally by fertilizing a test area of soil and checking the decomposition rate that results. Ecologists have carried out this kind of experiment in a range of different types of tundra, and decomposition is usually enhanced if nitrate fertilizer is added to the soil. The rate of decomposition in these experiments is measured by the amount of carbon dioxide produced by the soil, much of which is due to microbial respiration.

There are additional chemical processes taking place in the soil, one of which is nitrification (see sidebar on page 98), in which nitrogen from decaying proteins is eventually made available to the growing plants of the ecosystem. Nitrification in the tundra is slower than in most other ecosystems, attaining rates similar to those of nitrogen fixation. This means that even when proteins are deaminated by microbial enzymes, the conversion of ammonium ions to nitrate ions is slow and can hold back the nitrogen cycle. In this situation some plants, such as members of the heather family (Ericaceae), have developed techniques for absorbing nitrogen in the form of ammonium ions and thus shortcutting the need for nitrification. This is an extremely efficient mechanism that allows these plants to tap the nutrient resource ahead of their competitors.

Nitrate ions, unlike ammonium ions, are not retained very efficiently by soils, so unless they are rapidly taken up by plants they are in danger of being leached out of the ecosystem. In the waterlogged soils of the tundra there is an additional cause of nitrogen loss from the soil, namely *denitrification*. Oxygen is in short supply in the stagnant water of saturated soils, and some microbes, such as members of the genus *Clostridium*, use nitrate as an oxidizing agent in their respiration instead of oxygen. As they decompose organic materials they produce carbon dioxide

and nitrogen gas, thus depleting the soil yet further of its nitrogen resource. Slow decomposition and low rates of nitrification result in the entire nutrient cycle of the tundra being retarded, and slow recycling can hold back the rate of growth in this ecosystem.

■ ECOSYSTEM DEVELOPMENT AND STABILITY

Ecosystems grow almost like organisms. Bare rock and lifeless detritus left by a retreating glacier is gradually colonized by organisms, first by photosynthetic microbes, lichens, algae, and higher plants, closely followed by detritivores and decomposers, and then by herbivores and predators. The ecosystem thus gradually increases in its diversity and complexity. As the ecosystem develops, changes occur in both its energy-flow patterns and its nutrient-cycling systems.

A young ecosystem is dominated by its nonliving components, such as the fragmented rocks of its primitive soils and the atmosphere. As living organisms arrive, often carried in the air as spores and seeds or, in the case of mobile animals, flying, arriving by sea, or floating on transported detritus, so the living component of the ecosystem begins to expand and assume greater importance. Blue-green bacteria and lichens arrive as spores in the air, colonize rock and soil surfaces, and begin to take in solar energy and carbon dioxide from the atmosphere to form organic compounds. Bacteria and fungi also arrive by air and feed upon the dead remains of the photosynthetic microbes. Mosses and ferns are widely dispersed by air as spores and colonize primitive soils, especially where damp conditions prevail, and they add more energy to the expanding biomass of the ecosystem and more organic matter to the soil (see "Soil Formation and Maturation," pages 54–59). Invertebrate animals may also arrive by air, such as tiny spiders, which spin webs that act as parachutes and carry their living passengers for many miles. Some detrivores, such as springtails (Collembola), manage to disperse in the surface waters of the sea, enclosed in envelopes of air trapped around their hairy bodies. Others are carried on human garbage, such as plastic debris that is widely dispersed, floats, and resists decay.

Flowering plants can also invade new areas by sea transport and by air if they have light seeds with feathery dispersal mechanisms. Others may be carried into the young tundra habitats by migrating birds, such as geese and seed-eating snow buntings (*Plectrophenax nivalis*). Many seeds are destroyed and digested in the guts of birds, but some survive and are dispersed in the droppings. It is also possible that a bird may die while viable seeds remain in its crop, and these surviving seeds in time germinate and establish themselves in a new habitat. The flowering plants are often faster growing and more robust than mosses and lichens, so the rate of energy accumulation in biomass growth increases with their arrival. Some provide a home for nitrogen-fixing bacteria (see sidebar "Nitrogen Fixation," page 97), and this leads to increasing nitrogen accumulation in the growing ecosystem. Many lichens also contribute to this process.

The outcome of these invasive events is that the composition of the ecosystem changes. There is a buildup of energy residing within the living biomass in the form of organic chemicals. Biomass continues to grow until climate or some other factor, such as shallow soil, prevents further growth. The biomass then stabilizes at its maximum level, which may still be very low in the polar desert tundra but can be greater in the shrub- and tree-dominated vegetation of the shrub tundra and forest tundra (see "Primary Productivity in the Tundra," pages 88–91). This final, equilibrium state is called the *climax* of the succession. At this stage the energy budget of the ecosystem is balanced. All the energy fixed in primary production is being used for the respiration of the plants, animals, and microbes within the system, so there is no excess energy that can accumulate as additional biomass. Ecosystem growth has terminated.

Just as energy builds up in the living biomass during succession, so do nutrients. The expanding biomass can be considered a living reservoir of elements, especially carbon, nitrogen, phosphorus, calcium, and so on. At the start of the succession all the chemicals were contained in the nonliving (abiotic) part of the ecosystem, but at climax the biomass itself forms a nutrient reservoir. Locked up within living tissues, elements are less likely to be leached from the ecosystem and lost, but they are released gradually by decomposition after death, so they can be recycled. The nutrient budget of the entire ecosystem is now in equilibrium in the same way as the energy budget. The continued gains of nutrients to the ecosystem, by rock weathering, atmospheric deposition, and so on, is equaled by the losses from the ecosystem, especially by leaching. The ecosystem is now in nutrient balance unless conditions change and biomass starts to grow again, which is possible if there is a shift in climate, for example.

Because the climax ecosystem is in equilibrium, it is also regarded as being stable. Stability is an important ecological concept, especially in the modern world, in which human modification of the environment is so widespread and ecosystem conservation is a matter for concern, but the concept of stability is also difficult to define and demands careful consideration (see sidebar at left).

A stable ecosystem possesses both inertia and resilience, but what features of an ecosystem can supply these qualities? One helpful way of approaching this question is to use the analogy of a bank. Before depositing money in a bank an investor is well advised to check upon the assets of the institution. A large bank with extensive assets is more likely to be a safe place in which to deposit money. If there is a sudden demand upon its resources, it should be able to take the strain without injury, and any money invested remains safe. A small bank with low assets, on the other hand, could be severely damaged in the event of any sudden large demand on its resources. Invested money is therefore less secure. To summarize, the stability of a bank depends upon the size of its assets in relation to the demands placed upon it. The same reasoning can be applied to ecosystems. The assets of the ecosystem are its biomass and nutrient reserves, and

The Concept of Stability

Stability is difficult to define. If a situation exhibits no change over time, it can be considered stable. An ecosystem that is in equilibrium will remain in its current state and will not alter over the course of time unless there is a change in the prevailing conditions, such as climate. Such an ecosystem can be regarded as stable. But the word *stability* implies more than simply a lack of change. There are two additional ideas contained within the concept of stability: One is the ability to recover rapidly from any disturbance, and the other is the degree of resistance to alteration.

Rapid recovery from disturbance helps an ecosystem sustain itself in the face of modifying forces. Ecologists give this property the title *resilience*. A resilient ecosystem can cope with perturbation by returning to its original state without being altered by the experience. An ecosystem that is capable of healing itself in this way can clearly be regarded as stable.

Resistance to change is called *inertia*. All ecosystems are subjected to disturbing forces, such as extreme conditions of weather or human pollution events. Some ecosystems are easily damaged by such perturbations and can thus be considered relatively unstable, while others are more resistant, and the possession of such inertia protects them from harmful effects.

A stable ecosystem, therefore, is one that is strong and unmoved by disturbing pressures, that recovers rapidly and returns to its original state if it is damaged, and that therefore continues unaltered in the long term despite the forces that tend to deflect it from its condition.

the kind of pressures exerted on those reserves range from heavy grazing (by caribou, for example) to pollution events.

Tundra ecosystems are generally low in biomass. They exist in nutrient-poor environments and often grow upon nutrient-poor rocks. Their rates of productivity and nutrient cycling are relatively slow compared with those of other ecosystems. The tundra must therefore be regarded as potentially a very vulnerable ecosystem. At present it is relatively stable in the sense that rapid alterations are not apparent, but the pressures upon the biome are increasing, and the properties of the tundra suggest that it may not prove to be stable in the face of these pressures (see "Tundra Conservation," pages 229–232).

PRODUCTIVITY AND POPULATION CYCLES IN THE TUNDRA

Some changes take place in a cyclic fashion in the tundra, and this does not mean that the ecosystem is intrinsically unstable. Annual cycles are very pronounced because of the strong seasonality of the tundra climate. Primary productivity is confined to a relatively short period in the tundra summer, and for the rest of the year the vegetation biomass may actually shrink as dead leaves are shed and plants become dormant. Many animals leave the ecosystem or become inactive, so the winter ecosystem is very different from the summer ecosystem in its biomass, energy flow, and nutrient-cycling patterns. In the winter the tundra ecosystem still respires gently as the dormant organisms sustain their lives. So the energy resource of the ecosystem gradually declines, and carbon flows from living organisms to the atmosphere as the respiratory waste product carbon dioxide. A full study of the balance of the tundra ecosystem, therefore, must take into account both its summer and its winter condition. The fact that so many ecosystem functions vary so greatly between seasons is not an indication of instability because the cycles are repeated in a predictable fashion, and the long-term equilibrium of the ecosystem is maintained.

Some animal population cycles in the tundra are also cyclic in nature. Lemmings (see photograph at right) are perhaps the most well-known example. Every three to five years the lemming population of some tundra areas expands to exceptionally high densities. The population during one of these "explosions" can be 10 times that found following a population crash. When an expansion of this sort takes place, lemmings react by spreading out from the center of population and invading new areas. Sometimes this can be observed as crowds of lemmings moving together through the tundra sedge communities and shrub lands, seeking new feeding grounds. They do not commit mass suicide; this is an erroneous myth, but they may swim across creeks and waterways in their frantic journey. Precisely what causes the population cycle is not known. It does not appear to be directly linked to climatic fluctuations but may be related to general food supply in the form of herbage or perhaps the quality of the vegetation in terms of its nitrogen and phosphorus content.

Lemmings are not the only tundra animals that exhibit population cycles. Caribou populations fluctuate, and scientists have tried to identify the causes of their cycle. Caribou cycles are closely related to wolf numbers, and some workers have suggested that wolf predation causes the fluctuations in caribou numbers. It is more likely that wolf numbers simply reflect the caribou cycles. When caribou are abundant, the wolves also increase in number. A similar relationship has been found between snowshoe hares and their main predator, the lynx. Caribou populations in western Greenland have been extensively studied because some factors, such as human disturbance and complications due to emigration and immigration, can be eliminated there. The population cycle in western Greenland takes between 58 and 60 years to move from one peak to another, so the caribou cycle is evidently much longer in wavelength than the lemming cycle. Whether this cycle relates to climate or to food availability is still not known.

The Norway lemming (*Lemmus lemmus*), here emerging from its burrow among tundra vegetation, is distinctive because of the yellow patterns of fur upon its head. Lemmings are the most abundant of small tundra mammals. *(Sachiko Joko)*

Population cycles are well known in many different ecosystems apart from the tundra. Locust plagues in arid regions are often cyclic in nature and, as in the population cycles of the tundra, are often difficult to explain or predict. Cyclic changes of this sort in an ecosystem, however, do not necessarily indicate instability. A system can fluctuate and yet remain stable.

ALPINE TUNDRA ECOSYSTEMS

Alpine tundra climates (see "Alpine Climate," pages 21–26) have similar annual average temperatures to those of the polar regions but differ in many other respects, especially in their seasonal and daily fluctuations and in their patterns of precipitation. But the general structure of alpine tundra closely resembles that of polar tundra, and measurements of biomass, not surprisingly, are also very similar. Primary production has been studied in many alpine sites around the world, and, despite the climatic variations, the general picture of alpine productivity also resembles that of polar sites.

Niwot Ridge, Colorado, has been the focus of many ecological studies. The primary productivity of many different tundra habitats has been studied, and the results conform closely to those of the polar tundra (see the following table). They are also in line with studies from alpine locations in Europe, such as the Austrian Alps.

Resource allocation is also comparable to that of the polar tundra. The ratio of below-ground biomass to above-ground biomass varies between dry and wet locations, being approximately 1.0 in wet meadows and 2.3 in dry meadows. The alpine vegetation thus places much of its organic resources below ground, investing in roots that can seek out water and mineral nutrients. In extremely exposed situations on Niwot Ridge, the ratio is as high as 8, and precisely the same value has been obtained from studies in the European Alps. This means that almost 90 percent of the biomass in these ecosystems may be underground.

The most important limiting factor for productivity during the summer period is the availability of nitrogen. As in the case of polar tundra, nitrification is very slow in the soils because of low microbial activity during the cold winter, and the result is that available nitrates are scarce. As in the case of polar tundra, there are some alpine plants that are able to obtain their nitrogen as ammonium ions or as organic compounds, such as amino acids, and thus avoid the need for nitrification. Fertilization experiments in alpine tundra, like those in polar tundra, show that productivity can be increased and that there is a shift in the composition of vegetation when nutrients are added to the soil. Aerial pollution with nitrates can be a serious problem for the conservation of alpine tundra ecosystems when mountains are found close to sites of dense human occupation and sources of pollutants. The alpine tundra ecosystem, therefore, shares many of its processes and problems with the equivalent ecosystems in the polar tundra.

CONCLUSIONS

The concept of the ecosystem involves the living organisms within a habitat together with the nonliving components, such as atmosphere and soils. The tundra can be viewed as an ecosystem in which energy flows from high intensity sunlight to dissipated heat, passing through plants, consumer animals, and decomposer microbes on its passage. This flow of energy provides the power needed to circulate elements from inorganic to organic within the bodies of organisms, first the plants and then the consumers. In decomposition these elements are released back into the nonliving environment, where they are available for the next cycle.

Most energy fixation in an ecosystem takes place as a result of green plant photosynthesis, and in the case of the tundra ecosystems this results in a dry weight gain of less than one pound per square yard per year. In comparison with other global ecosystems this is a low level of produc-

Primary Production in Various Habitats from Alpine Tundra Sites
(Values for Productivity in Pounds per Square Foot [Kg M⁻²] Dry Weight)

SITE	VEGETATION	PRIMARY PRODUCTIVITY
Niwot Ridge, Colorado	Fellfield	0.05–0.07 (0.17–0.24)
	Alpine meadow	0.04–0.09 (0.13–0.30)
	Shrub tundra	0.09 (0.3)
Medicine Bow, Wyoming	Fellfield	0.04 (0.13)
	Alpine meadow	0.04–0.11 (0.14–0.38)
Mt. Patscherkofel, Austria	Dwarf shrub heath	0.03–0.12 (0.11–0.42)

tivity, similar to that of the hot deserts of the tropics. But in tundra ecosystems the growing season is very short, so most of the energy fixation takes place in the course of just two or three months. Consumer animals may move into the ecosystem from other regions to avail themselves of this short burst of food production. Some of the energy fixed by plants is used in their own metabolism, but what is left (the net primary production) is available for consumption by animals or may die and become part of the decomposition pathway of energy flow. Consumer animals occupy a series of different feeding levels, or trophic levels, including herbivorous grazers, predatory carnivores, and top carnivores. The community of animals may form a complex food web, but the number of links in any chain of energy movement is limited by the quantity of energy available. Since tundra ecosystems have low productivity, food chains tend to be short.

Chemical elements cycle within the ecosystem, first being absorbed from the soil by plants, then passed through the trophic levels, and finally liberated as a result of excretion or by death and decomposition. Unlike energy, nutrient elements can be reused in the ecosystem. Elements also enter and leave the tundra ecosystem, arriving by soil weathering of the underlying rock, or in rainfall, or by microbial fixation of nitrogen from the atmosphere. Tundra regions close to the oceans receive more chemicals in their rainfall than regions far from the sea. Oceanic regions may also receive an input of nutrients from the droppings of seabirds that nest or roost on land. Elements are lost by leaching and erosion of soils and by transport out of the ecosystem by migratory animals, such as caribou.

Microbes (bacteria and fungi) are responsible for the decomposition of dead plant and animal materials. Some of these are capable of maintaining metabolic activity even at very low temperatures, but the cold conditions through much of the year restrict the rate at which decomposition can take place. In some locations, especially at wet sites, this can lead to a buildup of organic matter in the soil. As the microbes operate on dead tissues, they make elements available for reuse by plants, as in the case of nitrification, whereby amino acids are oxidized to nitrates that plants are capable of absorbing.

Ecosystems develop over the course of time. Bare areas in the tundra, perhaps exposed by a retreating ice mass, are gradually colonized by plants, animals, and microbes. As the mass of living matter (the *biomass*) increases, so the environment becomes more complex as the structure of the vegetation affects microclimate. Increasing productivity during the course of succession supports a greater diversity of animal life. Eventually, the biomass growth is limited by the climatic conditions so that an equilibrium state is attained, called the *climax*. This is the most biodiverse and complex type of ecosystem that can be achieved within the harsh restrictions of climate, and ecologist believe it to be the most stable stage in the succession. But tundra ecosystems have such low productivity and biomass that they are easily damaged and are slow to recover, so they are very vulnerable in comparison to other types of ecosystems.

Equilibrium and stability do not mean that the ecosystem is static. Cyclic processes often occur within tundra ecosystems, such as the population cycles of lemmings and other mammals. The causes and controls of such cycles are still not fully understood, but they may relate to variations in productivity and food availability from year to year.

Alpine tundra ecosystems are very similar to those of the polar tundra in their properties. They have low levels of biomass and productivity and a high degree of seasonality in their primary production and in the presence of animals. When viewed as ecosystems, polar and alpine habitats are closely comparable.

5

Biology of the Tundra

The climate of the tundra is both harsh and bleak, so living in the tundra presents very considerable problems to all the organisms found in these extreme conditions. It is not surprising that the tundra contains a low diversity of plants, animals, and microbes. Relatively few organisms have developed the capacity to survive here, but those that do are of considerable ecological and evolutionary interest because of their high degree of specialization. Tundra plants and animals have to be tough to endure the problems that face them, and they have developed some unusual features that assist them to survive.

■ WRAPPING UP IN THE FREEZER

Cold is the most obvious problem in the tundra. It is not a constant cold; indeed, the temperature in summer on the ground can rise to more than 80°F (26°C), and the high angle of the Sun over tropical mountains can take the daytime temperature well above these levels. But the annual average temperature is low, so all permanent residents of the tundra need to be able to cope with very cold conditions for much of the year. All chemical reactions become slower at low temperatures, and this includes the biochemical reactions within living cells. Cell metabolism, that is the general working and functioning of the living cell, is largely controlled by the activities of specialized molecules, the *enzymes*.

Enzymes are proteins, large molecules built from chains of smaller units called *amino acids*. These chains are usually not straight but are twisted and curled, the folds and bends being held together by weak bonds, and their contorted structure is important for the actions they perform in the cell. Enzymes are chemical catalysts. They assist in chemical reactions but are themselves unchanged at the end of the reaction, as a result of which they can be reused time and time again in the living cell. An example of an enzyme is nitrogenase, which is used by some microbes for the fixation of atmospheric nitrogen (see sidebar, "Nitrogen Fixation," page 97). In this process the nitrogenase enzyme acts as a catalyst in the reduction of nitrogen gas to ammonium ions, which is a reaction that can be carried out in the absence of the enzyme only by applying very high temperature and pressure to this relatively inert gas. So enzymes facilitate chemical reactions that would otherwise be extremely difficult and energy consuming. They achieve this by forming a loose combination with the substance they operate on, called a *substrate,* forming an *enzyme-substrate complex.* When they combine in this way, they distort the substrate molecule, straining its chemical bonds and thus allowing it to undergo chemical changes more easily, and this distortion is partly due to the complex structure of the protein molecule.

Enzymes each have their own optimum working temperature, which varies from one enzyme to another and, indeed, among the same enzyme derived from different species or even different populations of organisms. The optimum temperature for activity is under genetic control and can be modified in the course of evolution. Most enzymes are inhibited at very high and very low temperatures because their structure begins to disintegrate, and they no longer function effectively as catalysts. Relatively few enzymes in the natural world operate effectively above a temperature of about 105°F (40°C) because at this temperature the protein molecule begins to unravel, and a *denatured* protein that has lost its structure can no longer operate as an effective catalyst. At low temperatures an enzyme also loses efficiency due to the fact that all chemical reactions are slower when the temperature drops. As a general rule the activity of a reaction roughly halves with every 18°F (10°C) fall in temperature. In the cold conditions of the tundra, therefore, chemical activities, from soil weathering to enzyme activity, take place more slowly than in lower latitudes and altitudes. All life processes are believed to cease when the temperature falls to –94°F (–70°C).

Plants generally assume the temperature of their surroundings, so when the air temperature falls, their shoot temperature is also likely to fall with it. If the Sun is shining, however, any dark pigments in the plant, including chlorophyll, absorb some of the radiant energy and can raise the temperature of leaves and stems. Invertebrate animals and some vertebrates, such as amphibians and reptiles, are unable to control their body temperature and, like plants, assume the same temperature as their surroundings. They are said to be *poikilothermic,* or cold-blooded. Like plants, however, they are able to take advantage of the radiant energy of sunshine, and they have the advantage of mobility, which means that they can seek out suitable microclimates, such as illuminated, sheltered spots where they can maximize their energy absorption. By deliberately raising its temperature in this way a snake or a butterfly can enhance its cell metabolism by encouraging enzyme activity, and this enables it to perform its functions, such as movement in hunting or escaping predators, more quickly and effectively. When overheated, poikilotherms can regulate their temperature to some degree by losing water and dissipating some energy as latent heat of vaporization.

Mammals and birds are warm-blooded, or *homoiothermic.* They control the temperature of their cells quite closely by maintaining a constant blood temperature. This might seem to be the answer to living in very cold conditions. It does indeed offer many advantages, but an organism that maintains its temperature well above its surroundings inevitably suffers rapid energy loss. Newton's law of cooling states that a body cools more rapidly if there is a large temperature difference between the body and its environment. So a warm-blooded animal needs to expend more energy to keep its higher temperature, and this means that it has to eat more to sustain these energy losses. Scientists have conducted some detailed experiments on the energy budgets of wading birds breeding in the Arctic, and they find that a bird foraging and breeding in the tundra loses approximately 50 percent more energy than the same species of bird breeding in the temperate zone. Birds foraging for food in the tundra expend energy at twice the rate of birds incubating eggs. The summer food supply in the tundra must be adequate to meet this high energy expenditure, or the birds would not succeed in their breeding efforts. The high demand for food, however, cannot be met in the Arctic winter, so the birds migrate to regions with a less challenging climate.

As an alternative solution to migration, a homoiotherm can enter a comatose state during the winter months by lowering its body temperature and consequently its energy consumption. This is the strategy adopted by animals that hibernate. Marmots, for example, hibernate beneath the ground in alpine tundra regions, and their body temperature falls to about 40°F (4°C) when they are deeply comatose and may even reach the freezing point without resulting in death. Their cell enzymes are not denatured at these low temperatures, but their chemical efficiency is clearly very low, so metabolic activity in the animal's body is strongly suppressed.

Humans are homoiotherms, and when subjected to extreme cold they usually respond by wrapping themselves in additional layers of clothing. In other words, they avoid excessive heat loss by resorting to insulation. Even before Sir Isaac Newton had formulated his law of cooling, the consequences of having a body temperature well above one's surroundings had led to the adoption of this behavioral technique on the part of humans to avoid rapid loss of heat. Insulation, however, is an idea that is not limited to

---

Hair

The word *hair* is usually applied to fine extensions of tissue situated on the surface of an organism. In plants hairs (*trichomes*) may be found on a range of different organs, including leaves, twigs, fruits, and roots, and can serve various functions. On roots their value is to expand the surface area of the roots and to act as absorptive organs, taking up water and nutrient elements from the soil. Above ground hairs can serve to reduce transpiration, to deter grazers, or to reflect sunlight from the surface of a leaf, protecting it from overheating. But plant hairs can also act as a means of insulation, keeping leaves and stems warm when the air temperature falls to dangerously low levels. Hairs on insects (*setae*) are hollow projections of the cuticle. They may serve to deter predators, as in some caterpillars, or to insulate these poikilotherms from heat loss. Mammals have more complex hairs than plants and insects. Each hair consists of numerous dead cells that have become *cornified,* that is, enriched in the protein keratin. They grow from hair follicles, which are hollows within the epidermis where active cell division is taking place, replacing damaged and lost hairs. All mammals, including humans, are covered with a layer of hair, and their main function is insulation.

Hairs are very effective insulators because they trap a layer of air close to the surface of the organism. Air is a poor conductor of heat, so as long as it is not removed by turbulence, it creates a protective cover, reducing the gradient of temperature between the organism and its environment and thus preventing rapid heat loss. The hollow hairs of insects and some mammals, such as polar bears, are particularly effective as insulators because they contain air within their structure, and even the most penetrating wind cannot disturb these tiny enclosed air masses.

people; it is actually found throughout the plant and animal kingdoms, from mosses to magnolias, and from bees to bison. The physical principle of insulation is the reduction of the temperature gradient between an object and its surroundings. Newton's law is concerned with temperature differences, and if the organism's body is surrounded by material that is intermediate in temperature between the body and the surrounding air, then the loss of energy will be impeded, especially if the intervening material has poor properties of heat transmission. The body will then cool more slowly. In the natural world the two most common insulating materials are hair (see sidebar on page 105) and fat.

Some mosses and lichens have hairs. They serve the dual purpose of protecting these delicate organisms from desiccation and also help them to retain warmth because the process of photosynthesis involves the activity of enzymes. Warm temperatures therefore need to be maintained. Among higher plants some Arctic willows are extremely hairy, such as the woolly willow (*Salix lanata*), which has leaves, twigs, and even buds and bracts covered with a dense, tangled mass of white hairs. Again, these may serve various functions, including the prevention of water loss and even protection from very high solar intensity at times, but in the long, windy winter the hairs provide protective insulation from the cold. Willow leaves, protected by a layer of hairs, can attain temperatures 20°F (11°C) above those of the surrounding air. The high winds of the Arctic winter often carry crystals of ice, which have an abrasive effect on any surface they collide with, whether rock, plant, or animal. A thick cover of hair thus reduces the impact of ice abrasion. Among alpine plants the reflective properties of hairs can be of particular importance because the intensity of damaging ultraviolet light is enhanced in high altitude situations, where the atmosphere is thin, so hairs protect leaves from radiation that could damage cell chromosomes. The seasonal development of a hole in the stratospheric ozone layer in polar regions has added to this source of biological stress for polar tundra organisms (see "Ozone Holes," page 226). Damaging radiation is also a problem on low-latitude mountains, where ultraviolet rays arrive at a high angle and are less efficiently filtered by the atmosphere. Frost damage is even possible on tropical mountains, such as those of East Africa, and some plants, such as the tree groundsels

Invertebrates, even hairy ones like spiders, need to take shelter when the temperature falls, but their presence is still evident, shown here by a spider's web covered with frost. *(Brandon, Yellowstone Digital Slide File, Yellowstone National Park)*

The snowshoe hare, like many other tundra mammals, has a thick white coat in winter, which gives it camouflage against a snowy landscape. *(Yellowstone Digital Slide File, Yellowstone National Park)*

(*Dendrosenecio* species), retain their dense leaf bases long after the leaf itself is dead to provide a fibrous insulation. The internal living cells of the shoots can be kept at a temperature of 36°F (2°C) even when the outside temperature falls below 25°F (–4°C).

Insects vary considerably in their degree of hairiness. Among the most hairy are the bumblebees, which combine an abundance of hair with large body size. They generate heat by their metabolic activity and by absorbing solar energy as they bask, and they conserve heat very efficiently by means of their hair. As a result of their energy conservation mechanisms, they can maintain activity at lower ambient temperatures than most other insects. Bumblebees also have an unusual technique for generating additional metabolic heat. They vibrate their wings rapidly, producing the characteristic buzzing noise, but do not actually take off. Using this mechanism, they can elevate the temperature of their nests to more than 85°F (30°C). Bees even select the flowers they visit by favoring those at the highest temperature. They use color as a guide for selecting the warmest flowers. Many caterpillars have dark hairs, which help absorb heat and retain it. Ecologists have checked this idea experimentally by shaving some caterpillars and comparing their temperatures with those that were left with their hairs intact. As expected, the shaved caterpillars lost heat much faster than their unshaved neighbors. Hairiness increases with altitude in many alpine insects, including flies (Diptera), and this is believed both to insulate them and to add to their ability to absorb solar heat. Some flies have particularly hairy faces, and it is possible that the face then acts as a heat sensor, leading the fly to warmer patches of rock in which to bask. Most spiders are hairy, which helps them to survive in the cold (see photograph on page 106).

Less-hairy insects are generally very sensitive to the cold, and all insects become inactive and dormant during the tundra winter. For example, the common housefly is fully active only between 59°F and 73°F (15 to 23°C). Its movements become feeble below 50°F (10°C), and at 42°F (6°C) it enters chill coma. When its temperature is lowered to 23°F (–5°C), the housefly dies within 40 minutes. Some insects are more tolerant of the cold than this. For example, tundra mosquitoes of the genus *Aedes* have been observed

Breeding in the Arctic tundra, the snowy owl (*Nyctea scandiaca*) migrates south for the winter. This bird was photographed in South Dakota. It has feathered legs, which aid in heat conservation. *(Denni Larson, Natural Resources Conservation Service, USDA)*

flying when the temperature is close to freezing. Some high-altitude members of the beetle family Tenebrionidae have flattened bodies with raised wing cases. The air-filled cavity acts as an insulating layer.

The mammals most in need of thermal insulation are those that remain active throughout the year and have to cope with the rigors of a long, dark winter. The arctic fox (*Alopex lagopus*), for example, is active in both winter and summer and retains a dense white coat throughout the year. In winter, however, the density of hair growth increases, providing additional insulation for the increasingly cold conditions. Even its paws are covered in fur, both on top and below. As a result of this thicker coat, the arctic fox does not need to increase its metabolic rate to provide extra heat until the air temperature falls to –40°F (–40°C). Without this adaptation the fox would need extra food resources at precisely the time

when they are most difficult to obtain. The animal is so well insulated that it can curl up and sleep on the open surface of the snow even when the air temperature is only –112°F (–80°C). No organism without effective insulation could possibly survive at this temperature. Herbivores have even greater energy balance problems than carnivores, because their food contains less fat and therefore is less calorific. Animals like the arctic hare (*Lepus arcticus*) and the snowshoe hare (*Lepus americanus*), as shown in the photograph on page 107, have much denser hair and more efficient insulation than their relatives in temperate latitudes. An autumn molt provides an even thicker coat for better insulation and can be accompanied by a change to white coloring to provide camouflage, as in the case of the snowshoe hare.

Birds use feathers rather than hair as their means of thermal insulation. Most birds cannot increase feather density when conditions become colder, but one resident of the tundra, the ptarmigan (*Lagopus mutus*), can do so. The alternative is to increase metabolic rate to produce more internal heat, and this is the response of most tundra birds, including waders. Even their chicks lack extra down, but they are *nidifugous,* which means that they leave the nest soon after hatching and actively search out their invertebrate food to keep up their energy supplies. Like the arctic fox, the ptarmigan has insulated legs and feet that are covered by a layer of feathers. The snowy owl (*Nyctea scandiaca*) is another polar tundra species with feathered legs, and the underside of its feet are covered by horny lobes that prevent loss of heat when the bird is standing upon the cold ground (see photograph on page 108).

Insulation, therefore, is a very widespread feature of tundra plants and animals. But why is extreme cold such a dangerous experience for living cells? To answer this question, it is necessary to examine the effects of low temperature on cell function.

■ COLD STRESS

Just as enzyme activity is destroyed at high temperature, it can also be severely curtailed at very low temperature. At low temperature enzymes operate progressively less efficiently and eventually lose their structure, becoming denatured. This means that the activity of the cell begins to slow down as temperature falls, but the relationship between temperature and enzyme activity varies greatly between different species of plant and animal. Some species are much more tolerant of low temperature than others, and this can relate to the temperature optima of their enzymes, which is in turn controlled by the genetic constitution of the cell. The cells of sensitive organisms, both plants and animals, may even die at low temperature before the freezing point of water is reached. They are said to demonstrate *chilling injury*.

Some tropical organisms begin to show signs of chilling injury if their temperature falls below 50°F (10°C), and many are affected below 41°F (5°C). In the case of aquatic organisms, which live in a less variable environment than terrestrial ones, chilling can occur at quite a high temperature. Some tropical fish, for example, may die if the temperature drops below 68°F (20°C). Close inspection of a cell that is undergoing chilling injury reveals a series of changes that eventually prove fatal. At low temperature the mobile cytoplasm of a cell becomes more static; cytoplasmic streaming is reduced. The respiration rate of the cell then increases due to the mitochondria that control the energy relations of the cell becoming uncoupled. The energy-regulating molecule of the cell, adenosine triphosphate (ATP), is not properly linked up to the oxidation of substrates, which is a process that takes place within the mitochondria. So the mitochondria run out of control. Membranes of chilled cells then begin to malfunction. Cell membranes are layered structures of proteins and phospholipids, and they are temperature sensitive. The lipid component, as its name implies, is a fat, and, like all fats and oils, it alters its state according to the temperature. As the temperature falls, the state of the membrane changes, becoming more stiff and solid, and it

no longer effectively controls the passage of chemicals into and out of the cell. In other words, the cell becomes leaky. At this stage the whole structure of the cell collapses, and the damage is irremediable. In some fungi that are cold tolerant (*psychrophilic*) the fats in the membranes are largely unsaturated, which results in their being more fluid even at low temperature.

The effect of cold upon the cell is greatest if the temperature drops suddenly. If a cell is gradually exposed to an increasing degree of cold, it may have a limited capacity for adjustment. This is called *cold acclimation*. Some plants are better able to cope with chilling if they are first exposed to drought. Perhaps they close their stomata and effectively enter a more dormant state, leaving themselves less susceptible to injury.

Once the temperature falls below the freezing point of water, cells become liable to frost damage, which is quite different from chilling injury. When water freezes, it increases in volume by about 10 percent. If the contents of a cell freeze, therefore, considerable disruption of the structure, particularly the membranes, is likely to result. Often this damage reveals itself not during the freezing but when the cell subsequently thaws, especially if the thaw is rapid. When a household water pipe becomes frozen and bursts, the resulting leak is not evident until the thaw comes, and so it is with a frozen cell. The presence of a solute lowers the freezing point of water, so cell contents, which are rich in solutes, are unlikely to freeze at 32°F (0°C). In plant tissues, however, relatively pure water does occur between the cells, outside the main cell wall. Ice crystals thus form and grow between cells, and their formation gradually extracts water from the living cells, causing them to desiccate. The plant may recover from this condition, but if ice crystals actually form within the cell, cell death is inevitable. Freezing injury within the cell is thus caused by membrane disruption, protein denaturation, and dehydration. Frost acclimation in plants is familiar to gardeners, and a plant exposed to progressively lower temperatures often develops the capacity to withstand frost more effectively. It becomes *frost hardy*.

Animal cells can survive freezing, especially if both the freezing and the subsequent thawing are rapid, which is quite different from plants. Animal cells do not have a vacuole, which tends to be the focus of freezing injury inside the plant cell, so they are better able to cope with low cell temperature. At low temperature the whole cell becomes solidified with amorphous ice.

One way in which a cell can protect itself from freezing is to increase the solute concentration, because this lowers the freezing point of the solution, but the choice of solutes is limited. Common salt, sodium chloride, is used in this way to keep roads from becoming frozen in winter, but this is not generally a suitable material for lowering the freezing point of cells. If it is present in any concentration in the cytoplasm, it interferes with enzyme activity and so is toxic. Plants are able to overcome this to some extent by placing the solute in the vacuole away from the living cytoplasm, but salt is used in this way mainly in maritime, salt-tolerant plants (*halophytes*), which also use it to increase the osmotic potential of the cells and thus extract water from saline surroundings. Sugars can be used as an antifreeze, especially among plants, many of which convert stored starch into sugars when subjected to chilling, and these are not harmful to enzymes or to cell organelles. Sugar alcohols, such as mannitol and glycerol, can also be effective and are used as protective solutes by a number of different organisms. Amino acids, the building blocks of proteins, may also be used to lower cell freezing points in both plants and animals. The amino acid proline has often been found in abundance in chilled organisms, ranging from cabbages to fish. In the case of cabbage, the proline content of leaves rises from about 2 percent to 60 percent when they are frost hardened. Similarly, goldfish accumulate proline in their cells when they are subjected to gradually lower temperatures. Thus, animals can also acclimate to frost over a period of time by changing the chemistry of their cells.

Most tundra invertebrates survive extreme cold by supercooling at subzero temperatures. Their cells are protected by antifreeze compounds, and they must empty their guts of all water-containing materials. Springtails, for example, using glycerol as an antifreeze, can be chilled to -40°F (–40°C) and still survive. Mites have been found on the nunataks (the high mountaintops emerging above the ice sheet) of the Antarctic continent that have survived temperatures of –60°F (–50°C). These are the extremes that can be tolerated by highly adapted invertebrates, but all inhabitants of the tundra must be able to cope with long exposure to relatively low temperatures. An alpine ground beetle, *Pelophila borealis,* for example, was kept in a laboratory for 13 months at a temperature of 25°F (–4°C) and survived. In another experiment Antarctic mites were kept at –4°F (–20°C) for more than three months and suffered only 20 percent mortality. So survival in the tundra requires not only the ability to cope with episodes of extreme cold, but also to emerge intact after prolonged but less extreme cold stress.

Tundra organisms, therefore, have adaptations for cold tolerance at the cellular level. There are two broad strategies for cell survival at low temperature. *Freeze avoiders* use antifreeze compounds to lower their cell freezing point and become supercooled when the temperature falls. *Freeze tolerators* undergo freezing but remain relatively undamaged by the experience. Most organisms of extreme conditions, either in the high latitudes of the polar tundra or in the high altitude regions of the alpine tundra, use freeze avoidance techniques. Freeze tolerance is usually associated with tundra boundary regions where the climate is less predictable and when freeze-thaw cycles occur. Under these conditions,

it is evidently more expedient to tolerate freezing rather than develop elaborate metabolic systems for avoiding it. True freeze avoiders can actually be badly damaged when they are subjected to alternating cold and warm, so they are best suited to sites where extreme winter cold is a regular feature.

■ PLANT FORMS IN THE TUNDRA

Tundra vegetation is very variable, ranging from a low carpet of mosses to relatively tall shrubs, depending upon the latitude, the climate, and the local microclimate. One very obvious feature of tundra vegetation, however, is that it is low in stature when compared with any of the other world biomes, with the possible exception of some arid deserts. The structure of vegetation in the tundra is extremely simple (see sidebar "Vegetation Structure," page 64). Botanists have tried to devise systems for describing the general types of plants that contribute to vegetation, classifying plants according to their form rather than their taxonomic and evolutionary relationships. One of the most successful and widely used systems for classification was constructed in 1909 by the Danish botanist Christen Raunkiaer (1860–1938) (see sidebar on page 112 and the diagram on page 111).

Using the classification of plants on the basis of their perennation characteristics, Raunkiaer found that different vegetation types could be described according to the proportions of the various life forms they contained. He called this their *biological spectrum* (see sidebar on page 113). The biological spectrum of the polar tundra consists largely of chamaephytes and hemicryptophytes, which hold their buds just above or at the soil surface, respectively. Phanerophytes are absent from the High Arctic, although nanophanerophytes may grow in the warmer conditions of the Low Arctic, where some shrubs, such as dwarf birches and willows, grow above the 10-inch (25-cm) defining limit for the chamaephyte.

In tundra habitats, therefore, there is clearly great advantage in keeping the vulnerable growing points close to the surface of the ground because the high winds in winter have the capacity to chill the developing buds and cause frost injury. The situation is made even more severe when crystals of ice are carried by the wind and have a corrosive effect on vegetation surfaces. Any shoot rising above the general low canopy of dwarf shrubs or cushion plants is rapidly shredded by ice blasting. Hemicryptophytes, with their buds actually at the surface of the soil, escape most of the impact of wind. The shoots grow in the spring as conditions improve and emerge above the canopy of vegetation during the warmer summer. This strategy has the advantage that taller shoots can grow when the weather permits, thus raising their heads above the surrounding cushion plants and gain-

Plants can be classified into life-forms according to where they locate their organs of perennation (buds, bulbs, corms, and so on). Phanerophytes are trees and shrubs that hold their buds well above the ground, which leaves them susceptible to damage by extreme climatic conditions. Chamaephytes are dwarf shrubs that have buds less than 10 inches (25 cm) above the soil surface. They are well adapted to extreme conditions of cold, drought, and wind. Hemicryptophytes (which means half-hidden) have their buds at ground level. They die back in winter and sprout in spring, which is an appropriate growth form for many temperate regions. Geophytes are plants that survive periods of drought beneath the soil surface as bulbs or corms. This is a useful adaptation for desert plants, but the frozen soil of the tundra is damaging to this strategy. Hydrophytes survive beneath water in the mud of marshes and ponds. Geophytes and hydrophytes are together classed as cryptophytes (meaning hidden) because their perennating organs are out of sight.

fact that both life forms survive in the tundra indicates that there are advantages and disadvantages to both strategies.

The cryptophytes of the tundra are largely aquatic hydrophytes. They hold their perennating organs in the mud beneath frozen bodies of water through the winter. Water loses its heat less quickly than soil, so these sealed-in muds are relatively safe environments in which to overwinter. The same is not true of the terrestrial soils. In the high-latitude polar tundra, permafrost lies a little way beneath the surface right through the summer, and the active layer of unfrozen soil soon solidifies as temperatures fall through the late summer. The cryptophyte form is therefore unsuitable for survival in the tundra soils, as it would be subjected to a pincer movement of ice that develops both from the surface and from below.

The absence of therophytes from tundra may seem surprising. Soils are constantly disturbed by freezing and thawing and by the erosion of surface water in the summer. Plant communities in the tundra are often "open" in the sense that the vegetation rarely forms a complete cover. In every type of vegetation there are gaps and openings revealing bare soil surfaces. Such locations in temperate latitudes would rapidly be colonized by annual plants, opportunistic "weeds" that take advantage of the low levels of competition and quickly complete their life cycles before additional stress or disturbance brings their brief period of success to an end, but this does not occur in the tundra. Very few annuals manage to survive there, such as the Iceland purslane (*Koenigia islandica*), a plant found in the polar tundra biome of both the Northern and Southern Hemispheres. It is said to be bipo-

ing competitive advantage in the better light conditions. But the growth of these taller shoots can take place only when the snow has melted and the temperature rises. Early growth can result in disaster if cold conditions return during the spring, killing the growing shoots. The chamaephytes have no such problems. Their growth is generally slower, but they start the season with the advantage of some height because their perennating buds are above the ground. They are also often evergreen, so they begin their photosynthesis as soon as light and warmth arrive. So there is evidently a trade-off between the hemicryptophyte and the chamaephyte. The

lar in its geographical distribution. But this unusual species is an exception to the general rule that therophytes simply cannot cope in the tundra. The reason for this is likely to be the lack of stress tolerance among such plants. Most annuals have very broad tolerance limits but are unable to cope with severe stress of any kind. Intense drought, soils with very poor nutrient content, or heavy metal pollution, for example, all take their toll on annual weeds. On the other hand, their rapid reproduction and short generation time can lead to a fast rate of evolutionary adaptation, as in some pollution-tolerant weeds. Therophytes have evidently not adapted to cope with the intense cold of the tundra winter, however, and they remain very rare or absent from most tundra regions. Perhaps the growing season is too short for even these fast-reproducing plants to be able to complete their full life cycle in the time available. Alternatively, per-

Plant Life-forms

For several centuries plant geographers have tried to describe and classify plants according to their structure in a way that would relate to their adaptations to climate. It has long been appreciated that the vegetation of the world can be divided into major types, such as tropical rain forest, savanna, deciduous woodland, and tundra, and that the general structure, or *physiognomy*, differed between these various types. Regardless of the taxonomic relationships, vegetation types could be described in terms of the general *life-form* of their components: Rain forest consists largely of trees, tundra of low-stature plants, and so on. The problem was how to classify them in a way that described this form and that expressed the way in which the form related to climate. One solution, which has proved very widely accepted and used, was devised by the Danish botanist Christen Raunkiaer, who felt that the most important feature of a plant as far as climatic adaptation was concerned is how it survives from one year to the next. For many plants this depends on how they manage to protect their delicate growing points, the buds. In a wet tropical environment survival is made relatively easy for the plant because there is little seasonal variation and no extremes of temperature. The buds of tropical plant species can therefore be held high in the air and do not even need protection from drought or frost by the development of bud scales. Some tropical environments have a dry season, and this results in stress to the plant's growing points, meaning that buds may not be held quite so far from the ground and have to be covered by protective scales. In deserts the situation is even more testing for plants, and many can only survive by becoming dormant as seeds.

Raunkiaer therefore developed a classification scheme for plants that bore no relationship to their evolutionary origins, but to their strategies for survival. He classified plants in particular on the basis of where they held their perennating organs.

Phanerophytes: plants holding their buds more than 10 inches (25 cm) above the soil surface, including all trees and shrubs. The group can be subdivided according to height into mega-, meso-, micro-, and nanophanerophyte, the last named having its buds between 10 inches and 6.6 feet (25 cm to 2 m) above the ground.

Chamaephytes: plants with their perennating buds less than 10 inches (25 cm) in the air, but above the soil surface. This group includes the dwarf shrubs and herbs that grow in dense cushions.

Hemicryptophytes: plants that hold their buds at the level of the soil surface. Many temperate herbs fall into this category, including many grasses, together with plants such as stinging nettles.

Cryptophytes: plants with their perennating organs below ground (*geophytes*) or beneath the surface of water (*hydrophytes*). This group includes plants with bulbs, corms, tubers, and rhizomes together with aquatic plants.

Therophytes: plants that survive through unfavorable periods of cold or drought as dormant seeds. These have no vegetative perennating organs and rely entirely on seeds for survival. The group includes annual plants together with ephemerals, plants that have such a short life history they can produce more than one generation in a growing season.

Raunkiaer found that the vegetation of the major biomes of the world differed in the proportions of these various types of plant life forms present within them. Climate, he felt, determines which of the life forms is most appropriate as a survival mechanism for each geographic region.

Biological Spectrum

Having defined the different types of life-forms found among the plants of the world, Raunkiaer analyzed the major vegetation types (now called *biomes*) to establish whether there was a pattern to the global distribution of life-forms and whether this pattern could be explained by variation in climate. He took the full list of the flora growing in selected sites within the different biomes and determined the proportions of each life-form within that flora. The outcome was a *biological spectrum* depicting the life-form assemblages of the various biomes. Some generalized results are shown in the table. The figures are the percentage of the flora that fall into each life-form category, with major contributors shown in bold type.

Predictably, the tropical rain forest flora is dominated by trees, but temperate deciduous forests and the coniferous boreal forests actually have a higher proportion of hemicryptophytes in their flora than they have trees. The grasslands of the world, both tropical (savanna) and temperate, have high proportions of hemicryptophytes but also have a large contingent of annual therophytes, which take advantage of open gaps in the turf and benefit from any disturbance by trampling animals. Deserts have an even greater proportion of these opportunistic therophytes, and the flora is dominated by them. The tundra, both polar and alpine, is dominated by hemicryptophytes and chamaephytes, having a higher proportion of chamaephytes in its flora than any other biome.

	TRF	SAV	DES	TDF	TG	BF	PT	AT
Phanerophytes	**69**	18	3	**36**	5	13	0	0
Chamaephytes	8	16	16	3	6	17	**23**	**25**
Hemicryptophytes	13	**27**	19	**42**	**56**	**50**	**66**	**68**
Cryptophytes	3	13	5	14	10	13	11	9
Therophytes	7	**26**	**57**	5	23	7	0	3

Key to Biomes: TRF = tropical rain forest; Sav = savanna; Des = desert; TDF = temperate deciduous forest; TG = temperate grassland; BF = boreal forest; PT = polar tundra; AT = alpine tundra

haps their seeds fail to survive the intense cold of the frozen tundra soils. This seems unlikely, however, because the seeds of the hemicryptophytes and chamaephytes survive well, and there are records of seeds germinating after 200 years of burial in Arctic soils.

The biological spectrum of the alpine tundra vegetation is very similar to that of the polar tundra. Chamaephytes and hemicryptophytes dominate, while phanerophytes and therophytes are absent. This similarity between the spectra of the two habitats confirms that the life-form assemblages result from adaptation to particular climatic conditions. In the life-form concept botanists have developed a meaningful system of ecological classification of plants.

■ BODY SIZE IN THE TUNDRA

Botanical studies have thus shown that there is a general trend of body size reduction with increasing latitude. Trees give way to shrubs, which in turn are replaced by dwarf shrubs, then herbs, and finally mosses and lichens as the High Arctic polar desert is approached. The body size of mammals in the tundra appears to follow very different rules. There are some very large terrestrial mammals in the polar regions, such as the polar bear and the musk ox, so the plant life form pattern is evidently not repeated here. In 1847 the zoologist Carl Bergmann (1814–65) set out one of the earliest biogeographical principles to be formulated, which later became known as Bergmann's rule. This states that for closely related groups of mammals and birds, the high-latitude representatives tend to be larger than their lower-latitude relatives. So if the rule is to be believed, these vertebrates seem to behave in quite the opposite way from the plants.

There are some observations that support the rule, but many biogeographers consider that the many exceptions that have been described have put the rule in doubt. There is some evidence that backs up Bergmann's rule in the tundra regions, such as the fact that the emperor penguin (*Aptenodytes forsteri*) in Antarctica is considerably larger

Chinstrap penguins (*Pygoscellis antarctica*) breed in colonies on some of the islands in the Southern Ocean and also on the Antarctic Peninsula.　*(Keith Grochow)*

than the Galapagos penguin (*Spheniscus mendiculus*) in the equatorial regions. But, on the other hand, the little blue penguin (*Eudyptula minor*) is found even farther south in Antarctica than the emperor penguin yet is much smaller, and the chinstrap penguin (*Pygoscelis antarctica*) shown in the photograph above, and the gentoo penguin (*Pygoscelis papua*) are large birds, but are mainly found farther north, around the oceanic islands of the Southern Ocean. Some biologists claim that Bergmann's rule is based upon some very carefully selected sets of data.

The rule does not seem to apply to cold-blooded poikilotherms. Ground beetles (Carabidae), for example, tend to be smaller at higher altitudes and in colder conditions. Indeed, the size of invertebrates in general is smaller in the tundra.

Bergmann based his rule not simply on observations but also some theoretical reasoning. Assuming that all endotherms lose an equivalent amount of heat from a given area of their body surfaces, then large-bodied animals should be better at heat conservation than smaller ones because they have less surface area per unit body volume than a small animal. This means that a small animal needs to take in more food in relation to its body size than a large animal, which is clearly a disadvantage when food is in short supply, as it is in the low-productivity tundra. A small animal is in danger of suffering from "cold starvation." This, Bergmann would claim, is why large animals such as polar bears and blue whales are able to survive in the extreme north. But one could argue that a large animal needs more food per

individual than a small animal, so small animals should be favored in a land where food is scarce.

Zoologists in Canada have carried out a survey of the musk ox (*Ovibos moschatus*) in different parts of Alaska, Arctic Canada, and Greenland and have checked whether there is any relationship between body size and latitude. They found no relationship over the range of latitudes 60°N to 83°N. There were changes in the dental structure, but these probably related to diet, which ranged from predominantly willows in the south to mainly grasses in the north. Some evidence from fossils, however, does appear to support Bergmann's rule. Horses (probably *Equus ferus*) were present in Alaska at the close of the last glaciation but became extinct around 12,500 years ago. Before they finally became extinct the fossil record shows that the population became progressively smaller in body size. This was taking place at the same time as the temperature was rapidly rising and glaciers were retreating, so it could be claimed that the horse population was responding in accordance with Bergmann's rule. Similar responses have been recorded in studies of woodrat (*Neotoma cinerea*) fossils, which also change in size in harmony with climatic changes. Arguments about the validity of Bergmann's rule will undoubtedly continue for some time, but the idea has proved a very attractive one, and there are indeed examples of animals that appear to obey it.

A very similar and perhaps less controversial rule was propounded by Joel A. Allen (1838–1921) in 1877 and became known as Allen's rule. This principle is concerned with the extremities rather than the entire body. Allen observed that populations of closely related animals tended to have smaller extremities at higher latitudes. Take ears, for example. The antelope jackrabbit of Arizona (*Lepus alleni*) has very long ears, whereas Oregon populations of the black-tailed jackrabbit (*L. californicus*) have shorter ears. In Canada the snowshoe hare (*L. americanus*) has even shorter ears, while the arctic hare (*L. arcticus*) in the northern tundra has the shortest ears of all. Similarly, the fennec fox (*Fennecus zerda*) of North Africa has very large ears in porportion to its head size, while the red fox (*Vulpes vulpes*) of Europe and North America has much shorter ears. The arctic fox (*Alopex lagopus*), however, has the shortest ears of all. Its rounded, bearlike ears hardly project beyond the thick fur of its neck. The rule does not apply to migratory animals, which escape cold conditions by moving into warmer climes, so this does suggest that the tendency for polar animals to have smaller extremities is linked to climate. Scientists have tested the theory experimentally by breeding mice at different temperatures, and those reared at low temperatures have shorter tails than those accustomed to greater warmth.

Like Bergmann's rule, Allen's rule can be explained physiologically by reference to energy conservation. Ears and tails are supplied with blood vessels and yet have little hair cover, so they radiate heat very effectively. Indeed, elephants use their ears to lose energy when they become overheated. Although large ears can be a great advantage to an animal like a hare, enabling it to hear the approach of a predator, there is a trade-off between the detection of an enemy and the loss of precious heat. Similarly, a predatory animal such as a fox uses its ears to detect the movement of small mammals in low vegetation, but large ears would be a liability when the temperature falls and icy winds blow across the tundra.

As in the case of Bergmann's rule, however, there are some biogeographers who doubt whether Allen's rule can be regarded as a universal principle. They argue that even in the tundra there may be a need for animals to dissipate heat. Caribou, when overheated, thermoregulate (control heat loss) through their long legs, beavers use their tails, and blubber-coated seals can lose heat through their flippers. So these doubters are not convinced that small ears and short tails are necessarily a general adaptation to cold environments. Both of these rules, therefore, could apply to a limited number of well-studied examples.

■ COLOR IN THE TUNDRA

For much of the year the prevailing color of the tundra landscape is white because it is covered by snow and ice, but strong winds ensure that even in winter the exposed ridges and hilltops remain clear of snow and reveal the underlying vegetation or rock. Lichen-covered rocks may be light in color, often pale gray or orange, but the vegetation in winter is dark colored, brown or dark green. Herbaceous vegetation mainly dies down in the winter apart from the dense cushions of saxifrages and other chamaephytes (see sidebar "Plant Life-forms," page 112). Many of the chamaephytes, both the herbs and dwarf shrubs, have evergreen leaves and retain these structures right through the winter. There are many possible advantages of being an evergreen in the tundra environment, as described in the sidebar on pages 116–117.

Plants with evergreen leaves save energy and nutrients in constructing and maintaining their leaves, are better able to take advantage of the improved conditions at the onset of spring, and are often more efficient at water conservation and herbivore deterrence. This raises the question why are there any deciduous species of plant in the tundra. Many of the hemicryptophytes (see sidebar on page 113) die down for the winter, losing all their leaves and becoming reduced to a ground-level growing point and an underground root or rhizome. These plants clearly require deciduous leaves. Some dwarf shrubs, such as certain members of the blueberry family (*Vaccinium* species) and the dwarf birch (*Betula nana*), also lose their leaves in winter. In doing so they lose the many advantages

that the evergreen habit provides, but they are less easily damaged by fierce winter winds bearing ice crystals. All above-ground vegetation that is not protected by a snow cover is liable to be subjected to the corrosive force of the wind, and leaves present through the winter are easily damaged. Bare twigs covered with a layer of bark are more resilient and resistant in the face of such stress, so several plants adopt this habit as a survival mechanism.

With the coming of spring the dark greens and browns of the tundra are replaced by brighter greens as both deciduous and evergreen species begin their vegetative growth and

Cloudberry (*Rubus chamaemorus*), growing here in Finland, is a widespread plant of peaty soils in tundra habitats. Its fruits are eaten by birds, such as ptarmigan, which are responsible for the plant's dispersal.   *(Peter D. Moore)*

Why Be an Evergreen?

Many plants in the tundra have evergreen leaves, including the bearberry (*Arctostaphyllos uva-ursi*), shown below. This does not mean that the leaves last forever; they may survive only two or three years, but they are not discarded at the end of each summer in the manner of deciduous leaves, so the plant has a green appearance even in winter. For most plants the main function of the leaf is to photosynthesize, trapping solar energy and using it to fix carbon dioxide from the atmosphere and produce sugars. In the polar tundra the days are short or even nonexistent during the winter, so the presence of leaves may seem entirely redundant. Indeed, living leaves are costly to a plant in dark conditions because they continue to respire but fail to photosynthesize, so they consume more energy than they produce. Why, then, is the evergreen habit such a frequent feature in the tundra? One possible reason is that the plant is ready to begin photosynthesis as soon as conditions become suitable. As days become longer, the temperature rises, the cover of snow melts, and the evergreen leaf is fully constructed and ready to begin its activity. A deciduous plant has to build new leaves, which means that it has to commence a high level of metabolic activity very early in the season when the temperature may still be quite low. Stored food reserves must be mobilized, new elements must be taken up from the soil, and the expensive business of leaf building must commence at a time when the plant has just survived the most stressful time of its year. An early start without the necessity for new leaf construction may be worth all the energy losses associated with the long winter nights.

One must also consider the total energy balance of the leaf over its entire lifetime. Building a leaf costs energy, so if it can be made to last longer, then it may prove better value. Renewing leaves each year is expensive in terms of energy investment, especially for plants that grow in the tundra where growing seasons are short and all energy must be used with parsimony and efficiency.

Bearberry (*Arctostaphylos uva-ursi*) is a dwarf shrub that grows abundantly in both polar and alpine tundra, such as here in Scotland. Its distribution pattern can therefore be described as arctic-alpine.   *(Peter D. Moore)*

Not only do leaves cost energy, they also require nutrient elements for their construction. Nitrogen is needed for protein building and phosphorus for membranes and nucleic acids, and these must be obtained from soils that are generally poor in these elements. If a leaf lasts longer, then less effort needs to be expended in extracting these elements from the soil, so the investment in an evergreen leaf may prove worthwhile. Evergreen leaves, in other words, are an economically sound option.

There are two additional possible advantages of evergreen leaves. Leaves of this kind are often tough, leathery, and covered on their upper surface by a thick waxy layer. Such leaves generally have low rates of transpiration, and water conservation can be an important adaptation in the tundra, even though water appears to be in abundant supply (see "Drought in the Tundra," pages 119–121). Because they last longer, evergreen leaves also have a greater opportunity to accumulate chemicals that deter grazers, so these structures can help to protect the plant from both the invertebrate and the vertebrate herbivores that gain their energy by consuming them.

the production of new leaves. The flush of new herbage is attractive to herbivores because these fresh leaves are rich in nutrient elements, are high in proteins, and contain relatively low concentrations of herbivore-deterrent chemicals (see "Grazing in the Tundra," pages 129–130). Very rapidly this period of vegetative growth turns into a time of flower production, because the season is short and the life cycle must be quickly completed.

Some tundra plants are pollinated by wind (see sidebar "Pollination," page 126), and their flowers are usually green and inconspicuous. These plants include sedges,

The tundra becomes colorful in the fall. These leaves of James' saxifrage (*Boykinia heucheriformis*) become bright red as they accumulate anthocyanin pigments. *(J. Schmidt, Yellowstone Digital Slide File, Yellowstone National Park)*

willows, and dwarf birches. But others are pollinated by insects, and these require attractive floral structures, often using color as a means of advertising their presence. Pollinating insects, however, may be in short supply, especially early in the short summer season, so many tundra plants, particularly those of the alpine tundra, have large and brightly colored flowers. The tundra in early summer can become a very colorful place.

Flower color in plants is mainly determined by pigments, especially *carotenoids* and *flavonoids*. Carotenoids consist of long chains of carbon atoms with a six-carbon ring at each end. Beta-carotene, for example, consists of two molecules of vitamin A joined together in the form of mirror images. Carotenoids are colored orange or yellow, and they are important in photosynthesis, acting as accessory pigments and gathering light energy, which they pass on to chlorophyll. If chlorophyll is removed from a leaf, a yellow color remains because of the residual carotene present in the plastids of the cells. Some yellow and orange flowers owe their color to the presence of carotenoids, such as those of the marsh marigold (*Caltha palustris*). Far more important in the determination of flower color, however, are the flavonoids. These are generally smaller molecules than the carotenoids but contain two or more six-carbon aromatic rings. The red and blue colors of flowers, such as those of the gentians (*Gentiana* species), often result from the presence of *anthocyanins*, which are water-soluble flavonoids that are located in the vacuoles of plant cells. One other group of flavonoids is the *flavonols*, which are widely distributed in leaves and flowers, and, although colorless, add to the whiteness of white-colored flowers.

The tundra is often made more colorful by fruits in the late summer, especially the bright red fruits of the bearberry (*Arctostaphylos uva-ursi*) (see photograph on page 116), and the pink of the cloudberry (*Rubus chamaemorus*), shown in flower in the picture on page 115. Red is a color that proves particularly attractive to birds, which have acute color vision, so its adoption by plants is often associated with bird dispersal (see "Reproduction and Dispersal in the Tundra," pages 123–129). Many of the tundra fruits are black or very dark blue, as in the case of the blueberries (*Vaccinium* spe-

Many tundra butterflies have underwing patterns that help them merge into the background of lichens and mosses. This is the case with the Uhler's arctic (*Oeneis uhleri*), which is found both in the polar tundra of the Northwest Territories and in the alpine tundra of the Rocky Mountains south to New Mexico.   *(Yellowstone Digital Slide File, Yellowstone National Park)*

cies), and they are often covered with a whitish waxy bloom. These fruits reflect ultraviolet light, and birds are more sensitive to radiation of this wavelength than humans, so the fruits may be more conspicuous to a foraging bird than they are to people.

The late summer is also a time when the vegetation itself changes color because the leaves of deciduous plants begin to die and take on a wide variety of colors. The colors of autumn leaves are caused by a number of biochemical changes. Chlorophyll degrades, so other pigments in the leaves become more apparent, such as the yellow carotenoids that have served as accessory pigments in photosynthesis. Flavonols, usually colorless, are converted into anthocyanins, and so contribute a range of red colors to the dying leaves, as in the leaves of the saxifrage, shown in the photograph on page 117. The outcome is a blaze of red and yellow colors in the tundra landscape, rivaling that of the deciduous forests of New England.

Animal coloration in the tundra is of two types. Pigmentation is present, as in the hairs of some mammals, especially in the summer, when dark colors provide camouflage and protection against predators. Many insects, including beetles and butterflies, however, achieve metallic coloration by using iridescence. The colors in this case are not produced by pigments but by the physical structure of the surface, which is often covered in dense papillae, or wartlike protrusions. These tiny projections are usually aligned in rows or dense masses so they split the incident light like tiny prisms, resulting in a colored appearance to an observer. The color may change when it is viewed from different angles, as in the change from green to blue or violet on a beetle wing case.

Animal coloration serves several purposes. Often it is *cryptic,* which means that it provides a camouflage, allowing the individual to merge into its background and not be noticed by its predators. Some of the animals that use cryptic coloration in the tundra change their colors with the different seasons. The willow ptarmigan (*Lagopus lagopus*), for example, is a grouse that is largely brown in summer, when it is inconspicuous at its nest on the ground among dwarf shrubs, but is almost entirely white in winter, when it is hard to detect against a white background of snow. The race of the willow ptarmigan that occurs farther south in Scotland, where winter snow is not quite so extensive or prolonged, does not change its color during the winter but remains brown all year. The greens and browns of many insects are also cryptic, but these animals either pupate or become dormant and hidden in the winter, so they do not need to adopt a white winter coloration. The bright colors of many butterflies are associated with attracting a mate rather than avoiding a predator, but the underwing patterns often serve to make a resting butterfly inconspicuous, as shown in the photo on page 118.

Crypsis, however, is a common reason for coloration among tundra animals, but it does not provide all the answers. Why, for example, are seals dark colored? They are very conspicuous when out of water on the ice and are more easily spotted by polar bears and orcas. They may, on the other hand, be less conspicuous when underwater and hunting for their aquatic prey. Their dark coloration may also serve to enhance their ability to absorb solar energy when they are basking in the sunshine on the ice. Polar bears, on the other hand, seem well adapted to be inconspicuous against the ice, but they have no real predators in the tundra, so why should they adopt cryptic coloration? The likely answer is that they are not colored white as a means of avoiding predation but to assist in their hunting. Basking seals, their main prey, are more likely to notice the approach of a black bear than a white one.

Insects at high altitude are often dark in color. Among the butterflies, for example, High Arctic and high-altitude species tend to be dark. The so-called alpines (*Erebia* species) are a good illustration of this tendency. All are very dark brown in color, with just a few decorative spots of black and white. The banded alpine (*E. rossii*) is found well into the High Arctic of Alaska, while the Colorado alpine (*E. callias*) inhabits the mountain summits of Colorado, Montana, and Wyoming. Dark coloration is called *melanism.*

The tundra, therefore, is a colorful place for a number of reasons. Some organisms, such as flowering plants, need to attract attention, while others, such as caterpillars, seek to avoid it. Much of the color of the tundra landscape, however, results from biochemical processes in the leaves of plants, especially those associated with the loss of deciduous leaves in the fall.

■ DROUGHT IN THE TUNDRA

Precipitation is generally low in the polar tundra (see "The Atmosphere and Climate," pages 10–11) and also at very high altitude in alpine tundra. But temperatures are also generally low at high latitude and altitude, so evaporation rates are slow. Water, either in liquid form or as ice and snow, is therefore not normally scarce in tundra habitats, but the lack of precipitation means that once the spring water supply in soils has drained into low-lying areas, it is not replenished. As a consequence, well drained ridges, especially among scree, or sandy and gravelly soils, can become extremely dry in the tundra summer. The poor cover of vegetation adds to the drought problem, because the soil surface is exposed to sunlight for up to 24 hours everyday, and water evaporates rapidly. The tundra habitat, therefore, has strong contrasts between its low-lying wet and raised, well-drained dry sites. Larger, mobile animals are not strongly affected by such local drought, but soil invertebrates and plants can become stressed. Living on a gravel ridge in the tundra summer is not very different from living in an arid desert, apart from the lower temperature.

The plants of dry habitats can be divided into two types, drought avoiders and drought tolerators. In hot deserts the most frequent types of drought avoiders are geophytes and therophytes (see sidebar "Plant Life-forms," page 112); they either resort to underground organs of survival, such as bulbs, or they become dormant as seeds and wait for better times. But neither of these life forms is found at all commonly in tundra habitats, so most plants have to tolerate the drought.

The main problem for a plant under drought conditions is that it loses water from its leaves whenever it is actively photosynthesizing, so it needs to replenish its water supplies from the soil. Water is thus lost by transpiration (see sidebar below), and maintaining water balance can become difficult if the soil is dry.

The tundra contains many plants adapted to tolerate drought. The chamaephyte life form, which is low-growing, close to the soil surface, has many advantages in terms of water conservation. It avoids wind, which means that it is not exposed to the drying effect of moving air to the same extent as plants that grow high above the soil surface. Most tundra plants have small leaves, often evergreen and leathery, with their stomata concentrated on the undersurface, where they are protected from air movements and from direct insolation. The upper surface of the leaf is often coated with a thick layer of wax (the *cuticle*), which prevents any evapora-

Transpiration, Water Potential, and Wilt

Photosynthesis involves trapping energy from sunlight and using that energy to reduce atmospheric carbon dioxide to organic molecules. Obtaining the carbon dioxide gas, however, means opening pores (*stomata*) in the surface of leaves so that the gas can enter the plant, and open pores allow the diffusion of water vapor out of the leaf and into the atmosphere. Inside the leaf there are small chambers below the pores surrounded by the cells that absorb the carbon dioxide gas, and these cells are constantly losing water when the pores are open. As the cell loses water, the cell contents become more concentrated in their solutes, and the walls of the cell may even be put under a negative pressure, all of which indicate a demand for water, or, to use the technical expression, a reduced water potential. A cell with a low water potential exerts a strong pull on the water in surrounding cells that have a more abundant supply of water (a higher water potential). They draw water from these cells, and the demand for water is thus conducted across the leaf to the vascular bundles, the veins through which water passes up the plant stems from the roots.

Water is thus removed from the vascular bundles of the leaf, which effectively pulls water in a continuous column from the roots, which act as absorptive organs in the soil. Water molecules have the property of cohesion, binding together so effectively that a very strong pulling force can be exerted without disrupting the water column. As water is pulled upward from the vascular tissues in the roots, a negative pressure is exerted on the surrounding cells, removing water from them and placing them under a water demand that then extends across the roots to the root hairs. When a root hair cell becomes short of water, it has only one remaining source, the water in the soil outside the plant. If the soil is moist, this water is relatively easily available because its water potential is high, but if the soil is dry, then water is held more tightly by the capillarity of the fine pores between the soil particles, and the water potential of the soil is thus low. In saline conditions the soil water potential is even lower because of the effect of the dissolved solutes, making it even more difficult for the plant root to extract water.

In prolonged drought conditions the soil water potential may fall to such a low level that plant roots are unable to extract the remaining tightly bound water. The entire plant continues to lose water without being able to compensate by root absorption, and eventually the turgor of the cells is lost and the plant begins to wilt. This shows itself visually by a drooping of the leaves and branches, no longer held rigid by water tension. In temperate latitudes, even in alpine tundra, there is some relief during the night, when photosynthesis ceases, stomata close, temperature falls, evaporation rates drop, and the water stress on the plant is reduced to some extent. In a polar summer there is not even this respite. If water arrives as precipitation or water flow from melting snow, for example, then wilting may not prove fatal, and recovery of the plant takes place. But if the drought persists, then the plant enters a condition called permanent wilt, and at this stage even a renewal of water supply cannot prevent plant death.

tion from the epidermal cells. Many members of the heather family (Ericaceae) are of this type. Water losses by transpiration are thus reduced, but there is a danger that reducing transpiration will have the side effect of reduced gaseous exchange in general, which means less carbon assimilation and slower growth. It is not surprising, therefore, that most tundra plants are slow growing. The other possible strategy for avoiding drought stress is establishing an extensive root system to forage for water in times of scarcity, but, although this is often found among desert plants, it is not common in tundra species. The permafrost lies close to the surface, and the gravel ridges where drought is most common often have very shallow soils, so root extension rarely solves the problem.

One other type of drought tolerance is to become dehydrated and yet survive. Botanists call plants with this survival strategy *poikilohydric*. Some mosses and lichens have developed the capacity to cope with extreme desiccation and yet retain the ability to grow once again when they are supplied with water. Mosses give the impression of being delicate structures, their thin leaves lacking any waxy cuticle, and easily damaged. But although they dry quickly, they can then survive in a dormant, desiccated state for long periods of time. Mosses that have been collected, dried, and stored in a museum collection for more than a century have often proved viable when rewetted and given a chance to grow. Some higher plants also have this property. The alpine plant *Ramonda myconi,* from the Pyrenees of Europe, for example, can survive for up to three years completely desiccated and yet be able to recover and grow when given water once again. In a desiccated condition this plant can tolerate temperatures of up to 130°F (55°C). The remarkable capacity of *Ramonda* to survive such extreme conditions indicates the stresses to which alpine plants at relatively low latitudes are exposed. It is a hemicryptophyte and holds its leaves in a flat rosette close to the ground surface. The leaves and the tap root die during drought, leaving a withered but viable center to the plant.

Small invertebrate animals can also suffer from desiccation in tundra regions. Small size is a disadvantage in water conservation because the animal has a very large surface area per unit body mass, so water is quickly lost from its surface. The exoskeleton of soil invertebrates is usually thin and poorly adapted to water retention, and those that lack an exoskeleton, such as flat worms (planarians) and nematode worms, are particularly sensitive to drought. Woodlice and centipedes have no cuticular wax layer on their exoskeletons, so they also avoid dry conditions and actively seek out locations of high humidity. Above the soil surface, plant bugs also demand humidity. Some, such as the plant hoppers, or spittle bugs, surround themselves with a layer of bubbles to prevent desiccation. Most small invertebrates avoid exposure to the sun unless they occupy a moist habitat where there is no danger of excessive drying.

■ WATERLOGGING AND OXYGEN SHORTAGE

As far as wetness is concerned, the tundra is a place of contrasts. While some well-drained ridges and screes may suffer from summer drought, there are other areas where water is present throughout the year, in a liquid form in summer and as ice in winter. Life in this situation generates a new set of problems for plants and animals.

All plants and animals need to respire. They obtain their energy by oxidizing organic molecules, releasing carbon dioxide and water in the process. Oxygen is a very reactive molecule, and it needs to be very carefully controlled within the cell. In the atmosphere it is a very abundant element, so plants and animals are not likely to run short of it, but in water this is not always the case. Oxygen dissolves in water, the form in which aquatic organisms obtain it, but it diffuses through water about 10,000 times more slowly than it moves through air, so organisms consuming oxygen can face a shortage when demand is high. Photosynthesis generates oxygen as a by-product, so a well illuminated water body containing green plants is constantly being renewed in its oxygen content, and the organisms within it are not totally dependent on atmospheric oxygen being dissolved at the surface. But in the dark conditions below the sediment surface there is no photosynthesis, and diffusion is the only source of oxygen. So the local environment often becomes *anaerobic* or *anoxic*. Plant roots, invertebrate animals, and microbes then suffer from oxygen starvation. Also, under such conditions some elements form toxic compounds in the soil. Iron and manganese can link up with sulfur and nitrogen to form sulfides and nitrites, respectively, and these compounds are toxic to plants. Some plants overcome this toxic effect from the soil chemicals by leaking oxygen from their roots and oxidizing the offending compounds before their toxicity is felt, but this mechanism leads to an additional strain upon the plant's oxygen supplies.

Tundra organisms that inhabit the wet regions have developed a wide range of techniques for dealing with the stress of anoxia. Some plants have structural features that help them obtain oxygen, such as hollow stems allowing air penetration to the roots. These may continue to function even when they are dead because they form a series of hollow tubes leading up from the sediments to the atmosphere above. In this way they can enhance the supply of air to the roots. It is possible to test in a laboratory precisely how a reduction in oxygen supply affects the function of a plant's roots. Even in a flood-tolerant plant, such as rice, reducing the oxygen supply from the normal atmospheric level of 21 percent to 10 percent reduces the rate of new root growth by about 60 percent. Many rushes (*Juncus* species) are killed by just a few days of anoxia. They survive in wet places only if the periods of inundation are of short duration, so they can cope with periodic floods and fluctuating water tables in the soil, but not permanent waterlogging.

Some plants, however, are true tolerators of anoxia. One of the main problems with living in anoxic conditions is that the respiration of a molecule such as glucose proceeds only as far as the generation of ethyl alcohol (ethanol). It is this reaction, of course, that is used in the commercial production of beers and wines, using the anaerobic respiration of the microbe yeast. Yeast colonies, however, can generate ethanol only to a concentration of about 15 percent because at higher levels it becomes toxic to the yeast cells. In precisely the same way, ethanol is toxic to plant and animal cells, so anaerobic respiration that results in ethanol generation in the cell can cause cell death. Some anoxia tolerators avoid producing excessive amounts of ethanol by slowing their metabolic activity and their growth. Others manage to export the ethanol into their environment. In the case of some water lilies, for example, oxygen can be taken down the hollow stalks of some floating leaves (usually the young ones), and the ethanol produced by anaerobic respiration diffuses up other hollow stalks (often the old ones) and is lost to the atmosphere. One species, the yellow water lily (*Nuphar lutea*), is sometimes called the "brandy bottle" because its leaves often smell alcoholic.

Many animals of wet conditions are able to absorb oxygen from water over their entire skin surface. Examples include worms, crustaceans, many insects, and some amphibians. Larger animals often have specialized gills over which they can pump water and absorb oxygen even from low concentrations. Arthropods have parts of their bodies rich in tracheae and covered by a thin, permeable cuticle so that oxygen can diffuse across. These may be concentrated in gill structures, as in the stoneflies (Plecoptera) and May flies (Ephemeroptera). Some aquatic insects have spiracles at the end of their bodies with a terminal siphon that can be extended so that an insect at the water surface can extrude the siphon into the air above. The siphon can even be used to penetrate the hollow tubular stems of aquatic plants and tap in on their air supply. The larva of the drone fly (*Eristalis tenax*) has a particularly extensible siphon. Although this larva is less than an inch (2 cm) in length, its breathing tube can be extended to almost five inches (12 cm), hence its popular name of "rat-tailed maggot."

Many invertebrates underwater maintain a thin layer of air over their surfaces and draw upon this for their oxygen supply. Hairs on the body surface help in this respect. Some have even developed efficient oxygen-transporting molecules within their bodies, including hemoglobin, the main oxygen transporter of mammals. Hemoglobin is found in some earthworms, mollusks, and midge larvae.

Waterlogging is not the only reason why some tundra habitats are relatively poor in oxygen. At high altitude in alpine tundra, low air pressure leads to a low availability of oxygen, which is why mountain climbers and airline passengers need enhanced oxygen supplies under such conditions. The gravitational pull of the Earth acts upon the gases of the atmosphere, pulling them toward itself. The weight of gas is greatest close to the Earth's surface, where the air is compressed and dense, but declines with altitude as the air becomes thinner. The atmospheric pressure at 16,400 feet (5,000 m) is only half that at sea level, so all the components of the atmosphere are less dense, including vital oxygen.

In mammals atmospheric oxygen dissolves in a thin layer of water lining the surface of the lungs and diffuses into the rich supply of blood vessels surrounding the alveoli, the air-filled vessels that make up the lung structure. Dissolved oxygen in the blood becomes attached to the iron-containing protein hemoglobin, located within the red blood cells, the erythrocytes. Every erythrocyte contains approximately 300 million molecules of hemoglobin, and each cubic inch of human blood contains about 80 billion erythrocytes (5 billion per cm^3). Blood is therefore an extremely efficient transporter of oxygen from the lungs to the body tissues, where it can be used for cell respiration. Mountain mammals have even higher hemoglobin concentrations to cope with the lack of oxygen in the atmosphere. They also accumulate higher levels of myoglobin in their muscles, providing them with greater muscular efficiency.

Under high-altitude, low-pressure conditions, the diffusion of oxygen from lungs to blood is impaired, so the blood supply to all parts of the body is reduced. Even at an altitude of 10,000 feet (3,000 m) the effects of oxygen shortage can be felt, causing dizziness, breathlessness, headaches, and general lack of energy, as anyone who has been skiing or mountain hiking will know. A rapid ascent to very high altitudes can result in altitude sickness, when fluid accumulates in the lungs and the brain (edema), which can be very serious. In 1968 the Olympic Games were held in Mexico City at an altitude of 7,400 feet (2,250 m), and even at this height there were serious concerns about the fitness of athletes to compete under low-pressure conditions. In the outcome, sprint events were unusually fast, probably because of the reduced air resistance and the short time involved in the race, which reduced the risk of tissue oxygen starvation. Long-distance events on the whole were slower than might have been expected because athletes suffered from the low oxygen concentration. The great success of long-distance runners from Ethiopia may well be a consequence of their acclimatization to low levels of oxygen in their high-altitude African homeland, resulting in the development of greater lung capacity.

Taking off from high altitude, however, does require extra effort and energy, and the highest record for a bird getting airborne from the ground belongs to a chough, a member of the crow family, at an altitude of 27,000 feet (8,235 m) on Mount Everest. Apparently it needed a long run before finally taking to the air.

Cold-blooded poikilotherms demand less metabolic energy than warm-blooded homiotherms because they do not need to maintain their body temperatures. Laboratory experiments with frogs have shown that they are able to cope with low pressure conditions much better than warm-blooded animals. Frogs and toads are found on high mountains, but they are limited by low temperature rather than by low oxygen availability. Invertebrates are also relatively unaffected by low atmospheric pressure. Spiders have been found at an altitude of 22,000 feet (6,710 m) in the Himalaya, and laboratory tests on insects have shown that they can survive much lower pressures, even below that experienced on the top of the world's highest mountain, Mount Everest, at 29,000 feet (8,845 m).

■ REPRODUCTION AND DISPERSAL IN THE TUNDRA

All plants and animals need to reproduce so that the current generation can be replaced as it ages and becomes less fit. Darwin pointed out that organisms tend to produce more offspring than can possibly survive and that mortality, particularly infant mortality, is therefore an inevitable fact of nature. Many plants, together with some simple animals and protists, are capable of vegetative reproduction, or fission, enabling them to produce clones of themselves that can spread into new habitats or replace the original organism. This can prove a very successful means of reproduction, but it has the disadvantage that the offspring are genetically identical to the parent. In a situation in which the parent is particularly well adapted to an environment, this need not be a disadvantage, and many organisms make use of this fact to expand populations rapidly when conditions are suitable. But in the long term there is a need for genetic diversity in populations so that the capacity exists for the species to adapt to new conditions, especially at times of rapid environmental change.

Sexual reproduction involves the fusion of two gametes, usually from separate individuals with slightly different genetic constitution. The gametes have to be sufficiently genetically similar to be compatible, so the gametes of different species are not normally able to fuse, but each individual in a population, apart from clones, has a distinct genetic makeup. When gametes form their chromosomes are also subjected to a shuffling that adds to the range of

Nepali woman in traditional costume. The people of Nepal live in the highest regions of the planet. *(The Nepal Trust)*

Oxygen shortage at high altitude undoubtedly limits the distribution of some animals, especially warm-blooded mammals and birds that demand a high oxygen input to maintain their body temperature and energetic activity. The highest human settlement in the world lies at 16,000 feet (4,880 m) in Tibet, but pastoral farmers take their domestic animals even higher, to around 18,000 feet (5,500 m). Wild sheep and ibex in Tibet have often been recorded at 19,000 feet (5,800 m), which is remarkable when one considers that the partial pressure of oxygen is less than half that of sea level. The people of Nepal and Tibet are physiologically adapted to high-altitude conditions. Birds can attain even higher altitudes and may have to do so when crossing mountain ridges while on migration. Condors and eagles conserve energy by soaring at high altitude, using thermals in the moving air so they can avoid using flight muscles that demand oxygen.

genetic combinations produced, so the fusion of gametes results in the production of a new individual with a completely new and distinctive genetic constitution. Sometimes the new individual is less well adapted to the environmental conditions than the parent. If polar bears produced a dark-colored cub, it would probably fail to hunt successfully because it would be highly conspicuous against the ice and snow. The likelihood is that it would starve before reaching adulthood. But often the new individual has some features that provide added advantages for life in the tundra. A plant may have larger flowers or a mammal thicker fur, both of which may give them advantages over their neighbors in the competition of life. They would therefore survive better and reproduce more successfully, leaving more of their progeny in future generations. The rigors of the environment would sift out the weaker genetic combinations and permit the survival of the more favorable genes. This is the process of natural selection that Charles Darwin first set out as the likely mechanism to drive evolution.

Reproduction, especially sexual reproduction, when combined with mortality among the less fit, thus lies at the heart of evolutionary adaptation. One might suppose, therefore, that it would benefit an organism to produce as many young as possible, but that is not always the case. A whole branch of the discipline of ecology, called population ecology, is devoted to the study of reproductive strategies, the most effective ways of expanding a population under different circumstances. Population ecologists have devised a classificatory scheme for reproductive strategies, dividing them into r-selection strategies, which concentrate on rapid population growth by maximizing reproduction rates, and K-selection strategies, which are concerned with the maintenance of population over the long term rather than rapid population expansion in the short term (see sidebar on page 125).

Both types of reproductive strategies have advantages in different locations and at different times. A species that can expand its population rapidly can take advantage of favorable conditions that may not last very long. If a tundra summer is particularly warm and plant productivity is high, grazing animals may be able to expand their populations to take advantage of the good conditions. Lemmings, for example, are capable of very rapid population expansion, having a gestation time of only three weeks, producing up to eight young in a litter, and reaching sexual maturity in just six weeks (see "Polar Tundra Mammals," pages 165–169). Lemmings can thus undergo population explosions when food is abundant and mortality is low. They are a good example of an organism with an r-selected reproductive strategy. Penguins lie at the other end of the reproductive spectrum. They lay just one or two eggs each breeding season and have a limited capacity to replace any lost or failed eggs. The emperor penguin, the largest of the penguins, lays one egg and has to incubate it for up to 64 days before the chick hatches. Not all of the adults breed, and the success rate for a colony is very variable. In one recorded season only one chick survived from a colony of 300 adults. Once fledged, the young penguin is not able to breed until it is five or six years old, but it then may survive for 20 to 50 years. So penguins are slow breeders but good survivors. They are an excellent example of organisms with K-selected reproductive strategies. From these two examples it is evident that both r- and K-selection operate in the tundra; each has certain advantages for particular groups of organisms.

About 80 percent of the plants living in the tundra habitat have the capacity to reproduce vegetatively, and many of them do so regularly. Given the genetic advantages of sexual reproduction, it is perhaps surprising that vegetative reproduction is so widespread. The reason may well lie in the energetic costs of sex. The production of a new stem that creeps over the surface of the ground, takes root, and eventually generates a new plant costs far less in energy investment than the production of a flower, the attraction of a pollinator, fruit and seed growth, dispersal, and the likelihood of considerable wastage in failed germination, predation, and inability to survive. One estimate claims that sexual reproduction is 10,000 times more expensive in energy terms than vegetative propagation. In an environment where energy is hard to acquire, such expense is difficult to justify, so many plants resort to the cheaper option.

The commonest method of vegetative growth in the tundra is by creeping stems, either above ground (*stolons*) or below ground (*rhizomes*). In an environment where lateral spread close to the ground is more likely to be successful than upward growth into the penetrating icy wind, ground-hugging growth is extremely appropriate. Studies of plants that spread by creeping stems have shown that these extending organs behave in a surprising way. When the stem encounters unproductive surfaces, such as very poor soil or rock, it actually grows faster, but when it arrives at a location that is more suitable for growth, its extension slows down and a new plant begins to form. The advantages of this type of "foraging" are evident. There is no advantage to be gained by growing slowly in areas that are unproductive, so the plant accelerates its extension until it reaches somewhere more suitable for root production and the establishment of a new cloned plant.

Some plants produce flowers but do not need pollination to generate embryos. The production of fruits and seeds without fertilization is called *apomixis* and is quite frequent among tundra plants. One familiar example of this process, found both in tundra and in temperate environments, is the dandelion (*Taraxacum* species). Unlike the stolon and rhizome types of vegetative reproduction, apomixis can lead to the widespread dispersal of cloned individuals. The parachute fruits of dandelions can spread over wide distances,

r- and K-Selection

The population growth curves of different organisms follow a similar basic plan. The simplest and fastest type of population growth can be observed among bacteria, and since they reproduce by simple cell splitting (binary fission), these organisms provide an ideal model for population growth studies. Suppose a bacterium divides in two every 20 minutes. If a single bacterial cell is introduced into an ideal growth medium and commences its process of division, within an hour there would be eight cells. In three hours the population would be 512, and in just six hours the population would have reached 262,000 cells. Mathematically, this is known as a logarithmic growth curve. If it continued indefinitely, then the Earth would be packed full of bacteria within a matter of weeks. In fact, there are limitations to population growth, which may take the form of resource limitation, predation, parasitism, and so on. In natural populations mortality resulting from such limitations keeps most populations under control for much of the time; only occasionally are there population explosions because of unusually high availability of food or lack of predation. A population growth curve in the course of time thus generally shows a rapid expansion in its early stages, accelerating to a maximum rate, and then gradually declining until it becomes level. In these latter stages the limitations to growth are holding back any further population growth, and the population has reached its maximum extent.

Population biologists use two terms to describe the particular shape of the population growth curve of an organism. The letter *K* represents the *carrying capacity,* which is given by the maximum number of individuals of the species that can be sustained by the environment. This is the density at which there is no additional population growth. The letter *r* represents the *initial rate of population increase,* and it relates to how rapidly a species can reproduce. A bacterium is very efficient at rapid reproduction, while an elephant is not. The bacterium therefore has a higher *r* value. Some organisms are very good at expanding their populations but do not survive very well when placed under competitive stress. They are typical of open, noncompetitive environments, where resources are not limited, such as plowed arable fields or gaps in a forest canopy created by a minor catastrophe. Such species are said to be *r-selected.* The evolutionary pressures upon them over many millennia have led to their adaptation as pioneers and invaders. They are the opportunists of the plant and animal worlds. *K-selected* species, on the other hand, are not very good at invading new habitats. They may take a long while to reach an appropriate location, but once they establish themselves they can build up their populations and outcompete many of the pioneer species that have preceded them. The K-selected species is thus best adapted for long-term survival in stable conditions.

so that a genetic race of dandelions may become very widespread. There are numerous races of the dandelion, some botanists believe over 2,000, which may account for the plant's very wide global distribution. Each race has its own environmental requirements and its particular characteristics that may prove successful in certain types of habitat. Thus, one race may be found in long-grass habitats, another in short; one in disturbed soils, another in stable soils; one in alpine tundra, another in polar tundra, and so on. One advantage of apomixis is that the mechanism preserves a favorable genetic constitution. The abundance of dandelions in their many forms provides evidence of the success of this type of vegetative reproduction.

Some plants hedge their bets and resort to both sexual and asexual reproduction. The alpine bistort (*Polygonum viviparum*), for example, has flowering shoots with small white flowers that have the usual components of anthers and carpels and are able to produce seeds by sexual reproduction. But some of the locations on the flowering shoot are occupied by nonsexual bulbils that fall from the plant on maturity and grow into new plants. If the sexual process fails, the vegetative option supplies an alternative strategy. Some grasses, including the viviparous fescue (*Festuca vivipara*) and the alpine meadow grass (*Poa alpina*) also produce vegetative bulbils in their inflorescences. Vivipary is found mainly in tundra habitats and also in arid deserts. Its advantage may be a consequence of the unpredictability of the success of sexual reproduction. Pollinators are relatively scarce, freak weather is a possibility, and a plant may fail to produce fruits. Both of these strategies are also valuable if a habitat is patchy and the chances of finding better conditions far from the parent plant are bleak. In that case the offspring may be better served by depositing them close to the parent, where they are assured of appropriate growing conditions even though they face more intersibling competition.

Sexual reproduction in plants can take a number of forms. In the lower plants, including mosses, liverworts, club mosses, and ferns, the male gamete is a motile sperm, so fertilization depends upon the gamete swimming through a film of water to reach the female egg cell. Although some tundra habitats are dry for part of the summer, most are wet in the early summer when snow and ice melts, so the availability of water on plant and soil surfaces is usually adequate for this form of reproduction. Lower plants, therefore, are a major component of the tundra flora. Scientists have examined the reproductive strategies of mosses in the Antarctic and have found that short-lived species expend a greater proportion of their energy on sexual reproduction than longer-lived species. Among mosses sexual reproduction creates the sporangia in which air-borne spores are produced, so mosses that rely on colonization of new areas rather than survival in their present site need to resort to the sexual process. In this case sexual reproduction and subsequent spore production is a form of r-selection for these lower plants. Among higher plants, including the flowering plants, sexual reproduction and the transfer of the male gamete involves the production of pollen grains, which are transferred from the anther of one flower to the stigma of another in a process termed *pollination* (see sidebar below).

Wind-pollinated plants are very common in tundra regions, which is not surprising since windy conditions prevail for much of the time. The flowers of wind pollinated species require no colorful or scented attractions but must simply protect the stamens and carpels until they are mature. The stamens then extend their flexible stalks (*filaments*) so that the pollen-containing anthers are held away from the flower and release their large quantities of pollen into the air. Willows, birches, grasses, and sedges are among the most common of wind-pollinated species of plant, but other herbs also use this mechanism, including the mountain sorrel (*Oxyria digyna*), a member of the dock family (Polygonaceae), which colonizes scree in both polar and alpine tundra. Wind pollination is a very chancy affair (see

Pollination

Gymnosperms (a plant group including the conifers) and angiosperms (flowering plants) transfer their male gametes from one plant or flower to another by means of pollen grains. The pollen grain is a means of enclosing the delicate male cells, or nuclei, and protecting them from damage and desiccation on their journey. The pollen grain wall is therefore made of tough material that resists drying yet is light and pliable. It is usually penetrated by one or more pores or slits through which recognition proteins are secreted when the grain arrives at its destination and which provide an opening for the germination and extension of the pollen tube.

The means of conveyance from one flower to another is very variable among plants, but the main mechanisms for pollen transfer in the tundra are wind and insects. There is no shortage of wind in tundra habitats, but this mechanism is extremely chancy. The statistical likelihood of male pollen being transferred by wind to a stigma of a female reproductive organ belonging to the same species is very low. This can be overcome only by the production of very large quantities of pollen by plants adopting this pollination mechanism, and this is extremely expensive in terms of the energy invested. A single anther of the wind-pollinated sorrel (*Rumex acetosa*), for example, contains approximately 30,000 pollen grains, so the output from one plant runs into many millions. Insect pollination is much more reliable because the insect visitor is very likely to visit more flowers on leaving the source

of pollen, and because of the behavior patterns of many insects it is quite possible that the next flower visited will be of the same species. So targeting is much more certain with insects but is by no means infallible, so many pollen grains are still produced. A single clover (*Trifolium* species) anther contains about 200 pollen grains, but clover flowers are clustered in dense inflorescences, so many anthers are found close together. Attracting an insect involves advertising, so flower structures have to be more elaborate, brightly colored, perfumed, and supplied with energy-rich rewards, such as nectar. So insect pollination is also energy expensive despite its more efficient focusing of resources.

Insect pollination leaves the plant dependent on another organism for completion of its reproductive cycle, whereas this is not the case for wind pollination. The insect pollinated plant must therefore delay its flowering until the animal vector is active, in other words until summer is well established. Wind pollinated plants, on the other hand, can begin their flowering and pollen production as soon as the temperature is high enough for these processes to begin. In an environment where the growing and flowering season is short, early pollination may mean a better chance of fruit maturation and the production of viable seed. The fact that both systems operate alongside one another in the tundra indicates that each strategy has certain advantages.

sidebar on page 126), so pollen is produced in large quantities from flowers that are often grouped into large clusters (inflorescences). In the case of willows and birches these inflorescences take the form of catkins, while in grasses they form small, dense spikes called *spikelets,* which themselves may be arranged in a variety of patterns.

Insect pollination is also common in both polar and alpine tundra. Insects are abundant and active for only a very limited summer season, so there is intense competition between plants to attract the attention of potential visitors. The main emphasis is upon large flowers with brightly colored petals or sepals. Scent is not a common feature of tundra plants because the esters and other organic chemicals used in flower scents volatilize best under hot conditions; the prevailing cold does not encourage the use of scents as insect attractants. Humans, however, have a poor sense of smell compared with most other animals, and even the apparently weak scents of tundra flowers may be sufficient as a guide for pollinators. Flowers and their insect vectors have evolved together over millions of years, adjusting to one another's requirements and resulting in a finely tuned symbiosis in which the insect receives energy-rich rewards (either nectar or some of the pollen) and the plant gains an assured means of transport for its genetic material. As a result of this *coevolution,* various types of flower have specialist visitors, and their color and structure often reflect such adaptation.

Flies (Diptera) have a highly developed sense of smell and also good vision. Flowers of the tundra plant sea mayweed (*Tripleurospermum maritimum*), which has a very weak scent, are attractive to flies that use both olfactory (scent) and visual means of locating them. The flowers are arranged in dense inflorescences with white outer ray florets and yellow inner disc florets, very much like a large daisy. The initial attraction is visual, when the fly detects the plant with its targetlike flowers, yellow discs surrounded by a white circle. As it approaches the insect is then able to detect the scent, which may indicate the age of the flower and the likelihood of reward. Flies have short tongues, so they have a strong preference for flowers that do not have a deep tubular structure. Daisy-type flowers are ideal for them because they have a flat inflorescence with very shallow tubular florets.

Butterflies (Lepidoptera) become abundant in tundra habitats in the brief summer, especially in alpine tundra. They have good color vision, and different species seek out particular colors in the flowers they visit. Some alpine entomologists have observed that butterflies tend to choose flowers with the same color as themselves and suggest that this is related to their behavior patterns in mate choice. Certainly butterflies often visit flowers that are yellow, orange, blue, and purple. Bees (Hymenoptera) are similarly attracted to flowers with these colors. Butterflies, moths, and bees have long tongues that they carry in a tight coil. They are able to extend these tongues deep inside the long tubular structures of some flowers, such as the yellow rattlebox (*Rhinanthus minor*), which is most frequently visited by bumble bees.

Entomologists have been able to show experimentally that bees are blind to red. When they are offered varying shades of gray mixed with red, they are unable to distinguish the two. On the other hand, they are able to distinguish yellow and blue very efficiently. Red light has a long wavelength, and bees are evidently insensitive at this end of the spectrum. At the other end of the spectrum, in the short wavelength region, they are capable of detecting blue, violet, and even very short wave ultraviolet radiation, which is beyond the human visual range. Purple and blue flowers are frequent in the tundra flora, including such well known alpine flowers as the gentians (*Gentiana* species). Some of these flowers have elaborate adaptations for encouraging and guiding the visiting insect, such as lines on the petals to indicate the direction in which the pollinator should go. Dark-colored flowers with bell-shaped structure, such as the gentians, become warmer in the sunshine and some bees seek out these warmer flowers.

Most plants have their male and female organs in the same flower, or sometimes as separate flowers on the same plant. Some plants, however, have separate male and female plants, each producing single-sex flowers. These plants are called *dioecious,* literally meaning "having two homes." Only about 4 percent of all known flowering plants have this system. This proportion is true of the tundra flora also, but in the tundra dioecious plants form one of the most widespread components of the flora. The wind-pollinated willows are among these plants with separate male and female plants, and analysis of willow populations reveals that about 60 percent of plants are female and 40 percent male. Perhaps this explains in part why dioecy is so abundant in the tundra, because females are the individuals that produce the new seeds, while males simply cast their pollen to the wind and then have no further function in the population cycle. Males are not indispensable, but they are not needed in great quantity.

All tundra plants have to cope with a very short summer season in which to complete the development of fruits and seeds. The time between pollination and seed maturation has to be brief. As a consequence seed size among tundra plants is generally small. There is no equivalent to the coconut palm in the Arctic. Many plants are therefore able to produce large numbers of seeds, so they would be classified as r-selected species (see sidebar "r- and K-Selection," page 125). The dandelions (*Taraxacum* species), for example, produce large quantities of small seeds that can be dispersed by air. The louseworts (*Pedicularis* species) are similarly prolific producers of tiny airborne seeds. In part seed size is determined by the dispersal mechanism. An airborne seed has to be small if it is to have any chance of long-distance dispersal.

Seeds that pass through the gut of an animal, however, can afford to be larger and heavier. The fruits of many grasses, sedges, members of the dock family (Polygonaceae), and the goosefoots (Chenopodiaceae) are eaten by small granivorous (seed-eating) birds. Many of these are ground up in the crop of the consumer and are digested, but some escape and are voided with other fecal material, providing an excellent means of dispersal. A bird may travel several miles between eating a seed and depositing it, so the transport mechanism is very effective even if wasteful in terms of seed losses to the bird. To ensure that birds consume fruits and seeds, therefore, they must contain stores of food materials, such as starch or oils that can be used by the seedling after germination but also act as incentives encouraging the birds to feed. Inevitably, therefore, bird-dispersed seeds need to be relatively large but not so big that the bird is unable to swallow them. Just as there is a close evolutionary relationship between flowers and their pollinators, so is there a correspondence between seed size and the gape size of the birds that consume them.

Some plants produce conspicuous fruits that attract the attention of birds or even fruit-eating mammals. Many of the evergreen dwarf shrubs, such as blueberries (*Vaccinium* species), bearberry (*Arctostaphylos uva-ursi*), and baked-apple berry (*Rubus chamaemorus*) produce quite large berries that are clearly visible to their consumers. Often these are colored black or red, reflecting the fact that birds especially are strongly attracted to red colors and spot them easily within a background of dark green. In the case of these fruits, the seeds are found deep within a fleshy layer that is the main object of the bird's dietary intentions. The seeds are small and coated with a protective layer (*testa*), allowing them to pass unharmed through the guts of their transporters. Protected seeds of this type have a much better chance of successful dispersal and germination than those eaten and digested by their vectors. As a result the plant does not need to produce as many seeds in order to be assured of persistence and spread of the species. It does, however, need to invest energy in providing the consumer with attractive, tasty, and nutritious outer layers in the form of fleshy berries.

Animals are also faced with the challenge of how much energy to invest in reproduction in an ecosystem where energy is scarce. In the case of birds, for example, how many eggs should be laid for maximum efficiency and survival? Clutch size is quite variable among tundra birds depending upon the postnatal chances of survival. This, in turn, is affected by such factors as food availability, predation, and parental care. Penguins are among the slowest breeders (see "Birds of the Southern Tundra," pages 164–165), usually laying only one or two eggs, and several tundra birds have similarly small clutch sizes. The arctic loon (*Gavia arctica*), for example, also lays just two eggs; most tundra waders, such as the red knot (*Calidris canutus*), lay four eggs;

the snow goose (*Chen caerulescens*) lays three to five eggs; and the rough-legged hawk (*Buteo lagopus*) lays two to five eggs. Ducks on the whole are more prolific in their egg-laying. Many temperate breeders, such as the redhead (*Aythya americana*), lay up to 14 eggs, but the Arctic breeders tend to lay fewer. The arctic nesting harlequin duck (*Histrionicus histrionicus*) lays six to eight, and the king eider (*Somateria spectabilis*) four to seven. So even among ducks, the clutch size is lower in the high latitudes. The same is true of game birds. The willow ptarmigan (*Lagopus lagopus*) of the tundra lays seven to 10 eggs, while the more southerly sharp-tailed grouse (*Pedioecetes phasianellus*) lays 10 to 14.

Among birds, therefore, the most frequent strategy for reproduction is the production of relatively few young in order to ensure a better level of food provision and parental care. Some birds modify their clutch size according to the availability of prey. The snowy owl, for example, lays up to nine eggs (average 6.3) in years of abundant prey, when the lemmings are enjoying a population explosion (see "Productivity and Population Cycles in the Tundra," pages 101–102), but only three to six (average 4.8) in years of food scarcity. This type of breeding strategy is extremely efficient because it allows populations to expand when conditions are favorable but does not waste energy in the production of young that are doomed to starvation in bad years.

Albatrosses, such as the black-browed albatross (*Diomedea melanophris*), shown in the photograph on page 129, are also extremely slow breeders. They lay a single egg and do not breed every year. The young birds do not return to the nesting colonies for three or four years and may not breed for several years after that. The result is that these birds are very sensitive to any catastrophe leading to a population fall because they take a long time to recover from any losses.

Mammals in the tundra exhibit a full range of breeding systems, from the highly productive r-selected system (see sidebar "r- and K-Selection," page 125), to the slow-breeding K-selected strategy. Lemmings are renowned for their population explosions, a consequence of their early attainment of sexual maturity, their short gestation period, and their large litter sizes. They represent the productive end of the spectrum of mammalian reproduction in the tundra. Among lagomorphs (rabbits and hares), both the arctic hare (*Lepus arcticus*) and the Alaska hare (*L. othus*) on average produce a litter of six young. Marmots usually produce four or five young. Both the gray wolf (*Canis lupus*) and the arctic fox (*Alopex lagopus*) can produce as many as 12 pups in a litter, but the litter size is related to food availability and the likelihood of offspring survival. The wolverine (*Gulo gulo*) produces just one to five young in a litter, while moose (*Alces alces*) and caribou (*Rangifer tarandus*) bear only one or two calves in their breeding cycle. Polar bears (*Ursus maritimus*) usually have two cubs, and the musk ox (*Ovibos moschatus*) has only one calf every two years.

The black-browed albatross (*Diomedea melanophris*) is probably the world's most abundant albatross species. Like other albatrosses, it spends much of its time on the wing, taking advantage of air turbulence above the ocean waves to keep it aloft with a minimum of energy expenditure. (*British Antarctic Survey*)

The tendency among tundra mammals, therefore, is for the number of young in a litter to decline with the size of the animal. Wolves and foxes sometimes break this trend by producing larger litters in years of abundant prey, but the tundra heavyweights, the polar bear and the musk ox, have small litter sizes and are thus vulnerable if conditions change and their populations come under stress. This has already had severe consequences for the musk ox, which was exterminated by human hunting in Eurasia.

Reproduction in the tundra, therefore, is a process that is finely tuned to the severity of the environment. Some species operate a boom-or-bust strategy, breeding rapidly when conditions are good and undergoing population crashes when things take a turn for the worse. These species often undergo regular population cycles. Others maintain a steady, slow level of reproduction, neither expanding their populations markedly in times of plenty nor overstretching resources when conditions are difficult. There is evidently room for both strategies in the tundra.

■ GRAZING IN THE TUNDRA

The short burst of primary productivity that takes place in the tundra each summer leads to very considerable changes in the functioning of the tundra ecosystem as herbivores take advantage of the new sources of energy, predators avail themselves of new supplies of prey, and the

detritivores and decomposers make a living of whatever is left (see "Food Webs in the Tundra," pages 91–93). Rapid expansions of populations can sometimes lead to instability, whereby a herbivore may exceed its carrying capacity (see sidebar "r- and K-Selection," page 125), as sometimes happens when moth caterpillar populations explode and strip birches of all their leaves. Grazing by various herbivores can occasionally become so intense that the vegetation is strongly affected.

Invertebrate herbivores are not abundant in the tundra. Only about 17 percent of the insects found in the polar tundra are herbivorous. So with the exception of occasional rapid expansions of populations, insect depredations on vegetation are unusual. Mammals, on the other hand, are capable of inflicting considerable damage. Small mammals are numerous, including various species of lemmings and voles, and these have regular population expansions that can leave a distinct mark on vegetation. In a temperate grassland ecosystem managed for sheep grazing, about 20 percent of the net primary production in any year is likely to find its way into the sheep. In the tundra during years of small mammal abundance, almost 100 percent of the production can be consumed by the grazers. The landscape can be severely altered as voles or lemmings strip grass-sedge tundra of virtually all its plant biomass. Once all the above-ground cover has been removed, the grazers dig up roots and rhizomes. It is under these conditions that the famed mass dispersion episodes of lemmings have been recorded.

Light grazing, on the other hand, can stimulate growth of herbs by removing older growth and encouraging basal shoots. Grazing can even alter the composition of the tundra by removing grazing-sensitive species and encouraging those plants that recover rapidly, such as the grasses and sedges. Woody plants, including the willows, dwarf birches, and mountain avens (*Dryas octopetala*), have growing points in the form of buds near the tips of the shoots, so when they are removed by grazers their growth is set back. Grasses and sedges, however, have leaves and shoots that grow up from the base, so removal of the upper parts of the plant does not suppress growth but allows additional growth from the base. Grazing on a community of mixed grasses, sedges, and woody plants, therefore, suppresses the woody shoots and encourages the herbs, leading to a change in the composition of the community.

Some plants in the tundra have adopted chemical defense mechanisms that deter grazers. Birch leaves, for example, contain tannins that are distasteful to some grazers, and Labrador tea (*Ledum groenlandicum*) contains aromatic compounds in its foliage that deter grazing by lemmings. Ecologists have discovered that several woody plants, including birches (*Betula* species), are less palatable to mammal grazers, such as hares, in the Arctic than they are farther south. This suggests that grazing pressures upon

Snow geese (*Chen caerulescens*) breed in the Arctic regions of Canada, then migrate south to winter in the southern parts of the United States and Mexico. As they gather prior to migration, they graze very heavily upon local areas of tundra vegetation.   *(Hal Korber, Pennsylvania Game Commission)*

these plants have been a long-term factor in their evolution and have encouraged the development of chemical defenses against the depredations of grazers. On the whole leaves and twigs that are chemically protected against grazing animals are also unpalatable to microbes, so they decompose more slowly in the soil. This aspect of plant adaptation thus has far reaching effects upon the function of an entire ecosystem, reducing grazing intensity and also slowing the process of nutrient cycling.

Large mammalian grazers, such as the caribou, can also have a marked impact on vegetation, especially where large numbers gather on migration or during calving. In the winter and early spring caribou feed mainly on lichens, which are abundant both in the tundra and in the winter quarters of boreal forest. An adult caribou can eat up to 15 pounds (7 kg) of lichens every day, and lichens generally have a very slow growth rate, so local impact on the lichen communities can be quite severe. On the calving grounds they turn to herbaceous food, especially the sedges and cotton grass

(*Eriophorum angustifolium*). Later in the summer they turn to grasses, particularly the meadow grass (*Poa pratensis*), and they also take many fungi. As winter settles in the caribou turn increasingly to dwarf shrubs, such as crowberry and willows as they begin to move south. There is thus a seasonal shift in grazing pressure upon different types of vegetation, which usually prevents overgrazing in one area.

Among the most serious offenders as grazers are geese. Snow geese, shown at left, gather in large flocks after breeding and before migration and graze both heavily and closely on sedge meadows and salt marshes. The marshes around the shores of Hudson Bay, for example, are often badly damaged by snow goose grazing.

Grazing, therefore, can result in damage to the tundra ecosystem, but usually this is local and seasonal. If overgrazing were sustained, then either the population of the grazer would crash or it would be forced to move on to other areas or other sources of food.

■ BEHAVIORAL ADAPTATIONS: HIBERNATION OR MIGRATION?

When faced with the prospect of an extremely cold, dark season in which food will be very scarce or absent, organisms are faced with two choices: remain in place and cope with the problems or move somewhere less taxing. Plants have no option. They must stay in place and enter a phase of dormancy in which new growth ceases and all unprotected organs must be shed or risk damage. But mobile animals do have a choice, and different species have taken separate paths in their behavioral adaptations to life in the tundra.

Some animals, such as the arctic fox and the musk ox, remain active through the tundra winter, but many, including both warm- and cold-blooded examples, enter a dormant state called *hibernation*. The word is derived from the Latin of the Roman armies, where the verb *hibernare* meant to spend the winter usually confined to quarters because of unfavorable conditions. Amphibians and reptiles as well as mammals often resort to spending cold winters in this way, usually beneath the surface of the ground or snow, where the temperature is a little warmer than on the surface. The behavior pattern is not restricted to the polar and alpine tundra regions but is also quite common in the higher latitudes of the temperate zone. In forested regions animals may find suitable shelter in hollow trees or dense vegetation as well as beneath the ground. In the lower latitudes hibernation is rarely necessary, but when summers become very hot and dry then the equivalent summer dormancy is called *aestivation*. Many animals, from earthworms to fish and alligators, aestivate through summer heat and drought.

Snow does not conduct heat well, especially if it is lightly compacted and contains many air spaces. Buried beneath the snow, both plants and animals can remain protected from the worst of the winter cold. Some mammals make nests, including the marmots of the alpine tundra, and collect vegetation to act as bedding and to provide additional insulation below the ground and thus conserve heat.

But warmth is relative. The temperature in the hibernation burrow is still too low to be able to maintain full metabolic activity, and the body temperature of the hibernating animals falls. Ecologists have studied marmot physiology during hibernation, and they have found that the body temperature can drop to 40°F (4°C). Some mice have even been recorded with body temperatures as low as 35°F (2°F). Blood pressure falls, the heartbeat slows, and endocrine gland secretions decrease, leading to the entire body entering a kind of coma in which metabolic activity is low. The animal is said to be *torpid*. In a hibernating marmot the heart beats once every two or three minutes, and the animal takes a breath only every 10 minutes. The consequence of this very low activity is that energy consumption is low and is maintained by drawing upon the stored fat reserves built up during the summer. At the same time waste products are reduced so that the animal does not need to urinate or defecate frequently during the period of hibernation. The marmot wakes about every three weeks to urinate. Since it is not eating, there is no waste material in its gut, and defecation is not necessary. By the end of its hibernation period the marmot has lost half its body mass.

The alternative behavior pattern to remaining in place for the winter is to move to a site with a more equable climate. This activity is called *migration*. The Latin verb *migrare* actually means to change one's abode, so the word *migration* can be applied to almost any movement from one place to another. Members of the plankton in a lake migrate upward during the day and downward at night, and sand hoppers on a beach migrate up and down the strand with the movements of tides. Most often, however, the word migration is reserved for regular seasonal activities of animals, such as the movement of caribou (*Rangifer tarandus*) herds and insect-feeding birds such as the barn swallow (*Hirundo rustica*). It does not include the normal processes of dispersal when organisms move into new territories, extending their range, even when such dispersals attain the scales of irruptions of large numbers, as sometimes happens with lemmings and snowy owls (*Nytea scandiaca*). Migration can involve very-long-distance movements, but in the alpine tundra it may simply involve the shifting of a population from high to lower altitude during the winter months.

Hibernation is found mainly among small animals, the marmot being one of the largest hibernators. Birds and large mammals either remain active at the same site or migrate. Caribou (called reindeer in Eurasia) live in large herds, often approaching a quarter million animals. Many of the caribou herds in North America and Russia migrate between winter quarters and summer calving grounds (see map on page 166), but some herds are not regularly migratory. The herds that move south into the boreal forest for the winter may travel up to 620 miles (1,000 km) and then face the return journey in the spring. The pattern of these seasonal migrations has remained unaltered for at least 150 years, since they were first formally recorded. But it is likely that they have been taking place for thousands of years, and the precise routes are deeply embedded in caribou behavior. In early March small groups of caribou begin their northward movement, and by late March and early April vast herds of animals are on the move. They gather in calving grounds for the communal birthing of the young, then often move to higher ground if it is available to avoid mosquito harassment. Then in early September as the first snow begins to fall, the herds start to move south once more.

Seasonal migrations in alpine tundra are also related to changing temperature and the winter extension of snow cover in the mountains, but the return to the uplands may be affected by changes in the quality of food during early spring. As the snow recedes at higher altitudes and the vegetation begins to grow, there is a flush of proteins into the young leaves. Most herbage is relatively low in proteins compared with animal tissues, so a protein-rich leaf or bud is very attractive to grazers. Elk (*Cervus elaphus*) migrate up mountains in the spring at a pace that corresponds with the surge of proteins in the vegetation and thus maximize their intake of nutritious food.

Many birds migrate over even larger distances than elk or caribou. Flight enables them to travel in a way that is less restricted by geography, for even mountain ranges and oceans do not necessarily impede their movements. Many small seed-eating birds that breed in the tundra move south in the winter. The snow bunting (*Plectrophenax nivalis*) breeds in the Arctic and winters as far south as Georgia and California, and the Lapland longspur (*Calcarius lapponicus*) also breeds north of the tree line and winters in Canada and the United States south to Texas and Virginia. Wading birds also move south as winter sets in. The least sandpiper (*Calidris minutilla*), for example, although less that six inches (15 cm) long, migrates from Arctic Canada and Alaska to northern South America for the winter. Similarly, the American golden plover flies over a long stretch of the Atlantic Ocean as it migrates between the tundra of northeastern Canada and eastern South America, some 2,500 miles (4,000 km). Ducks and geese also migrate over considerable distances to breed in the Arctic and then spend the winter in less severe conditions (see "Birds of the Northern Tundra," pages 154–161).

Migration is often very costly both in terms of energy and mortality. Migrating mammals and birds often spend time building up layers of fat prior to migration so that they will have enough fuel to complete their journeys. But too much fat, especially for a bird, leaves it less fit to indulge in vigorous activity such as long-distance flight. This means that stopover locations play an important part in migratory behavior, allowing the organism to refuel for the next stage in the journey. A nonmigratory bird usually has only 5 percent of its body weight as fat, and those that migrate over fairly short distances (up to 500 miles [750 km]) may contain 25 percent fat. But a long distance migrant that needs to cover 2,500 miles (4,000 km) needs to have fat equivalent to half its body weight. The red knot increases its rate of fat accumulation by 10 times immediately prior to leaving the tundra on its fall migration.

Predators often take advantage of large numbers of prey, weakened by days or nights of travel and following predictable routes. Many fall exhausted on the journey, and others fall prey to their enemies. Caribou and elk are subject to harassment and predation by wolves, as seen in the picture below. But the costs of leaving the breeding grounds rather than overwintering are evidently regarded as worthwhile for many birds and some mammals. Ornithologists have often raised the question why birds should leave the amenable, low-latitude climate of their winter ground to enter the bleak polar regions. The answer seems to be breeding

success. Long days in the high latitudes allow birds to raise more young than would be possible if they remained in temperate or tropical areas. Nest predators and parasites are also more common in the Tropics, so the tundra offers a means of escape from many of these.

Some birds migrate by day. These are usually the birds that use soaring mechanisms in their flight, taking advantage of thermals and air currents to carry them upward and forward with a minimum of muscular effort. Many raptors are of this kind. The nighttime migrants include many smaller birds as well as ducks, geese, and swans. One advantage of night flight is that the birds are less likely to become overheated as they indulge in maintained physical effort. The decision to begin migrating is stimulated by a number of factors. Many observers have reported restlessness among birds when their time for migration nears. Changes in the weather, such as increasing warmth or cold, depending upon the direction of the migration, can stimulate the birds

Wolves surround a bull elk. The wolf is the top predator over many Arctic tundra areas, especially in the transition zone with the forest. By adopting pack hunting strategies it is able to prey upon animals far bigger than itself. *(Christine Smith, Yellowstone Digital Slide File, Yellowstone National Park)*

into movement. There is some evidence to suggest that both birds and mammals can detect changes in daylength that stimulate migratory movement, perhaps subconsciously detected by the pineal gland that lies between the hemispheres of the brain.

It is important that birds arrive at their breeding grounds at approximately the same time so that pairs can be established, and this means that the timing of migration must be carefully controlled. The black-tailed godwit (*Limosa limosa*) has been subjected to careful scrutiny in its migratory and nesting habits, especially the timing of its movements. Some populations of this long-legged wading bird nest in the tundra of Iceland and spend the winter in western Europe between the British Isles and Spain. Ornithologists have been studying the species for many years and have now placed bands on the legs of many of these birds so they can follow the life history of individuals. The godwits develop long-term pair bonds, and they meet up on the breeding grounds, having spent their winters apart. The question arises, exactly how do the male and female of a pair manage to synchronize their arrival in Iceland for nesting? The answer is still not clear, but the male and female of a pair usually arrive within three days of each other. The only "divorces" that were observed occurred when the individual birds of a pair arrived more than eight days apart. Scattered over the mudflats of Europe, these birds are able to perceive some environmental signal that sets them on their course for Iceland, even though the pair may not have seen each other for six months. Understanding the environmental cues that initiate and control the rates of migration remains a major biological challenge.

Bird migration, like that of caribou, follows regular routes, or *flyways*. Often these pass along major river valleys, such as the Mississippi-Missouri. Undoubtedly, some of the birds' ability to keep on track results from visual identification of landmarks, especially when they finally settle at their destinations, but the fact that birds can navigate in the dark and even in foggy conditions indicates that there are other mechanisms at work. Some young birds even migrate without the guidance of adults, suggesting that there is some in-built guidance system that is genetically controlled rather than taught by experienced adults. Geneticists have made considerable advances in the understanding of the genetic control of migratory behavior. Two factors are involved, first the impulse to migrate at a particular time, the cue for migration, and second the navigation along the route. Both have a genetic base, and this aspect of genetic control is evidently subject to rapid evolutionary development, which allows birds to develop new behavior patterns in the event of climatic change. This is an aspect of bird biology that may prove particularly useful to migrant species in our changing world.

Birds can certainly use the Sun and the stars as a means of orientation, but this is itself a complex process because the turning of the Earth means that the relative positions of both the Sun and stars change in the course of a day, so a bird must also be aware of the time of day in order to interpret the positions of celestial bodies. Sunlight becomes polarized by the atmosphere in one part of the sky, lying at 90° to the position of the Sun. Birds are able to locate this plane of polarization, so they are able to detect the position of the Sun even in cloudy or foggy conditions. Night migrants can pick up this polarized light for a short while after sunset, and they may well orientate themselves on this basis for their nocturnal movements. The North Star, if it is visible, is then a fixed point in the heavens that can provide guidance for the birds after the sunlight has finally been lost. But evidence is mounting suggesting that birds contain an in-built compass capable of detecting the Earth's magnetic field.

The complexity of a bird's navigation system is illustrated by the fact that they do not always migrate in straight lines. Sometimes they reach a particular point in their journey and then adjust their flight direction to take up a new flight path. They seem to be able to detect their latitudinal position and use this as a cue to change their flight orientation. Ornithologists have put forward several ideas to explain a bird's ability to calculate its latitudinal position. The Earth's magnetic field varies in strength with latitude; perhaps birds can detect this. The dip of the magnetic field also changes with latitude. At the equator the magnetic field runs roughly parallel with the Earth's surface, while at the poles it operates at right angles to the surface. Between the two, the angle of dip changes with distance from the equator. The magnetic pole lies 870 miles (1,400 km) from the geographic pole. As the bird enters higher latitudes it may be able to detect the increasing angle between magnetic and geographical poles. It is difficult to explain how this could be achieved because the long days of the polar summer mean that even the stars are not available for determining the position of the geographic pole.

There is much to be learned about the navigational systems of birds and mammals, but it is evidently a system that has evolved over a very long period and has been adjusted and fine-tuned through the ages as climate, geography, and sea levels have changed. The Greenland white-fronted goose (see map on page 159), for example, migrates from western Greenland to the western coasts of Europe for the winter, when it would seem much simpler to fly down the east coast of North America. Perhaps this behavior pattern developed during the last glaciation 20,000 years ago, when an icecap over eastern Canada blocked that route.

The inhospitable nature of the tundra winter has led to the development of some remarkable behavioral adaptations among birds and mammals. Both hibernation and migration are very costly mechanisms adopted to enjoy the benefits of the polar summer and yet escape the rigors of the polar winter.

■ CONCLUSIONS

The harsh climate of the tundra has led to a wide range of adaptations among the plants and animals found there. By far the most serious problem that must be faced by all permanent dwellers and many visitors to the tundra is low temperature. Plants and cold-blooded animals (poikilotherms) assume the temperature of their surroundings unless they are exposed to solar radiation, when they may absorb the heat and raise their temperature. Birds and mammals are warm-blooded (homoiotherms), and they maintain a steady temperature usually higher than that of their tundra environment. This has many advantages in terms of maintaining an active life, but it is expensive in terms of energy.

Many tundra organisms have insulating covers to protect them from the worst of the cold, and often these coats take the form of hair. Plants, insects, and mammals have various kinds of hair, all of which trap a layer of air, which is a poor heat conductor, close to the skin to slow down the rate of heat loss. Birds have feathers that perform the same function.

The constituents of living cells are affected by cold, especially the enzymes that control cell activities. These proteins operate more slowly and less efficiently at low temperature and are inactivated by extreme cold. The fats in cell membranes solidify in the cold, leading to leaky cells, but the greatest danger is from cell freezing, particularly in plant cells, which contain watery solutions in their vacuoles. As water freezes it expands and tears cell membranes. Animal cells, which have no vacuoles, are more tolerant of rapid freezing and can become supercooled because of the solutes contained within the cells. Such solutes as glycerol and proline can act as antifreeze agents and allow the cell temperature to fall below the freezing point of water without damage.

Plants in the tundra adopt a distinctive range of life forms. Tall woody plants (phanerophytes) are virtually absent, as are annual plants (therophytes), and the vegetation is dominated by species of low stature (chamaephytes) or that hold their perennating buds at ground level (hemicryptophytes). Geophytes, which overwinter below ground, are absent because the soil becomes frozen, but there are some hydrophytes that inhabit the mud of pools. This biological spectrum is well suited to the climate, with its harsh, dark, cold, and wind-blasted winters.

Biogeographers have observed that mammals tend to be larger in the tundra than their temperate and tropical equivalents, and this has been formulated in Bergmann's rule. There are many examples that hold true both for present and past tundra faunas, but some critics believe this rule is an overgeneralization based on a small number of examples. The same is sometimes said of Allen's rule, which claims that tundra animals have smaller extremities, especially ears, because these organs are exposed to cold winds.

The plants of the tundra are a mixture of evergreen and deciduous species, so evidently both leaf forms can be advantageous. The evergreen is able to start photosynthesis early in the season without having to rebuild leaves, and it is also more efficient in energy and nutrient terms to build a leaf that will last several years. On the other hand deciduous leaves avoid the damaging and desiccating winds of winter and can dispose of pathogens, parasites, and small grazers on the leaf surfaces. The colors generated by dying leaves in fall make the landscape spectacular, the reds and yellows of deciduous leaves contrasting with the dark green of evergreens. Red berries, attractive to birds, add to these colors.

Animals in the tundra may be colored as a form of camouflage, or crypsis. Often they adopt the colors of the vegetation or soil that surrounds them to avoid predators. Some animals and birds become increasingly white with the onset of winter and the arrival of snow. Polar bears, however, have no problems with predators apart from humans. They are white because they wish to remain unseen as they stalk their prey on the ice. Some insects are dark colored because its allows them to absorb more solar heat and raise their temperature above that of their surroundings.

Although the tundra has many moist and wetland habitats, there is little precipitation, and well-drained ridges of shingle can become very dry in summer. Many plants have adaptations for drought tolerance, such as waxy cuticles. Invertebrates often hide in the soil or beneath the vegetation cover to avoid desiccation.

Wet tundra regions present a very different problem to plants and animals. Under waterlogged conditions oxygen is in short supply, and many organisms need to resort to anaerobic respiration. This can lead to an accumulation of toxic end products, such as ethanol, and flood tolerance involves developing biochemical techniques to avoid this. Oxygen is also scarce in high-altitude alpine tundra because the air is less dense. Mountain mammals have higher concentrations of hemoglobin in their blood than their lowland counterparts.

Reproductive strategies can be divided into two main types, r-selected, in which the emphasis lies on maximizing reproductive output, often associated with short life history and opportunistic invasion of habitats, and K-selected, in which long-term survival is the main concern and reproduction involves the generation of few offspring. Both types are found in the tundra, as illustrated by lemmings and penguins, respectively. Many plants resort to vegetative reproduction, which is an effective method of local persistence and spread. Flowers in the tundra vary according to their pollination mechanism. Wind pollination is common among shrubs, grasses, and sedges, and these flowers lack attractive colors and are often in the form of catkins or spikelets.

Insect-pollinated species, however, generally produce large and conspicuous flowers, especially in the alpine tundra. Competition for the limited number of insect pollinators is fierce, so each plant needs to draw attention to itself. The attraction is visual rather than by scent because cold air is not conducive to scent transmission. Sexual reproduction involves genetic mixing and is therefore effective in diversifying the species, but sometimes plants benefit by retaining certain genetic combinations, and this is favored by vegetative reproduction, or apomixis, in which seeds are produced without resorting to sexual transfers. This is common in tundra vegetation.

Birds in the tundra generally have small clutch sizes, but some, such as the snowy owl, increase their productivity when their prey of lemmings and voles is in abundant supply. The same applies to tundra mammals, such as the arctic fox, which also has larger litters in times of plenty. But large tundra mammals, including caribou and musk ox, have relatively low reproductive outputs.

Grazing can be intensive and can locally lead to the devastation of tundra vegetation. Small mammals, including lemmings, have periodic eruptions in their populations, and the effects on vegetation are clearly seen, often resulting in the complete stripping of the plant biomass. Caribou grazing can also have marked local effect, as can the grazing of snow geese on salt marshes.

The tundra winter presents animals with a stark choice. Either they must remain in the tundra, or they must migrate elsewhere. If they remain, they can continue actively seeking food, as in the case of the arctic fox, or they can hibernate, as in the case of the marmot in alpine tundra. Hibernation involves the animal assuming a torpid state when body temperature, blood pressure, heartbeat, respiration, and waste production all fall to a very low level. The alternative is migration, which is equally hazardous and requires the accumulation of sufficient fat reserves to carry the migrant on its travels. In the case of birds this is a time of high mortality and does assure the migrant of winter food supplies.

The mechanisms of navigation and orientation have been the subject of much research, yet they are still not fully understood. Birds seem to be able to detect the Earth's magnetic field, and from this they can orientate themselves and estimate their latitudinal position. Other tools are at their disposal, however, including solar and stellar navigation, together with simple visual recognition of landmarks during the journey. Future studies of bird migration may well concentrate on the polar tundra because of the complexity of magnetic fields as the poles are approached.

The tundra is not rich in plant and animal species compared with other parts of the world, but those able to inhabit this biome are remarkable in many respects. Over millions of years these organisms have adapted to life in some of the most inhospitable places on Earth. This has involved specialization in structure, physiology, reproduction, and behavior. The biology of the tundra provides one of this planet's clearest lessons in the adaptability of living things.

6

Biodiversity of the Tundra

It is entirely possible, perhaps even likely, that there is life elsewhere in the universe, but at present the only known living things reside on the planet Earth. The vastness of space and all the other stars and planets it contains are largely, perhaps entirely, a bleak and lifeless desert, while the Earth teems with life. Life makes the Earth special, and it is reasonable that people should have due regard and respect for the richness of living things that exist here, the Earth's biodiversity.

■ THE MEANING OF BIODIVERSITY

The word *biodiversity* is currently much used both by scientists and journalists, but it is not easy to define. The main component of biodiversity, and the one that immediately comes to mind, is the richness of species present in an area. Assembling species lists is therefore the first step in determining the biodiversity of an area, but diversity involves more than the number of species present. Consider a collection of 100 colored balls, and suppose that they come in 10 different colors. The collection could have 91 balls of one color and one each of the remaining nine colors, or it could have 10 balls of each color so that there was an even distribution of colors among the balls. The same could be true of a natural ecosystem with 10 species and 100 individuals. The individuals may be evenly spread among the different species, or one species might dominate the system and make up the bulk of the community. Faced with this situation, the ecosystem with the more even and equitable distribution of individuals among the species should be regarded as more diverse than the ecosystem that is dominated by one species. Ecologists therefore make a distinction between richness (number of species in an area) and diversity (a combination of richness and evenness).

Biodiversity, however, is yet more complex than this. All the individuals in a population, unless they are vegetative clones, differ slightly in their genetic constitution. In human societies only identical twins have exactly the same genetic makeup. All others differ to some extent. This variety makes society much more interesting and flexible; the thought of identical people behaving in precisely the same way is the basis for disturbing science fiction, as in Aldous Huxley's *Brave New World*. Sexual reproduction involves the shuffling and recombination of the genetic attributes of parents, leading to totally new arrangements in the offspring and making them distinct and unique individuals. This is true of humans and plants, mammals and microbes.

The outcome of genetic recombination is genetic diversity. A population of organisms in which breeding takes place between unrelated individuals is said to exhibit *outbreeding*. New genetic characters are constantly being added to the population, and new combinations lead to greater genetic variability. This, in turn, equips the population to cope with new challenges, including adaptation to changing climate, pollution pressures, predation, disease, and so on. A small population in which breeding takes place between related individuals is deprived of new genetic input and is said to be *inbred*. There is a danger that genes leading to physical weakness, normally recessive to more dominant healthy genes, will accumulate and make their presence felt in the population. Hemophilia in humans is an example of the type of recessive gene that shows itself under such circumstances. An organism that has outbreeding populations and a diverse genetic constitution is therefore more likely to survive in a changing world. It is said to have a wide *gene pool.*

Biodiversity is a concept that includes the breadth of the gene pool within the populations of organisms that constitute its species lists. Genetic diversity is a part of biodiversity because it is a component of the range of variation found within a particular habitat or ecosystem.

Biodiversity also includes the wealth of different microhabitats found within a landscape. This is closely related to species diversity because a diverse landscape consisting of patches of many different habitats and microhabitats will contain more species than a uniform landscape that lacks variety. Varied geology and topography in a region develop a range of different soil types, which in turn may support a variety of vegetation types with different species compositions and physical structures. Within these vegetation types many different animals and microbes are able to survive, leading to high biodiversity.

Conservationists lay much stress on biodiversity. They regard each species as an irreplaceable product of genetic material, sorted by the long and painful process of natural selection and honed to a high degree of fitness. Regions rich in biodiversity are therefore ranked highly in terms of conservation priority. Some conservationists lay stress upon the "hotspots of biodiversity" around the world, mainly concentrated in the Tropics, where large concentrations of species reside. It is reasonable to rate such sites highly, but it is dangerous to neglect less diverse areas that nevertheless contain species of plants and animals found nowhere else in the world. The tundra is such a region. It is not rich in species, but the organisms found there are remarkable for their capacity to survive in a harsh environment. Low biodiversity cannot be equated with low conservation value.

■ PATTERNS OF BIODIVERSITY

Biodiversity is not evenly spread over the face of the Earth. The Tropics contain a much higher diversity of species than either the temperate or the polar regions, and the same pattern applies to altitude. This is illustrated in the accompanying diagram on page 138. Alaska contains 40 breeding species of mammal, while British Columbia holds 70 species, California 100 species, and Costa Rica 140 species. At latitude 70°N in North America only four species of swallowtail butterfly are found, while there are 21 species at latitude 40°N and 80 species at the equator in northern South America. This relationship holds for almost all the major groups of species, so in terms of species richness the tundra is poor.

Biogeographers have expended much effort and thought into explaining the latitudinal gradient in species richness. Climate seems the most obvious cause of the pattern, but how might climate operate in controlling biodiversity? Climate has an immediate effect upon ecosystems by determining their productivity (see "Primary Productivity in the Tundra," pages 88–91). Vegetation is most productive when it is supplied with abundant water and warm temperature. If either of these two requirements is lacking, productivity falls. Overall primary productivity in North America, for example, is greatest in the southeastern region, including Florida, Alabama, Georgia, and South Carolina. To the west and north of that area productivity falls. Altitude also has a marked effect, with low productivity in the Rocky Mountain chain that rises again on the West Coast, but the general trend is a declining productivity away from the Southeast. Primary productivity lies at the base of almost all food webs, so a low productivity results in proportionally shorter food webs and fewer trophic levels. The consequence is that the ecosystem can support fewer species.

Productivity is certainly an important factor in explaining gradients of diversity, but there are other considerations. Warm, moist climates are the most suitable for the growth of trees, for example, and vegetation structure, that is, the complexity of its architecture, affects microclimates and the variety of microhabitats available to all kinds of organisms, including mosses, invertebrates, epiphytes (plants that use other species for support), birds, microbes, and mammals. A complex structure provides many opportunities for different organisms to survive and make a living. North America is again a good example of how this operates. The Southeast region contains 180 species of trees and shrubs of 10 feet (3 m) in height or more. New England and British Columbia each have about 60 species, declining steadily to zero in the High Arctic. As forest richness declines with increasing latitude, so does the complexity of habitats available to other species.

There is also the possibility that evolution progresses faster in the Tropics, so this is where new species are more rapidly generated. It is difficult to test this idea experimentally, but assembling fossil data can be of assistance. In the case of flowering plants, for instance, their fossils first appeared in the Tropics in late Jurassic times, more than 140 million years ago. They spread out from the Tropics, but it took at least 30 million years before they reached the high latitudes, and the total numbers have always remained lower there than close to the equator. It is clear that the latitudinal gradient in diversity has always existed, but this could still have been controlled by climate rather than rates of evolution.

The climate of the Earth has not been stable over the course of time, and the high latitudes have periodically been subjected to the advance of ice sheets, burying much of the land and part of the oceans beneath deep layers of ice. This has eliminated all living things from parts of the high latitudes and from high altitudes at various times in the past but particularly in the last 2 million years (see "Glacial History of the Earth," pages 176–178). The Tropics, apart from scattered high mountains, have not been glaciated in this way, but they have suffered from climate change, so they have not totally escaped the effects of climatic instability. Much of the temperate zone has been covered by ice or converted to tundra during the ice ages, so the plants and animals of this region have been partially eradicated or even brought to extinction in some cases. It can be argued, therefore, that the high latitudes have suffered most from the fluctuating climate of recent times and that this accounts for the low biodiversity. But this does not explain the fossil evidence, which suggests that the high latitudes have always been less rich in species than the low latitudes, even before the onset of the recent ice ages.

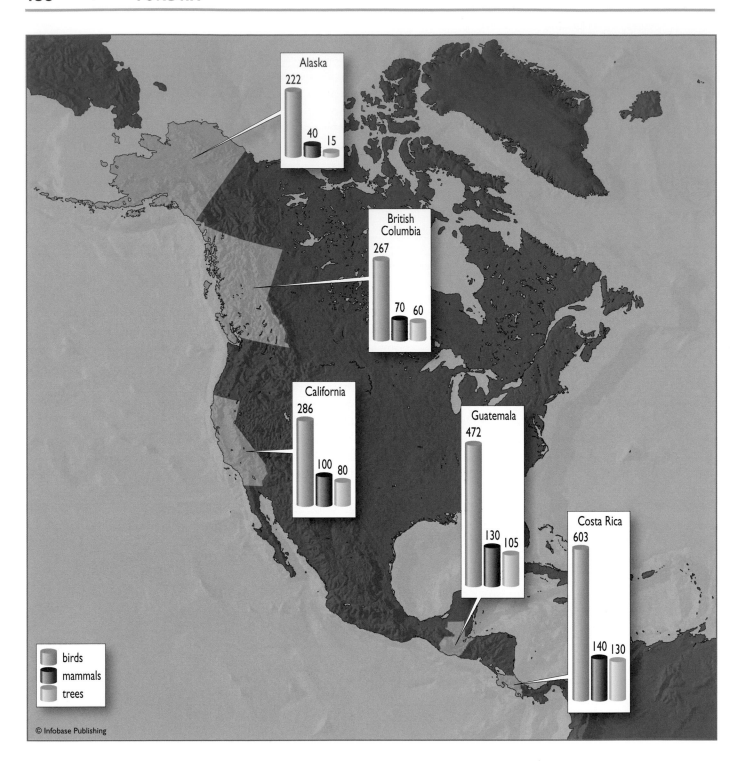

A gradient in species diversity is also found at different altitudes. A study of mountains on the Faroe Islands in the North Atlantic revealed that the vascular plant richness falls from 65 species at 1,000 feet (300 m) to 38 at 2,500 feet (750 m). A similar decline is found in more southerly mountains, such as the Himalaya Mountains in Nepal. At an altitude of 3,300 feet (1,000 m) there are approximately 1,000 species of vascular plants, but at 16,400 feet (5,000 m) this falls to about 200 species. The decline is quite gentle up to 13,000 feet (4,000 m) but then becomes very steep, reflecting the harshness of conditions in the high-altitude tundra biome.

Overall, the most likely explanation of why the tundra is low in diversity is the climate and its implications for ecosystem productivity. An ecosystem with low energy input will inevitably sustain fewer species than one with high primary productivity. The tundra is thus geographically predisposed to low diversity, but this should not be taken to imply that

(opposite page) Many types of organisms display latitudinal gradients of diversity. The low, tropical latitudes are usually much richer in species than the high latitudes. In the case of breeding birds, mammals, and tree species in North and Central America, all three have their highest diversity at the lowest latitudes in Central America. In comparison Alaska is very poor in species.

the biome is therefore less worthy of conservation, because the species that reside there, apart from the migrants, are found in no other of the Earth's biomes.

■ BIODIVERSITY IN SUCCESSION

Biodiversity is not static but is constantly changing over the course of time in any given location. Ecosystems develop and grow, eventually reaching maturity, but even the final stages often pass through a constant cycle of disturbance and renewal (see "Ecosystem Development and Stability," pages 99–101). During ecosystem development different species replace one another in a fairly predictable manner in a process that ecologists term succession. Succession is driven by the interaction of species leading to successive replacement of one set of organisms by another until eventually some degree of stability is attained (see sidebar on page 140).

Tundra habitats have proved very suitable for the study of succession because the retreat of glaciers during times of climatic warmth leaves the ground devoid of virtually all life, creating just the open and unoccupied conditions that pioneer organisms require. So biodiversity at the start of such a succession is zero. The arrival of pioneer mosses, lichens, ferns, flowering plants, and microbes soon establishes a functional ecosystem and initiates succession and the accumulation of biodiversity. Facilitation is most easily observed in plants, which provide an energy supply to invading animals and microbes, but also change the microclimate by shading the soil with their leaves and supplying the soil with organic matter as those leaves die and fall as litter. Such changes allow other organisms to invade, including dwarf shrubs such as willows and dwarf birch, or bearberry, and these have an even greater effect upon litter production, energy supply, microclimatic alteration, and soil stabilization. By casting a heavier shade, however, they usually prove too competitive for the pioneer plants, which soon succumb and fail to maintain their local populations.

The study of succession and changes in biodiversity, however, is hampered by the fact that the entire process, from initial invasion to stabilization, takes a long time, often more than 100 years. This means that observation by scientists takes more than one generation, and although ecologists are generally fairly patient people, this kind of long-term devotion to the detailed recording of research is unlikely. Other approaches to the study of tundra succession are therefore required. This is where the study of glacial retreat has proved particularly valuable, because it is often possible to estimate the time that has elapsed since a particular area of ground was left bare. Glaciers advance and retreat in response to changes in climate (see "Causes of Glaciation," pages 183–186), and the last 200 years have generally been ones of warming global climate and consequently general glacial retreat. Maps, historical records, and even old photographs are very useful in providing a chronology for the process of retreat and can indicate precisely where the glacial ice was positioned at any time in the past two centuries. The ecosystems that occupy these locations can then be examined and the course of succession inferred.

Some general patterns have emerged from these studies of tundra successions. At first communities are "open," which means that there is an incomplete covering of vegetation. As pioneer plants invade the ground gradually becomes fully occupied, and no bare soil or rock is visible. Increasing amounts of vegetation lead to the development of a physical structure leading to the overlapping of leaf layers, so that the ground is actually covered by several strata of leaves. Biomass thus increases during the course of succession. In the early stages there is little interaction between plant species because there is plenty of room for all to live in relative isolation, but interactions become more frequent with time. These interactions can be either negative or positive. A negative interaction is one in which two species require the same resources of light, water, mineral elements, pollinators, and so on. This means that they are in competition, and the more efficient species may well exclude the weaker competitor over the course of time. A positive interaction occurs when one species is aided by the presence of another, perhaps taking advantage of its shade and the microclimate it creates, or, in the case of epiphytes, growing upon the branches of the larger species, which it uses as support. These interactions can be studied using careful spatial analyses and testing the data statistically. In one Norwegian study, for example, a location that had been free of ice for 37 years showed just 13 species showing positive interactions among its component plants. A nearby site that had been exposed for 90 years had 23 species with positive interactions, and a more mature location that had been ice-free for 220 years had 37 plant species with positive interactions. This shows that the communities are becoming more complex and that their component species are developing more sophisticated and subtle relationships with one another. It also reflects the fact that more species are present within the communities.

It is difficult to study the changes in total biodiversity through a succession because identifying all the different microbes, plants, and animals within a developing ecosystem

Succession

Bare habitats, such as mudflats, sand dunes, or the lava left by a volcanic eruption, do not remain in that condition for long. Organisms with efficient powers of dispersal soon arrive at the site and establish themselves, as seen in the colonization of an area following glacial retreat, illustrated below. Although there is little competition for space at this stage, conditions for pioneers are often difficult. There may be poor water retention in the soil, or large variations in the daily temperature, or lack of shade from the Sun. The pioneer, therefore, must be capable of growing in a severe environment but does not have to be an aggressive competitor to settle in the area. The presence of an organism in a habitat, however, has an effect on the physical and chemical nature of its surroundings, so as soon as the pioneer has established itself the environment inevitably begins to change. The growth of a reed in a pond slows the flow of water and leads to higher rates of sedimentation, the growth of a grass on a sand dune slows the movement of air and leads to the deposition of airborne sand grains, and the establishment of lichen on a rock creates a source of organic matter for the soil. As it inadvertently changes its surroundings, the pioneer makes a habitat less severe and creates conditions in which other species are able to invade. This process is called *facilitation* and is one of the major driving forces of succession.

Facilitation does not imply that pioneer organisms are behaving in an altruistic way toward other species. The changes they make to an environment are inevitable consequences of their presence, and the arrival of new species is rarely beneficial to them. As new species arrive, the next stage in the process begins, and competition becomes an important force in the succession. Later arrivals may prove more competitive once established than the original pioneers, and the outcome of the competition may well be the loss of the pioneer species. In this way one species replaces another, and its continued presence alters the environment further, leading to new invasions, more competition, and additional local extinc-

Debris from glacial retreat near Lech in Austria has been colonized by herbaceous vegetation. Some shrubs are beginning to invade the developing soils.   *(Peter D. Moore)*

tions. Even when a final stability is attained, the equilibrium is a dynamic one. At any given location there is a turnover of species as individuals grow old and die. The death of a tree, for example, can lead to a cycle of successional events on a miniature scale as invaders once again struggle for supremacy in the gap.

Diversity changes in the course of succession. On the whole, more mature ecosystems contain more species of plants, animals, and microbes than immature ones in the course of succession, especially when the final, or climax, ecosystem has matured into a mosaic of patches each in different stages of recovery and renewal. Disturbance, however, can enhance biodiversity because it offers an opportunity for pioneers and early successional species to commence the process again, so that these species are not totally lost to an area. Fire, flood, and frost-heaving on a local scale can diversify a landscape and create just the kind of complex arrangement of different types of habitat that ensures maximum diversity.

would require a very large team of biological experts, and this is rarely available. As a consequence, most information on diversity change is based on studies of particular groups of organisms. Studies of colonization of glacial debris in Russia by flowering plants have shown that it can take more than 250 years to accumulate just 20 species. This is clearly slower than the Norwegian study of plant interactions, which probably relates to the overall climatic conditions. Harsher conditions, such as colder winters in this case, lead to a slower rate of biodiversity increase and ecosystem development.

Sometimes the pattern of species richness does not follow a steady course of increase but reaches an early peak and then settles down to a lower value. Some observations from the Austrian Alps in Europe, for example, indicate that the number of flowering plants in a glacial retreat succession reaches a maximum of 30 species between 30 and 70 years after being left ice free, but this number decreases to about 15 after 200 years. This pattern of "overshooting" can be found in a wide range of successions in different biomes and in various groups of organisms. It indicates that the initial invasion of pioneer species creates a kind of supersaturation, in which the ecosystem temporarily contains more species than its resources can support. Intense competition ensues, and some species are eliminated as the community settles to an equilibrium state.

The pattern of diversity in succession, therefore, is one of general overall increase, but early peaks in species numbers can occur prior to the establishment of balance in the community. Pioneer species are usually adept at colonization, and this involves a capacity to disperse widely and rapidly into new locations and the ability to survive under conditions where there is little vegetation cover and the microclimate may prove challenging. The sequence of species in the course of a succession is thus only partially explained by the operation of facilitation and competition (see sidebar on pages 140–141). It is also controlled to some extent by how rapidly a species can arrive at a site, which is determined by its dispersal ability. In the tundra environment, where vegetation is sparse and conditions severe for much of the year, the availability of plant and animal pioneers may well control the course of succession. As a consequence, tundra successions are very variable and are dependent on local conditions.

■ MICROBES IN THE TUNDRA

Biologists divide the living organisms of the world into five kingdoms. One of these kingdoms, the bacteria, is distinct from all other kingdoms by lacking a true nucleus in the single cell that forms the body of its members. These are called *prokaryotic* cells. All the other four kingdoms contain organisms with nuclei in their cells, called *eukaryotic* cells. These four kingdoms consist of animals, plants, fungi, and protoctista (or protists). The first three of these are familiar to most people, and the fourth contains an assortment of organisms that have variously been regarded as plants and animals in earlier systems of classification. The protoctista consists of algae (red, green, and brown), various unicellular photosynthetic organisms (including diatoms, chrysomonads, dinoflagellates, and euglenids), slime molds, amoebas, and various other organisms formerly grouped within the protozoa. The term *microbe* is usually used to cover the bacteria and the fungi, so it encompasses two kingdoms of organisms.

In all ecosystems microbes are most abundant in the soil. An ounce of soil can contain as many as 3 million bacteria (100,000 per gram). The identification of bacteria, however, poses immense difficulties, so it is extremely difficult to estimate the richness of species in soil. Individual bacteria need to be cultured and developed into large colonies if they are to be identified effectively, and at present up to 95 percent of

Cyanobacteria

The blue-green bacteria were once regarded as algae, and their system of photosynthesis is very similar to that of algae and higher plants. They generate ATP and a reducing agent by using the energy of sunlight, which they absorb with the blue-green pigment phycocyanin together with chlorophyll a. Most are aerobic, and, like higher plants, use water as an electron donor, releasing oxygen as a waste product, but some are able to operate by using hydrogen sulfide as an electron donor, releasing sulfur instead of oxygen. Many species are filamentous, forming long chains of individual cells linked together at their ends. These bacteria were the main photosynthetic organisms of the early Earth, dominating the planet from 2,500 million years ago to 600 million years ago. They formed dense mounds called *stromatolites,* which can still be found in the shallow seas of the west coast of Mexico, the Bahamas, Australia, the Persian Gulf, and even beneath the ice in Antarctica.

Apart from their ability to photosynthesize, the cyanobacteria are remarkable biologically because they can fix atmospheric nitrogen. One of the photosynthetic cells in a chain enlarges and loses its pigments, forming a *heterocyst* that houses the enzyme nitrogenase. Most cyanobacteria are terrestrial or freshwater, though some have been found in seawater. They are particularly abundant in tropical wetlands, including agricultural systems, such as rice paddies.

low temperature and has its growth optimum below 68°F (20°C) is called a *psychrophile.*

Some bacteria are photosynthetic and need light for long-term survival. The most abundant of the photosynthetic microbes in tundra environments are the blue-green bacteria, the cyanobacteria (see sidebar at left).

Tundra soils are often rich in cyanobacteria, which may cover their entire surface. On the island of Svalbard in the Arctic tundra the cyanobacteria are the initial colonists of bare soils and form a dominant ground cover of around 30 percent of the surface for the first 60 years of the succession. During this time they fix nitrogen from the atmosphere and donate it to the soil as they die, thus enriching the soil and increasing the potential productivity of the ecosystem. In this way they facilitate the invasion of more demanding species (see sidebar "Succession," pages 140–141), including many higher plants. They decline only when they become shaded by the growth of more robust photosynthetic species.

Fungi are also present in tundra soils. Fungi germinate from very small spores, often less than 0.0002 inches (0.0005 cm) wide, so tiny that they can be transported like dust particles in the air for many hundreds of miles. When they germinate they produce a fine thread, or mycelium, that branches and extends through the soil, secreting enzymes and digesting and assimilating organic materials. The mycelium may develop fruiting bodies that produce more spores, and these can result from sexual or asexual processes. The fungi found in tundra soils are mainly *psychrotrophs* rather than psychrophiles. A psychrotroph is an organism that has its growth optimum at temperatures above 68°F (20°C) but is still capable of growth and metabolic activity at the freezing point. This suggests that tundra fungi are actually well suited to warm conditions, but they are able to cope with cold when called upon to do so. It is likely that they originated in the temperate zone and that they are best suited to decomposition in the warm summer, when the air temperature is raised and the soil can become heated to 60°F (15°C) or above as a consequence of absorbing radiant heat from the Sun.

Many fungi can survive the tundra winter because they have the capacity to dehydrate, thus reducing the likelihood of cell freezing, and also accumulate antifreeze compounds, including sugars, glycerol, and mannitol. Their enzymes are capable of activity at low temperature, as in the case of the mold *Sclerotinia borealis,* which can operate its enzyme polygalacturonase at 30 percent efficiency at 41°F (5°C) but has its optimum operative temperature at 100–120°F (40–50°C). This fungus is therefore capable of growth when the temperature falls near the freezing point but is clearly much better suited to the high temperature of summer. Some mycologists (those who study fungi) believe that tundra fungal communities are enhanced by a fresh input of spores from lower latitudes each spring. This is entirely likely because of

soil bacteria are regarded as unculturable, so they defy identification. In recent years microbiologists have approached the problem by extracting DNA from soils and amplifying this by a culturing system. But the DNA of bacteria is variable, and the bulk of the DNA extracted from a soil cannot be assigned to particular species. So researchers have to be content with arbitrary labels for the "species" they extract.

Bacteria reproduce by fission, and the rate at which a colony can grow is strongly temperature dependent. In tundra soils bacteria are mainly active during the warmth of summer, when the water in the surface soil melts, and become dormant during the long winter. There are some, however, that manage to maintain their metabolic activity and reproduce even at very low temperatures, for example Bacillus TA41, extracted from saline Antarctic waters, which is active at 14°F (−10°C). A microbe that can operate at very

the ease with which fungal spores are carried over long distances. Survival through the winter becomes less of a problem for a species if it can reinvade whenever conditions are favorable. Tundra researchers have trapped and counted the fungal spores in the air through the spring on the island of Svalbard and have found that the deposition of immigrant spores begins in late April and peaks in August. The density of spores is not high, 33 spores of *Cladosporium* per cubic yard (44 spores/m³) at the beginning of August, but this would still prove adequate for the colonization of tundra soils and the rapid development of a fungal community. The total content of spores in the air in the Arctic can reach 490 spores per cubic yard (640 spores/m³). Ornithologists in Antarctica have taken swabs from the legs and feet of seabirds and found fungal spores, so it is possible that birds act as vectors for fungal dispersal in the tundra as well as air currents.

One hazard that must be faced by fungi and other microbes in the tundra is the high intensity of ultraviolet radiation (see "Ozone Holes," page 226). Fungi are generally sensitive to UV radiation, though the spores tend to be the most resistant stage in the life cycle, especially if they are equipped with thick, protective coats.

Microbes in the tundra play an important part in nutrient cycling, releasing the elements bound in organic matter and returning them to an inorganic state. Plants are then able to reuse these elements as they take them up from the soil. Microbes also return carbon to the atmosphere as they decompose dead organic matter, thus regulating the carbon cycle at high latitudes. Microbes thus play a vital role in the functioning of the tundra ecosystem.

■ TUNDRA LICHENS

Lichens are an extremely important and very widespread group of organisms in tundra habitats. They are classified as organisms and given scientific names like species of plants

Lichens

A lichen consists of an association of a fungus with an alga or a cyanobacterium, but the closeness of that association is quite variable and can simply consist of the two organisms living alongside one another. Such an association would not normally be called a lichen. One definition of a lichen is a stable, persistent, and self-supporting association of a *mycobiont* (the fungal component) with a *photobiont* (the photosynthetic partner) in which the mycobiont forms a protective cover for the photobiont. Delicate photosynthetic cells of a microalga or cyanobacterium are enclosed within a sheath of fungal mycelium. These cells are protected by the surrounding mycelium, and although they desiccate quickly under dry conditions, they are able to recover remarkably well. Light penetrates the mycelium, allowing the photobiont to photosynthesize, and the fungus derives its nutrition from this source. Both partners thus benefit from the association, and these benefits allow the combined "organism" to occupy positions that neither component could achieve on its own. They are able to grow on rock surfaces, the surface of stony soils, or on the branches of trees and shrubs as epiphytes, able to cope with desiccation even in dry and exposed conditions.

The structure of the combined thallus is quite complicated and often takes on a distinctive and recognizable form, which is why lichens can be classified and identified visually. The "lichenized" thallus sometimes produces distinctive chemicals, secondary products of metabolism that protect it from grazing animals by making it unpalatable. Lichens growing on rocks may even secrete acidic compounds that slowly dissolve the rock surface, assisting in the process of weathering (see "Geology and Rock Weathering," pages 52–54). Reproduction in lichens can take the form of small particles containing both components, becoming detached from the main thallus and carried away by wind or water. Alternatively, the fungal component may produce fruiting bodies and release airborne spores that become dispersed and land on surfaces where they may encounter the appropriate alga and form a new lichen thallus.

The algal or cyanobacterial component of lichens are well known as independent species that can grow without the aid of the fungus, though usually in a more restricted range of habitats. It was once believed that the fungal component survived only in combination with its symbiont, but recent studies using molecular techniques have shown that the fungi of lichens do grow independently of the photobiont but take on such a different form that their involvement in the lichen thallus had never been suspected. This poses a real problem for taxonomists involved in classifying organisms and raises the question once more as to whether lichens should be regarded as true species.

and animals, but it could be argued that they are not true species at all because they consist of two distinct organisms living in a symbiotic union (see sidebar on page 143).

Lichens are often the dominant component of the tundra ecosystem, as shown in the photograph below, forming a complete cover over the surface of the rocks or the soil. Lichens may even penetrate into the pores of coarse-grained rocks, such as sandstones, and form a well-protected photosynthetic layer beneath the rock surface. Lichens that form a flat crust on the surface of rocks are called *crustose,* whereas those that have a thicker thallus curling up at the edges are called *foliose.* These types of lichen are also found on the soil surface, where they are joined by upright, branching lichens that are said to have a *fruticose* form.

In Antarctica there are 350 species of lichen, while there are only two species of flowering plant, thus demonstrating how much more successful this apparently primitive form of life can be in an extreme environment. Seven species of lichen manage to survive within 4° of latitude from the South Pole. In the Russian polar desert approximately 40 percent of the species of photosynthetic organisms are lichens, and a further 42 percent are bryophytes (mosses and liverworts).

Lichens contribute to the primary productivity of an ecosystem because of the photosynthetic activity of the photobiont, but their growth rates are very low. A circular crustose lichen may extend its radius by only 0.004 inch (0.01 cm) in a year. The primary productivity of lichen-dominated ecosystems in the polar desert region of the tundra is thus very low, only about 0.01 pounds per square foot per year (0.03 kg m^{-2} y^{-1}) or less. Photosynthesis takes place only when the thallus is wet, so, although it can survive very effectively in a desiccated state, the lichen does not grow in that condition and is dormant. Its water content may fall to just 2 percent of its weight, and it can survive in this desiccated form for several decades if necessary and can then be resuscitated simply by wetting. It is thus a poikilohydric form of life (see "Drought in the Tundra," pages 119–121). Several species of lichen can survive buried by snow for six months of the year.

High-latitude and high-altitude populations of lichens have a lower temperature optimum for their photosynthetic activities than those from closer to the equator or from lower altitude. There is evidently some evolutionary selection for low-temperature operation under these colder conditions. Some photosynthetic activity has been

Much of the bare soil and rock of the tundra lands is covered by a crust of lichens.   *(Peter D. Moore)*

recorded in the lichen *Umbilicaria aprina* at a temperature as low as 2°F (–17°C). Genetically distinct races within a species that are adapted to specific ecological conditions are called *ecotypes*.

Lichens absorb most of their water from the atmosphere, especially those that grow upon rocks rather than soil, so they are sensitive to atmospheric pollution. Many respond negatively to sulfur compounds in the atmosphere, such as sulfur dioxide produced by industrial processes and carried into the tundra regions by air movements. Lichens also accumulate heavy metals from the soil and the atmosphere, and the lichens of northern Europe absorbed a considerable quantity of cesium 137, the radioactive pollutant produced by the accident involving the Russian nuclear power station at Chernobyl. Some large mammalian grazers, including caribou, depend heavily upon lichens for their fodder, and harmful elements can thus accumulate in food chains under such circumstances of aerial pollution.

The nitrogen-fixing capacity of lichens contributes significantly to the nitrogen input of tundra ecosystems and also makes the lichen thallus attractive to grazing animals, which require nitrogen for protein production. Caribou are particularly fond of the "reindeer moss," *Cladonia rangiferina,* and other *Cladonia* species. Caribou scrape off superficial snow with their hooves in order to gain access to the lichens beneath, but the slow growth of lichens means that recovery from intense grazing is slow, sometimes taking 10 years to make up the losses caused by just one grazing animal. The semidomesticated flocks of caribou in northern Europe are constantly moved because grazing for three or four years in an area can severely damage the lichen cover. Damaged areas sometimes need to be abandoned for 15 years in order to allow complete recovery.

Some lichens benefit from the presence of animals, such as those that absorb additional inputs of nitrogen and phosphorus from fecal material. The orange and yellow lichen *Xanthoria parietina,* for example, is often associated with the nesting colonies and roosting places of cliff-inhabiting sea birds. In a similar way the green alga *Prasiola crispa* is found in Antarctica mainly around the nests of penguins, which are enriched by their droppings.

Lichens are used by various birds, including sandpipers and plovers, as nesting material. The polar tundra has little to offer in the way of building materials for nests, but lichens may well prove an effective insulating material, protecting eggs and chicks from the cold ground beneath. Lichens thus form an important and influential component of tundra vegetation, being highly adapted to the cold and dry conditions of the tundra, especially in the high-latitude polar deserts.

■ TUNDRA MOSSES AND LIVERWORTS

Together with the lichens, mosses and liverworts contribute the bulk of the plant diversity in the polar desert tundra, and they are very important features of the vegetation throughout the tundra. Arctic Alaska, for example, has 415 species of moss and 135 species of liverwort. These are photosynthetic organisms that are considered primitive in the sense that they evolved relatively early in the history of life on Earth, between 500 and 400 million years ago. The life cycles of mosses and liverworts are similar. Both have a dominant generation in which the tissues of the plant are haploid (*gametophyte*), in other words, each cell has only one set of chromosomes. They therefore produce motile sperm cells and egg cells without any reduction division (meiosis). These fuse to produce a relatively short-lived diploid generation (*sporophyte*) that is entirely dependent on the gametophyte and that gives rise to haploid spores by meiosis. These spores are the main means of dispersal into new habitats.

Both mosses and liverworts are small in stature and look fragile. The moss gametophyte produces leafy stems, while liverworts come in two main forms. One has fine, creeping stems with small leaves, and the other has a platelike, flat thallus. All lack any cuticle, so they are easily desiccated, but many have a capacity to withstand such desiccation and emerge unharmed when they become wet once more, even after centuries in some cases. Like lichens, they are poikilohydric. Thus, although they are primitive in an evolutionary sense, they are relatively robust in the face of extreme environments, performing better than most higher plants in tundra conditions. They vary in their ecology, all needing water for the process of sexual reproduction and also requiring a humid atmosphere if they are to photosynthesize and grow. Many can withstand shade conditions and therefore grow beneath a canopy of taller plant species, but some require open sunny conditions and do not tolerate shade. Some of these, such as *Andreaea rupestris* and *Dicranoweisia crispula,* form small cushions on open rocks in the tundra scree of Iceland, where they are exposed to the full drying effect of long periods of sunlight.

Many are able to cope with long-term burial by snow, which is a frequent experience in the tundra, and a few can survive even where they experience only a very short growing period. Snow beds and snow patches lie late into the summer where they are protected from the low angle of the Sun by their aspect or by shading cliffs (see "Microclimate," pages 18–21). They gradually retreat as they melt at their edges, exposing the dark-colored liverworts and mosses that survive beneath them. One of the most frequent snow bed bryophytes of the Arctic is the liverwort *Anthelia juratz-*

kana, which forms dark carpets revealed by the retreating snow. This liverwort also occupies similar habitats in alpine tundra of the lower-latitude mountains, so it is a true Arctic-alpine species. *Polytrichum sexangulare* is the equivalent moss, often growing with *Anthelia juratzkana* in snow beds both in the Arctic and in alpine locations throughout North America and Europe. This is a dark green, upright moss with stiff, spiky leaves that may be uncovered for only two or three months a year.

Open, stony ground and scree can be very unstable, so any plants growing there must be able to recover rapidly from the catastrophe of disturbance and burial. The moss *Racomitrium lanuginosum* is one of the most successful in this respect. It is sometimes called the "gray hair moss" because its leaves terminate in long white hairs that give the bottle-green moss a gray overall appearance. When it dries these hairs wrap around the stems like a corkscrew, and the gray-white color reflects the bright sunlight, preventing overheating (see sidebar "Hair," page 105). The plant is thus highly resistant to desiccation, but it is only able to photosynthesize if the relative humidity of the atmosphere is greater than 95 percent, so it alternates between periods of rapid growth and dormancy. It grows in tussocks, which easily fragment and become established wherever the moss portions eventually lodge, so it ideally suited to a mobile scree. The mobility of the habitat ensures that few other plants manage to establish themselves with any degree of permanence, so it avoids the competition and the shading it would suffer in a more stable location. Like the snow bed species, this is also an Arctic-alpine in its biogeographical distribution and is found throughout the Arctic and over many mountains and moorlands farther south, especially where the rocks are acidic.

The flora of the Antarctic continent, as has been seen, is dominated by lichens, mosses, and liverworts, with only two higher plants managing to survive there. The mosses form a compact turf between rocks and stones, together with small cushions in the more favorable sites, reaching four inches (10 cm) in height. It is possible to detect the annual growth of the stems in such cushions, and a moss with this stature can be more than 100 years old, which illustrates how slow is the growth of these Antarctic species. One Antarctic moss that is likely to be familiar to city dwellers in North America and Europe is *Bryum argenteum.* This is best known as a very small, silvery moss that grows between the paving stones in urban sidewalks and paths. It is a remarkably adaptable species that is prepared to grow in mountain rock crevices, tundra soils, or city walls and pavements. It illustrates the fact that the tundra environment is well suited to species that can reproduce rapidly, disperse efficiently, endure drought, and effectively colonize harsh, rocky habitats. Within the mosses there are different reproductive strategies. Some have short lives and invest much of their energy in the production of spores, ensuring that they can spread rapidly but do not need to persist. Others are slow growing and long lived, and these do not need to produce spores as frequently or as abundantly. In other words, both r- and K-selected strategies are present within the bryophytes (see sidebar "r- and K-Selection," page 125).

In the subantarctic islands mosses play an even more important role in the vegetation cover, having more than 400 species present together with about 300 species of lichen. The high rainfall in these oceanic islands is ideal for bryophyte growth, and the occurrence of high winds restricts the height of other plant species, giving more opportunities to those of low stature.

Bryophytes play an important part in tundra ecosystems by stabilizing the surface of the soil and preventing the erosion that would otherwise result from the runoff of surface water in spring as the snow melts. The moss cover also retains moisture in the surface soil layers and is vital for the germination and establishment of many higher plant seeds, which find warm, moist conditions in the moss carpet. For invertebrates the moss layers and cushions form a complex, structured environment in which to live, and the maintenance of high humidity close to the ground is vital for the survival of most ground invertebrates. But the moss layer is never very deep, and when large herbivores graze upon it and trample the moss carpet underfoot they reduce the complexity of its structure, which in turn can influence the effectiveness of the layer as soil insulation. Excessive trampling, whether by caribou or humans, can thus have considerable and unforeseen effects upon the tundra ecosystem. It is possible that grazing may also encourage the growth of grasses at the expense of mosses because grasses recover much more rapidly following grazing, and this can lead to the replacement of moss tundra by grass tundra if the grazing is maintained.

Many tundra wetlands are dominated by a very distinctive group of mosses, the bog mosses, genus *Sphagnum.* There are about 300 species of *Sphagnum* in the world. They occur in a wide range of wetland habitats, but they are most abundant in the boreal and subarctic regions of the Northern Hemisphere, particularly Canada and Russia. The mosses of the genus *Sphagnum* probably dominate a larger area of the Earth's surface than any other type of bryophyte. The bog mosses are also found in alpine wetlands throughout the world, including the Chilean Andes; the Snowy Mountains of Australia; the mountains of Malaysia, China, and Japan; and the Ruwenzori Mountains of East Africa. There are 15 *Sphagnum* species found around the hot volcanic springs of Iceland.

Sphagnum is constructed as an upright stem with clusters of side branches at intervals and a dense terminal head of short branches and branch buds. It grows in masses, either forming even carpets or hemispherical tussocks, with the heads of the individual stems interlocking to give a uniform surface layer. Many of the cells in the leaves and the stems are dead and empty but become filled with water and thus act as a sponge. A fully saturated clump of *Sphagnum* can hold up to 20 times its dry weight in water. The growing tussock of bog moss thus carries a water supply with it as it grows. In boreal climates it can form elevated raised bogs, but these do not develop in the tundra. The cell walls of *Sphagnum* are rich in polyuronic acids, which have the chemical capacity to adsorb inorganic ions from the surrounding water, replacing them with hydrogen ions and thus making the water more acidic. This is a highly effective competitive mechanism that restricts the ability of more robust plants to germinate and establish themselves on the *Sphagnum* carpet, creating unwanted shade. Under these acidic and waterlogged conditions the process of decomposition is impaired, so the dead parts of the moss remain partially intact and build up beneath the growing layer, gradually becoming compacted into peat as the litter continues to accumulate.

The upper layers of the *Sphagnum* carpet or hummock are not fully saturated and usually contain air spaces and gaps through which water can drain. In these layers there is an active community of invertebrates, mainly detritivores and carnivores that feed upon them. This layer also forms an insulating blanket over the surface of the ground, preventing heat penetration in summer and leading to the persistence of the permafrost in the lower soil. This may contribute to the development of palsa mires in the Arctic (see "Tundra Wetlands," pages 74–78).

Mosses and liverworts thus play a key role in the ecology of the tundra, both polar and alpine. Under the extreme conditions of these habitats relatively few flowering plants can survive, so the bryophytes come into their own.

■ THE TUNDRA SOIL FAUNA

Lichens, mosses, liverworts, and higher plants provide energy-rich supplies of dead organic matter that falls upon the surface of the soil as litter, and this resource is exploited by the detritivores, the heterotrophic organisms that are unable to fix solar energy themselves. The protists and invertebrate animals feeding in this way provide a source of food to predators, both invertebrate and vertebrate, resulting in the development of complex food webs within the soil (see "Food Webs in the Tundra," pages 91–93).

The moist conditions of the moss carpet create a home for an abundance of microscopic life. Bacteria feeding on the dead plant material are consumed by unicellular amoebas (Rhizopoda), including some that are naked blobs of cytoplasm and others enclosed in a protective shell, the testate amoebas. The shells, or cysts, protect the testate amoebas from being digested by other organisms in the litter. The naked amoebas can change their shape without restriction, but the testate forms enclose much of the organism, simply allowing the armlike pseudopodia to extend through an aperture in the shell to collect food. Amoebas are not alone in harvesting bacteria and other organic fragments from the moist habitats among the litter. Ciliates are another type of protist involved in this activity. Ciliates (Ciliophora) gather their food by means of fine hairs, or cilia, that create currents in water, driving suspended fragments into the mouthlike opening through which they feed. Some protists are photosynthetic, including the diatoms (Bacillariophyta), single-celled organisms with stiff silica cases called frustules. Other unicellular photosynthetic protists were once regarded as algae, among these the so-called "snow algae," which are able to live in the upper layers of snow. The snow algae possess a wide range of pigments besides green chlorophyll, including red anthocyanins, and when the protists are present in abundance they may give a pink color to extensive beds of snow.

The rotifers (Rotifera) include many mobile predators of the microscopic life in the moist layers of the soil. They also possess cilia, but these are arranged in a circular crown, which appears to rotate as waves of motion pass along the rows of hairs. As in the case of ciliates, the swirling cilia create currents that direct suspended matter into a mouth, but in the case of the rotifers, which are larger than ciliates, there is a specialized, multicelluar gut, including stomach, intestine, and anus, through which the food passes in the course of its digestion using gastric enzymes. The rotifers are thus true animals, in contrast to the protists, all of which are unicellular. Most rotifers swim freely through films of water, gathering their unicellular prey as they go, but some are permanently attached to a substrate and strain food from the water as it passes by. Small crustaceans (Crustacea) also swim in the water surrounding moss leaves, especially *Sphagnum*. These are yet more complex in structure, having sensory antennae, like microscopic shrimps, and are coated with a chitinous exoskeleton like the crabs and lobsters to which they are related. The water held by the mosses close to the soil surface is thus alive with a microcosm of producers, detritivores, and microcarnivores, forming a complex ecosystem in its own right.

Within the mineral soil below this mossy layer of living things lies a dark world where dead organic matter rains from above and forms the energy resource for another group of detritivores, including worms. There are two main groups of worms in the tundra soils, the unsegmented nematode worms (Nematoda) and the segmented annelid worms (Annelida). The nematode worms are extremely abundant, perhaps the most abundant of all animals on Earth, and very diverse. At present about 80,000 species are known to science, but this is undoubtedly a small fraction of the total. Some biologists believe that there are at least a million species of nematode worms on the planet, inhabiting almost all conceivable habitats, including the soils of the tundra. They have cylindrical bodies, tapering to each pointed end, and are usually less than 0.04 inch (0.1 cm) in length. They have a mouth at one end, an intestine that stretches through the length of the body, and an anus at the other. They consume a wide range of foods, some eating rotifers and others eating plant roots, shoots, and tubers, while some are parasites of larger animals, including mammals. The annelid worms include the earthworms (oligochaets), which are common in most soils, though not in podzols (see "Soil Formation and Maturation," pages 54–59). Most earthworm species require calcium, so they are infrequent in acid soils, but some called enchytraeid worms are less fastidious about the pH of their surroundings and are common in more acidic, peaty conditions. During the tundra summer, when the soil is unfrozen, the earthworms constantly burrow up to the surface, collect organic materials such as dead leaves, and carry it deeper into the soil. They effectively plow soil, passing it through their guts and voiding the materials in the form of casts upon the surface. This churning, sometimes called bioturbation, leads to the mixing of soils and plays an important part in nutrient movement within the ecosystem. Earthworms have even been found in the snowbeds of Mount Kilimanjaro in Tanzania at an altitude of 12,500 feet (3,840 m).

Tardigrades, or water bears (Tardigrada), are less than 0.02 inch (0.05 cm) in length but have a remarkable resemblance to bears when viewed from head on under a binocular microscope, hence their popular name of water bears. They vary in color from red to blue to brown, especially if they feed upon pigmented lichens, whose color they assume. Their tiny bodies are covered by chitinous armour plating, taking on the form of a minute armadillo, and they walk on four pairs of jointed legs, each ending in long, bearlike claws. Their lumbering gait adds to their resemblance to bears. They feed mainly upon dead plant material, and they are frequent in the soils of the tundra. Among the spider allies of the soil (Arachnida), the most frequent are mites. These, like the tardigrades, are detritivores and occur even in the waterlogged soils of the tundra wetlands, living among the bog mosses.

The larvae of some insects inhabit the soil, including those of the crane flies (Tipulidae), which are sometimes called leatherjackets. These spend their larval existence feeding upon plant roots and eventually pupate and emerge as long-legged flies, which are well known in temperate regions as well as in the tundra. Tipulid larvae and annelid worms provide a food resource for many of the wading birds that breed in the tundra. The long bills of these birds are ideal for probing the soil, especially the softer organic soil, to extract the burrowing invertebrates found within it.

Mollusks (Mollusca) include both shell-less slugs and shelled snails. The latter require calcium carbonate to build their shells, so they are rare in acidic conditions, where calcium is scarce. Several species are found on high mountains, including species of *Vitrina,* found at an altitude of 14,500 feet (4,400 m) on Mount Kilimanjaro, Tanzania. Snow cover provides the required insulation for these species in the cold nighttime conditions.

One other soil invertebrate that requires special mention is the springtail (Collembola). Springtails are minute, segmented insects that have no wings but have a long, flexible organ at their tail end that allows them to leap into the air when they are disturbed, hence the name springtail. They often live in large numbers in the surface layers of soils. They are usually less than 0.2 inch (0.5 cm) in length, and they feed on organic detritus, protists, pollen grains, fungi, and leaf litter. Springtails are covered with short hairs, and this makes them virtually unwettable because of the surface tension of water. They are able, therefore, to skate over the surface of water without sinking. Some springtails have a distinctly disjunct distribution, meaning that isolated patches of a species occur at great distances from one another, and biogeographers have contemplated how such small and flightless creatures could disperse over long distances, especially in the hostile conditions of the tundra. In one experiment, a zoologist found that springtails can survive for up to 16 days in agitated seawater because of their hydrophobic, unwettable surface, and this could permit travel by sea over considerable distances. Add to this the fact that some springtails can survive for four years at a temperature of –8°F (–22°C) and the possibility of lengthy journeys by sea or in floating ice becomes quite acceptable. The springtail may be small, but it is one of the most remarkable of tundra animals and is also one of the most abundant. Among lichens on the High Arctic island of Svalbard, springtails sometimes achieve densities of 23,000 per square foot (243,000 m^{-2}).

The soils of continental Antarctica are one of the most harsh environments on the face of the Earth. Away from the oceanic western coastal fringe there are no flowering plants and few mosses and lichens, but the soils contain 24 species of springtail, 21 mites, 14 crustaceans, 41 rotifers, six tardigrades, and 10 nematodes. No annelid worms, flies, or mollusks have been recorded. So the soil fauna remains relatively diverse even under such severe conditions.

The microscopic living creatures of the tundra soils are vital for the healthy functioning of the ecosystem. Detritivores, including the springtails, break up organic matter as it passes through their guts, a process called *comminution.* This increases the surface area of the material in relation to its volume, and when the detritus is voided as fecal pellets fungi and bacteria find it easier to attack the remaining organic matter and decompose it completely.

Detritivores can be considered as the first stage in the process of decomposition, which is an important step in the recycling of scarce elements, such as phosphorus, in the tundra nutrient cycle. The biodiversity of the soil is easily overlooked, but upon it depends the maintenance of all the above-ground components of the tundra ecosystem, from the mosses to the caribou. This unseen part of the tundra may be downtrodden beneath many hooves, but it is a support system in more senses than one.

■ VASCULAR PLANTS IN THE TUNDRA

Vascular plants are those that have specialized tissues in their roots, stems, and leaves that carry water from the ground to the upper parts of the plant and also carry the products of photosynthesis from the leaves in the opposite direction. The tissues responsible for these transport systems are called *xylem* and *phloem,* respectively, and they are gathered together in *vascular bundles,* seen in the leaves as veins. Lichens, mosses, and liverworts do not contain such specialized tissues, although some of the taller mosses, such as the genus *Polytrichum,* do have less highly modified tissues that serve the function of water conductance. Ferns (pteridophytes), conifers and their allies (gymnosperms), and flowering plants (angiosperms) all fall within the definition of vascular plants.

The tundra biome is poor in vascular plants compared with other biomes. The continent of Antarctica contains only two vascular plant species, one a grass, *Deschampsia antarctica,* and the other a pearlwort, *Colobanthus quitensis,* belonging to the same family as chickweeds, Caryophyllaceae. Even they are restricted to the western maritime edge of the Antarctic Peninsula, though beyond the Antarctic Circle many more vascular plant species, including other species belonging to these two genera, are found on the subantarctic islands and in Tierra del Fuego, the southernmost tip of South America. Tierra del Fuego has 417 species of vascular plant, of which 386 are flowering plants. The two Antarctic species are both small, cushion- or tussock-forming species that manage to survive in small pockets of soil and in rock crevices beyond the edge of the Antarctic ice sheet.

Ferns (Pteridophyta) as a group are poorly adapted to life in extreme cold. No species of fern grows on the continental mainland of Antarctica, though there are 16 species growing on the subantarctic islands. The young leaf fronds of most ferns are held tightly coiled to the main stem, the rhizome. Usually they lie close to the surface of the ground and are protected only by a mass of scales, which is of little assistance during hard frost. Some ferns survive well in alpine tundra, such as the parsley ferns (*Cryptogamma* species), which are nevertheless restricted to relatively oceanic mountain ranges and are found mainly where there is a deep snow cover during the winter, providing insulation for the delicate new fronds. Some ferns have rhizomes situated deeper in the soil, such as the bracken fern (*Pteridium aquilinum*), which is perhaps one of the most widespread and successful ferns in the world, but this strategy is not appropriate for the tundra biome, where permafrost limits the depth of rooting.

Among the flowering plants few families have been more successful in the tundra biome than the saxifrages (Saxifragaceae). Most of these are herbaceous with the dense cushion form of a chamaephyte (see "Plant Life-forms," page 112). A typical example is the tufted saxifrage (*Saxifraga caespitosa*), which is circumpolar in distribution. Its tufts are only two to five inches tall (2–12 cm), but it has a robust tap root, so it can grow in very dry locations among rocks. It is rare as an alpine species farther south but does extend as far as the San Francisco Mountains in northern Arizona at an altitude of 12,850 feet (3857 m). The purple saxifrage (*Saxifraga oppositifolia*) is another circumpolar Arctic species, but it is equally abundant as an alpine species on the mountains of the temperate zone. This species is of genetic interest because it grows in two distinct forms, a tufted cushion form on dry, exposed ridges and a trailing carpet form on flat surfaces, especially those close to the Arctic shore. The tufted form is more drought tolerant but is slow growing, whereas the trailing form grows rapidly and can extend quickly over the ground when conditions are favorable. The two forms retain their growth characteristics even when they are grown together in experimental gardens, so the differences must be genetically determined rather than a consequence of the different environmental conditions where they are found. They should perhaps be regarded as subspecies. This type of genetic diversity is widespread among tundra plants, where different microhabitats place varied stresses upon the vegetation. The biodiversity of tundra plants is therefore larger than one might suppose by simply counting species. Genetic diversity may well prove valuable to the tundra flora if it has to cope with changing climatic conditions (see "Biological Responses to Climate Change," pages 224–226).

In alpine situations populations of species, both plant and animal, are isolated from one another, and this can lead to genetic divergence and rapid evolution of different forms of a species on different mountains. This is shown by the various species of *Dendrosenecio* occupying the high mountains of tropical East Africa (see "Tropical Alpine Habitats," pages 82–85). In the Andes Mountains of Colombia, just north of the equator in tropical South America, a similar group of plants has evolved belonging to the genus *Espeletia,* belonging to the dandelion family

The spring gentian (*Gentiana verna*) is an early-flowering plant of the alpine tundra. Its conspicuous flowers attract the attention of the first bees and butterflies to emerge after the long, hard winter. *(Peter D. Moore)*

(Asteraceae). There are about 130 species of this genus in the northern Andes, and they differ widely in their architecture and branching patterns. Unlike *Dendrosenecio,* which probably evolved from herbaceous plants and developed into stunted treelike forms, *Espeletia* evolved from tropical trees that were exposed to increasing cold, aridity, and local isolation during the last glaciation of the Pleistocene (see "The Pleistocene Glaciations," pages 178–183). Smaller, stunted forms emerged, and the isolation led to genetic differentiation because of the lack of exchange of genes between populations. The outcome was a diverse assemblage of compact alpine species of *Espeletia,* superficially resembling the pachycaul *Dendrosenecio* species of the African mountains but having a very different origin. This is an excellent example of convergent evolution. It also illustrates the high evolutionary capacity of alpine plants when subjected to biogeographical fragmentation of populations.

The mountain avens (*Dryas octopetala*) is another widespread arctic-alpine plant that displays a highly diverse genetic constitution. There are many variants of this plant that occupy a range of microhabitats, from fellfields to snow beds, and these races are genetically determined and so can be called *ecotypes*. Like many of the plants of the Arctic and of high mountains, it has relatively large flowers that attract potential insect pollinators (see "Reproduction and

Dispersal in the Tundra," pages 123–129). The relative scarcity of insects and the very short flowering season lead to intense competition between flowers for the attention of pollinators, and many adaptations have evolved to add to the comforts and rewards for insect visitors. Large, brightly colored flowers are frequent, the spring gentian (*Gentiana verna*) shown in the photograph above being an example.

One common feature is the parabolic form of many flowers. Considered in cross-section, the petals of the alpine pasque flower (*Pulsatilla alpina*), as shown in the photograph on page 151, take on a form similar to that of an astronomical telescope mirror, or a satellite dish for television reception. As with these mechanical constructions, the form ensures that diffuse waves are collected and focused upon the central sensor. In the case of parabolic flowers, sunlight is focused upon the carpels and stamens in the center of the flower and causes a rise in the temperature of these organs together with the surrounding air. Visiting insects find this warming very helpful. It allows these cold-blooded invertebrates to become more active and gives them energy to collect pollen and move on to neighboring flowers. The warmth may also aid the maturation of the plant's reproductive organs and subsequent seed development.

Many arctic-alpine plants have large parabolic flowers, including the purple saxifrage and the alpine pasque flower. In the purple saxifrage the size of the flowers increases with latitude, northern Arctic polar desert forms having larger flowers than those of the subarctic region. Competition for pollinators evidently becomes even more intense in the cool, short summers of the far north, and those with the largest flowers prove most successful in producing seeds and persisting in this harsh environment. Ecologists have conducted experiments in alpine habitats to determine the overall effectiveness of pollination at different altitudes. Using the common harebell (*Campanula rotundifolia*) they found that although there were fewer pollinators making fewer visits to high-altitude flowers, the success of pollination and seed production was comparable at different altitudes. Among bees and wasps, bumble bees predominate under tundra conditions, and these seem to prove particularly efficient in their pollinating activity, so the plants do not lose out because of fewer visits.

Parabolic flowers work only if they are directed toward the Sun, and this changes over the course of the day, so there is a great advantage to be gained by moving the position of a flower in relation to the Sun. Such diurnal solar tracking is found in many arctic-alpine species, including *Dryas*, Arctic poppy (*Papaver radicatum*), and the glacier crowfoot (*Ranunculus glacialis*). Botanists are still not sure how this solar tracking, which they term *heliotropism*, works. The surface epidermis of petals is often covered with fine projections, called papillae, and these may serve as sensors, detecting the direction from which sunlight comes. They could then send signals to cells in the flower stalk that would initiate growth or the redistribution of water, causing a leaning of the entire flower. The scientific investigation of cell signaling is still in its infancy, but this behavior of tundra flowers may provide a useful experimental model to work upon.

The hairy catkins of some willows and the white-haired heads of the cotton grasses (*Eriophorum* species) use a different method for warming their fruits. The translucent hairs allow light to penetrate to the developing fruits, but

In the cold alpine climate some flowers, such as this alpine pasque flower (*Pulsatilla alpina*) in Switzerland, track the Sun across the sky and gather warmth by adopting a parabolic shape that focuses the rays on the flower's center. *(Peter D. Moore)*

the presence of the layer of air held by the hairs acts as a thermal blanket, retaining heat in precisely the same way as the glass in a greenhouse. Cotton grasses have thus been using the greenhouse effect to good advantage long before humans hit upon the idea.

The vascular plants of the tundra may be relatively few in species number, but they make up for this in their genetic versatility and their wide range of adaptations to this extreme environment. Many of these tundra species are found in both polar and alpine habitats, but some, especially the alpine species, may be restricted to one or more mountain sites, isolated from surrounding regions. These endemic species are particularly important from the point of view of biodiversity because they could easily be lost as a result of catastrophe within their very limited range, and they could not then be replaced from other locations. Tundra plants illustrate the problems that face conservationists when they try to establish criteria on which to select priorities for conservation effort. Biodiversity often ranks highly, but here is a biome in which biodiversity is relatively low yet which stands in need of protection in a changing world (see "Tundra Conservation," pages 229–232).

TUNDRA ARTHROPODS

One of the problems in trying to assess the insect biodiversity of a region is that many species, perhaps most, have still not been described and named by scientists. The figures currently available for the biodiversity of the tundra biome are therefore undoubtedly underestimates. This is particularly true of the arthropods, a group that includes crustaceans, arachnids (spiders and mites), and insects. The word *arthropod* is derived from Greek and refers to the jointed limbs possessed by these animals. According to current estimates there are about 33,000 species of arthropods in Canada, but only 2,000 of these have been recorded north of the timberline in the tundra regions. Although arthropods are much more diverse than plants, they still show reduced biodiversity with increasing latitude. The Queen Elizabeth Islands in the far north of Canada have only 553 arthropod species, and the island of Svalbard in the North Atlantic has only 244 species. Continental Antarctica has an estimated 77 species of arthropods.

Mites and springtails are mainly confined to the soil habitats of the tundra (see "The Tundra Soil Fauna," pages 147–149), and many of the tundra insects spend the larval and pupal stages in their life history either in the soil or beneath the water surface in wetland habitats. Some mites are parasitic on other animals, such as beetles and bumblebees. They are ectoparasites, meaning they cling to the outside of the animal rather than entering its gut. Most Arctic bumblebees carry a load of tiny mites that extract

haemolymph (insect equivalent of blood) by penetrating the thin tissues linking the main chitinous body segments. Lice are also common ectoparasites of mammals and birds in the tundra.

Mosquitoes are among the most familiar insects that plague all visitors to the tundra during the summer. The larvae live in pools, where they usually congregate in the warmest part, such as shallow areas unshaded from the sunshine. This location may change during the course of the day, in which case the mosquito larvae swim to the newly warmed area as the position of the Sun alters. Females of the mosquito species *Aëdes nigripes* lay their eggs in the middle of the day in the warmest sites they can find. The eggs remain in place through the winter. The chosen location is likely to become snow-free soonest in the spring because of its sunny location, and the larvae therefore have an elongated season in which to mature, pupate, hatch into adult insects, mate, and lay their eggs before the next winter season arrives. The two-year life cycle is typical of most tundra insects, and some even require three years for completion. Some spiders take seven years, but the record for an arthropod is held by an Arctic moth, *Gyneophora groenlandica*, which took 14 years to complete its life cycle.

Mosquitoes are not the only blood-feeding insects of the tundra. Biting midges (Diptera: Ceratogoponidae) and blackflies also plague the mammals of the region, including humans. Midges are very small biting flies with a wingspan of less than a tenth of an inch (0.25 cm), and, as in the case of mosquitoes, only the females take blood. The female lays around 80 eggs, and the larvae are aquatic, living in the acid pools of the tundra. They are shaped like minute eels and swim around the pond feeding on vegetable detritus and bacteria. They often occur in very high density, some nutrient-rich ponds having as many as 2,500 mite larvae per square foot (27,000 m^{-2}). They pass through four developmental stages (*instars*) before pupating and emerging as fully mature insects. Most midges are *crepuscular*, which means that they fly and feed at dusk, but polar tundra species have no option but to bite in the daytime. The females require a blood meal prior to mating and egg laying, and they have a preference for large mammals, such as caribou and people, and do not seem to attack birds. The midges have a very acute sensory system for detecting odors situated in the antennae, and the females can locate their hosts from considerable distances. Males use the same sensory system to home in on females for mating. Mammals release chemicals in their breath, sweat, and urine that act as *kairomones,* the term given to chemical signals used by blood-feeding insects, which include many types of organic acids, including lactic acid. Even the carbon dioxide emitted in respiration has a strong attraction for the hungry females. Entomologists working on the behavior of biting midges have found that some individuals are far more attractive as

Large butterflies of the alpine tundra need to be strong flyers to avoid being carried away by high winds. The anise swallowtail (*Papilio zelicaon*) is robust enough to cope with the mountaintops of the Rocky Mountains. *(Yellowstone Digital Slide File, Yellowstone National Park)*

hosts than others, presumably because of differences in the chemical secretions of individual mammals.

Beetles (Coleoptera) are among the most diverse of all animal groups, and there are undoubtedly several million species that have not yet been described. Tundra beetles tend to be small in comparison to those of lower latitudes, so they are a clear exception to Bergmann's rule (see "Body Size in the Tundra," pages 113–115), which claims that high-latitude animals are larger than their low latitude counterparts. Studies of alpine beetles show precisely the reverse trend. For example, the beetle genus *Bembidion* in the Himalaya Mountains has an average body length of 0.4 inches (1 cm) at 1,640 feet (500 m) and only 0.16 inches (0.4 cm) at 13,100 feet (4,000 m). The small body size for arthropods is beneficial when their survival depends upon finding microclimatic situations where they can escape the winter cold, such as beneath the cover of dwarf shrub vegetation. Large cold-blooded organisms would be hard pressed to find sufficient cover for their survival.

As in the case of plants, some beetles have very restricted distributions, especially those found on low-latitude mountains. One species that has become well known to geologists is the dung beetle *Aphodius holdereri*, which is now confined to a very limited area of the high plateau of Tibet, at between 9,800 and 16,400 feet (3,000 to 5,000 m). Fossil remains of this species dating from 40,000 to 25,000 years ago (the height of the last ice age) have been found in sites throughout southern Britain. The conclusion is that this beetle was very widespread over Asia and Europe when ice sheets covered the northern parts of these continents but became increasingly confined as the climate warmed and the ice retreated. Its status in Tibet, therefore, is one of a *relict*, greatly reduced from its former wide range.

Butterflies are relatively common at the height of the tundra summer, especially in some alpine situations. Like the beetles, their body size tends to decrease with latitude and altitude, again throwing doubt on the general application of Bergmann's rule. Both polar and alpine tundra experience high winds, so insects with large wings are in danger of damage or even of being swept away from their habitats, so large size is clearly a disadvantage. Butterflies of the genus *Phulia* in the mountains of the high Andes, for example, have particularly small wings. They are usually found taking shelter from the wind in low scrub vegetation of the alpine tundra, where bigger wings would prove a liability. In the course of their evolution many alpine species have lost the capacity for flight, rather like the birds of some oceanic islands. Being blown away from a mountaintop into the unfamiliar terrain of the lowlands would prove as serious a problem as being blown out to sea for a land bird. In the Himalaya Mountains above 13,000 feet (4,000 m) 60 percent of the insects are flightless. There are some large butterflies found in alpine environments, and these overcome the problem of being blown away by having very powerful flight. An example is the anise swallowtail (*Papilio zelicaon*), shown in the photograph on page 153. It is found throughout the Rocky Mountains and is able to cope with strong winds.

The problem with loss of wings and consequently the power of flight is that the population then becomes isolated from other populations, and the chances of genetic exchange are reduced. Also, if catastrophe strikes a population, reinvasion is unlikely. Some moths in the tundra are migratory, and the ability to fly is therefore essential if they are to move between different latitudes and altitudes according to the season. The diamondback moth (*Plutella xylostella*) is an example of a migratory insect, which periodically immigrates into the island archipelagos of the High Arctic, including the Svalbard group in the North Atlantic. These islands are 500 miles (800 km) north of Norway, lying between Scandinavia and Greenland at a latitude of almost 75°N. No other flying Lepidoptera (butterflies and moths) occurs on these High Arctic islands, so the arrival of parties of the diamondback moth is a conspicuous event. Meteorologists have studied the conditions under which migration occurs and have found that it coincides with warm air currents sweeping into the North Atlantic from eastern Scandinavia and Russia. It is entirely possible that other insects migrate into the polar desert under these conditions, but bees, wasps, and flies would be less noticeable, so they have not yet been recorded. Like most insects, the diamondback moth requires warm temperatures for flight and needs a minimum of 65°F (18°C), preferring 73°F (23°C). Because air temperature decreases with altitude (see sidebar "Lapse Rate," page 22), the migrating moths must remain relatively low to keep within the warm air mass. They fly actively and do not depend entirely on the speed of air movement, but it must take them at leas 48 hours in the air to cover the distance from the mainland to Svalbard. Once they arrive, they find sources of nectar and food plants for their larvae, so breeding is possible, but the season is short. Immigration often takes place quite late in the year, so the establishment of a permanent population has not proved possible. Changing climate in the future may well alter this.

Insects are unable to regulate their body temperatures internally, but they raise their body temperatures by basking in sunlight. Butterflies, grasshoppers, and flies all indulge in this activity, usually choosing a sheltered sunny spot and sitting motionless to absorb energy. Some have dark bodies or wings to enhance their capacity for heat absorption. Most butterflies bask with their wings spread open, revealing the back of their bodies, but some bask with their wings closed, absorbing heat on their undersides. Mead's sulphur butterfly (*Colias meadii*), for example, is a stubby-winged species of the Rocky Mountains in North America found in alpine tundra habitats from British Columbia to New Mexico. It is orange above and olive green beneath and basks with its wings closed, revealing its absorptive green undersides. It needs to raise its body temperature to 86°F (30°C) for it to be able to fly.

Spiders also bask to raise their body temperatures, and they are particularly prone to basking when they are carrying egg sacks that need warm incubation. In *Sphagnum* bogs some spiders, such as *Pirata piraticus,* build tubes that lead from the sheltered zone beneath the bog moss up onto the surface of the bog. Females with egg cocoons attached to their rear ends sit in these tubes with their tails and cocoons in the sunlight on the moss surface while their bodies remain in the cooler, more humid conditions below the canopy.

Spiders secrete a fibrous protein (fibroin) that is silky in texture, water resistant, and almost as tensile as nylon. It is extremely elastic and can be stretched by 30 percent before breaking. Some types of spider silk are the strongest known natural fiber. The silk is used for weaving webs to catch flying insect prey, and this enables spiders to invade ecosystems in their very early successional stages of development (see sidebar "Succession," pages 140–141). Following glacial retreat, for example, spiders can set up their traps among the rocks and scree and catch the insects blown into the site from other areas. The silk even enables them to invade effectively because the young spiders spin a thread that they use as a parachute, which provides them with aerial transport into new locations for establishing their traps.

The tundra is not an ideal place for arthropods, and only those that are highly adapted can survive. But wherever there is a means to make a living, there is likely to be an arthropod that can exploit the situation, and this is certainly true of the tundra. A remarkable range of insects and their allies manages to survive in this unlikely habitat.

■ BIRDS OF THE NORTHERN TUNDRA

Birds are warm-blooded, well insulated by feathers, and highly mobile, so they are in a good position to take advantage of the short but productive season in the tundra. In all of these respects they are better adapted to tundra conditions than reptiles and amphibians. In fact, there are very few reptiles and amphibians that reach the edges of the tundra biome. The northern leopard frog (*Rana pipiens*), the wood frog (*R. sylvatica*), and the mink frog (*R. septentrionalis*) all extend to the northern edge of the boreal forests, reaching the southern fringes of Hudson Bay, where the true tundra begins. Alpine habitats are slightly more favorable to amphibians because the daytime temperature can be high. The mountain yellow-legged frog (*R. muscosa*) of the Sierra Nevada of California lives and breeds at up to 12,000 feet (3,600 m). Its poor powers of dispersal, however, mean that the species lives in isolated populations on the mountains, including San Bernadino, San Jacinto (near Palm Springs), and Mount Palomar. The Yosemite toad (*Bufo canorus*) has an even more restricted distribution in the alpine habitats of the Sierra Nevadas, being found only in the Yosemite region up to an altitude of 11,000 feet (3,300 m).

Birds, on the other hand, can cope with tundra conditions better, either as permanent or temporary residents. The total number of bird species breeding in the Arctic tundra is 183, with 136 species breeding in Russia, 113 in Alaska, 105 in Canada, 61 in Greenland, and 31 on Svalbard. Birds are therefore far more diverse in the tundra than amphibians, but this is still a low total compared with California (286 species), Guatemala (472 species), Costa Rica (603 species), and Colombia (1,721 species). Of the 9,672 bird species currently known to science, therefore, less than 2 percent breed in the Arctic tundra, illustrating the latitudinal decline in diversity toward the poles (see "Patterns of Biodiversity," pages 137–139).

The birds of the Arctic can be divided into a number of groups: sea birds (gulls, terns, jaegers, murres, loons), waders (plovers, sandpipers, phalaropes), wildfowl (ducks, geese, swans), game birds (ptarmigan, grouse), predators and carrion feeders (falcons, owls, raven), and song birds and perching birds (buntings, redpolls, warblers, larks). Although the diversity of breeding birds may be low in

---

The breeding range and the migration routes of the arctic tern. It uses the long days of the Arctic summer for its breeding, then migrates to the Southern Ocean, where it spends the remainder of its year. This bird travels farther and spends more of its time in daylight in the course of a year than any other.

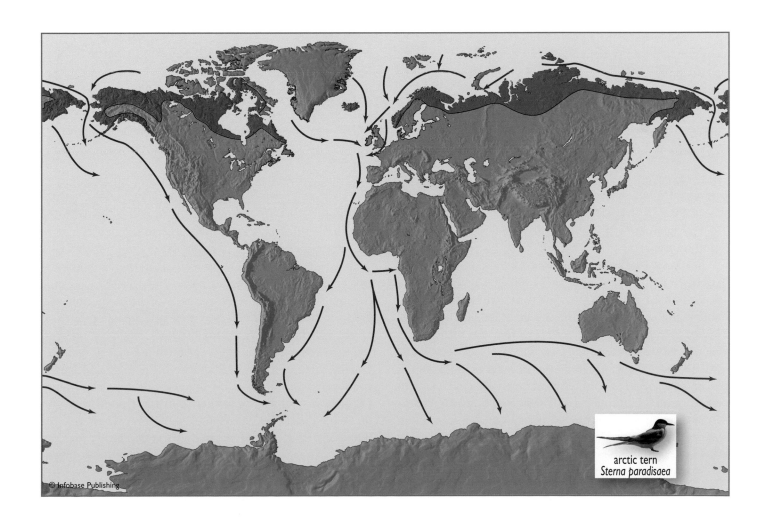

arctic tern
Sterna paradisaea

© Infobase Publishing

Ross's Gull

The bird that best personifies the icy wastes of the Arctic is Ross's gull, a small and elegant gull with a black ring encircling its face and a rosy blush to its breast and head. It was named after the famous Arctic explorer James Clark Ross (1800–62), who took the first specimens of the bird near Baffin Island in Canada while seeking the elusive Northwest Passage between the Atlantic and Pacific Oceans (see "Arctic Exploration," pages 203–206). Throughout the 19th century only a few sightings of the gull took place, and the species became much sought after by collectors of skins. Usually it was seen flying over the ice, often far from land, and its place of residence and breeding remained a mystery until 1905, when a Russian ornithologist, Sergei Alexandrovich Buturlin (1872–1938), launched an expedition with the aim of solving the mystery of Ross's gull. He left St. Petersburg in Russia in January accompanied by two assistants and journeyed to the Kolyma River in Siberia using teams of horses and dogs and covering the 3,400 miles (5,500 km) in three months. The ice on the river did not break until late May, and he was then able to travel north toward the Arctic Ocean. On his journey he came across several dozen Ross's gulls, some of which were breeding in the shrub tundra, the first scientist ever to observe their nests. The colony still exists, currently supporting 23 pairs of gulls, having grown from the 12 pairs noted by Buturlin in 1905. The nests are built on small islands on the edge of water-filled polygons (see "Periglacial Features," pages 59–62) and usually contain two or three eggs laid upon a bed of *Sphagnum* moss. They are laid in early June and hatch late in the month, when the surrounding vegetation has begun to sprout, providing cover for the chicks. Mortality is high, with less than 40 percent of chicks surviving even in a good year, and no breeding takes place when the weather conditions in June are poor. Despite the precarious status of this rare gull, which must have been clear even to Buturlin, he collected 38 specimens of Ross's gull together with 36 eggs, each of which fetched $8 (equivalent to about $500 in modern money) in the egg collecting market. It is remarkable that the Kolyma colony survived the subsequent onslaught of collectors.

Since the discovery of this breeding site, many other small colonies have been located in Siberia, and current estimates suggest that the world population may be as high as 45,000 to 55,000 birds. There have been isolated attempts at breeding elsewhere, such as western Greenland, and Churchill, Manitoba, some of which have been successful, so the situation looks hopeful for this attractive gull. But one mystery remains unsolved: Where does Ross's gull spend the winter months? Do these birds spend the long Arctic night flying over the frozen pack ice of the polar ocean? Winter sightings of the gull are extremely rare in the Atlantic or the Pacific, so Ross's gull still preserves its mysterious image.

---

comparison with other biomes, the density can be high. One survey of the coastal strip of northern Alaska provided an estimate of 11 million birds, including more than 5 million waders.

Sea birds feed in the ocean rather than on the tundra, but they need the dry land for breeding. Many species of sea bird breed in the Arctic and Antarctic tundra biome because the oceans bordering the land are rich in marine life and therefore a reliable source of food for their young. Colonies of cliff-nesting gulls, such as the black-legged kittiwake (*Rissa tridactyla*), and murres, such as the common murre (*Uria aalge*) and thick-billed murre (*U. lomvia*), often contain several hundred pairs of birds, and the cliffs become splattered with their droppings, phosphate-rich guano that fertilizes any local ecosystem, such as cliff grasslands. But other sea birds are less communal, including the black guillemot, which nests among rocks and boulders along the seashore.

Some gulls breed only in the tundra, including the glaucous gull (*Larus hyperboreus*), Iceland gull (*L. glaucoides*), and Thayer's gull (*L. thayeri*). The glaucous gull is typical of the Arctic gulls, having a circumpolar distribution and very flexible feeding habits. It takes seafood, particularly mussels and crabs from the intertidal zone, but will also consume carrion, such as dead seal carcasses, or adopt the role of a predator, taking eggs and young of other nesting birds. It breeds on cliff tops and promontories, singly or in small colonies, often constructing a nest from a pile of seaweed. Usually three eggs are laid, and both the male and female contribute to incubation. At the end of the polar summer the glaucous gull moves farther south, reaching Oregon on the West Coast and Maryland on the East Coast and the Greenland and Iceland populations spreading to western Europe as far south as the British Isles. Most tundra-breeding gulls move south for the winter, only the pure white ivory gull (*Pagophila eburnean*) and the pink-breasted Ross's gull

(*Rhodostethia rosea*) spending their entire lives in the Arctic. Biogeographers have expended much energy trying to trace the breeding grounds and the winter quarters of the rare and elusive Ross's gull, and many mysteries still surround this beautiful tundra gull (see sidebar on page 156).

Terns are closely related to gulls but are smaller with narrower wings, and many have long, forked tails, earning them the name of "sea-swallows." They are expert fliers and feed upon small fish, diving from considerable heights into the water. The tern most typical of the Arctic tundra is the arctic tern (*Sterna paradisea*). Like many other tundra birds, it is circumboreal in its distribution and nests on open ground, often on shingle ridges, where its speckled eggs are difficult to detect among similarly colored pebbles. Mortality is high, mainly because of predation by gulls and other birds together with arctic foxes. In a study of arctic tern nests in Spitsbergen ornithologists traced the history of 832 eggs and found that 68 percent of the eggs survived long enough to hatch, but only half the chicks survived after two weeks. So only one egg in three is likely to result in a mature fledged bird. The most remarkable feature in the arctic tern's life history, however, is the extent of its travel. Once fledged, the young birds disperse in various directions but eventually turn south and take to the open sea, crossing the equator

The arctic tern (*Sterna paradisea*) migrates annually between the Arctic Ocean and the Southern Ocean, making it one of the best-traveled birds in the world.   *(Lauri Dammert)*

into the Southern Hemisphere and eventually arriving in the Southern Ocean surrounding Antarctica. The accompanying map shows the extent of this journey. No other bird travels so far in the course of a year or spends so much of its life in perpetual daylight. Ornithologists study the patterns of movement of migrant birds by placing bands around the legs of young birds and hoping that some of these will be recovered when the bird dies. The method is chancy, but it has led to improved understanding of the directions and speed of migration movements. One arctic tern banded in Greenland, for example, was recovered three months later in South Africa, having traveled 11,200 miles (18,000 km). The likely pattern of arctic tern movements determined by banding methods is shown in the map on page 155. The total distance covered by an arctic tern in a year is thus at least 22,000 miles (36,000 km).

Jaegers (*Stercorarius* species) are predatory polar tundra sea birds with long, pointed wings and often have elongated central feathers in their tails. They are skillful fliers and are capable of tight turns and complex aerial maneuvers when in pursuit of their prey, usually gulls and terns. They do not usually kill their prey but harass them until in their fright they regurgitate the contents of their stomachs or drop the fish they are carrying in their bills. The jaeger then swoops and catches the food before it reaches the water. The parasitic jaeger (*S. parasiticus*) breeds throughout the Arctic tundra region, and, like the arctic tern, it migrates into the Southern Hemisphere for the winter. Birds from eastern Canada, Greenland, Iceland, and Arctic Europe migrate south through the Atlantic and winter in Argentina or South Africa, while birds from Alaska and Siberia move south through the Pacific and make their way to Chile, Indonesia, or even New Zealand. In the Southern Ocean the parasitic jaeger is likely to meet up with its Southern Hemisphere equivalent, the antarctic jaeger (*Catharacta antarctica*), which nests on the bleak, boulder-covered coastal regions of Antarctica and is a general predator of smaller birds, including penguins.

There are five species of loon (*Gavia* species), all of which are found in North America. All of them breed in the tundra, but most of them also extend southward into the boreal zone. The exception is the yellow-billed loon (*G. adamsii*), which is truly a tundra species, breeding only on the coastal marshy tundra of the far north. It prefers freshwater and avoids mountain regions. The loons are large diving birds with spearlike bills and with legs set well back on the body. The positioning of the legs makes the bird extremely ungainly on land but very agile in water, where it spends most of its time. Loons generally avoid flying when possible, although most have no alternative when they migrate south for the winter. The yellow-billed loon often flies from its nesting site in search of food. All loons are migratory, and even the yellow-billed loon leaves the Arctic in winter and migrates along the Pacific coasts and in the Atlantic south along the coast of Scandinavia, spending its winter at sea.

The tundra forms a focal region for breeding among the world's waders. The term *waders* is apt for many sandpipers, curlews, and phalaropes but is not entirely accurate for some other shorebirds, such as plovers and turnstones, usually included in this broad grouping. The waders generally have long legs and long bills, enabling them to wade in shallow water and probe the mud for their invertebrate prey. Many plovers and turnstones and some other "waders," such as stints and sanderlings, have relatively short legs and bills, and prefer foraging along the water's edge. So *shorebirds* is perhaps a better generic term for this wide range of maritime birds. There are 50 species of shorebird that breed in the Arctic tundra, and the region of the Arctic most favored by these birds is Beringia, the area at the northern tip of the Pacific Ocean surrounded by eastern Siberia, Alaska, and the Aleutian Islands. The northern coasts of Russia and Canada are also rich in shorebird species, but there are fewer species that breed in the North Atlantic region, including northeastern Canada, Greenland, and northwestern Europe. Biogeographers have analyzed possible reasons for this pattern and currently believe that it is related to the diversity of migration flyways in the Pacific region coupled with the higher productivity of vegetation in the tundra of Beringia.

There is some evidence to suggest that shorebirds with long legs (the true waders) lose more energy in the form of heat than those with short legs because their body is held higher above the ground and is more liable to lose heat when exposed to the wind. This could account for the fact that the shorebirds nesting in the High Arctic and the cold moss tundra are mainly the relatively short-legged species. Long-legged species prefer the taller vegetation of the sedge and shrub tundra at slightly lower latitudes, where their bodies receive more shelter from the penetrating tundra winds. This supports Allen's rule (see "Body Size in the Tundra," pages 113–115), which states that the size of an animal's extremities are reduced with increasing latitude and coldness of climate. The rule applies only to the legs of the shorebirds, however, not to wing ratios, which are not affected by latitude or climate.

Ornithologists have often wondered how shorebirds manage to begin laying eggs soon after arriving in the tundra despite the large amounts of energy they have to expend in getting there. Does the bird bring enough energy stored in its body fat to start laying, or is it dependent on feeding at its breeding location before it can begin breeding? Studies in the molecular biology of their feathers and eggs have supplied the answer to this question. Carbon, the fundamental element of all organic matter, comes in a number of forms, or isotopes, and these can be distinguished by mass spectroscopy. The organic matter derived from feeding in a temperate estuary differs in its ratio of carbon isotopes from that derived from tundra invertebrates. When they examined breeding tundra shorebirds, researchers found that their feathers contained organic matter from the temperate zone, but eggs contained carbon isotopes indicating a local tundra origin. The arriving migratory shorebirds need to feed and build up their reserves to enable them to begin the energetically expensive process of egg laying.

Some shorebirds rival the terns in their migration distances. The sanderling (*Calidris alba*), for example, breeds in the polar tundra of the High Arctic and moves south through North America, Europe, and Asia to winter in the Gulf of Mexico, California, and Central America. But some individuals continue into South America as far south as Chile, or down the west coast of Africa to Namibia, or down the western Pacific to Indonesia. These are very long seasonal journeys for a bird that is only eight inches (20 cm) in length. Before departing on this journey, however, the birds double their weight by accumulating fat. One of the risks of long-distance migration is predation along the way, and human predation has proved particularly intense for some shorebirds. The American golden plover (*Pluvialis dominica*) is an example of a shorebird that has been excessively hunted for its meat. In the 19th century it was shot by the wagonload in the southern states of North America, especially when the passenger pigeon became extinct, and attention shifted to this unfortunate bird. In New Orleans in a single day in the spring of 1821, 200 hunters managed to kill 48,000 golden plovers. Clearly, no bird can sustain slaughter at this level, and the species declined rapidly. It was not until it received protection in the early 20th century that its depleted numbers began to increase once more.

The golden plover is a good example of the genetic diversity of tundra organisms. There are three main species of golden plover, the American, the Pacific (*P. fulva*), and the Eurasian (*P. apricaria*) golden plovers, all of which have very similar plumages and are very difficult to separate in the field. The breeding male has a golden brown back, black face and underparts, and varying amounts of white separating these two areas of the plumage, providing the best means of identification. In addition, there is the black-bellied plover (*P. squatarola*), which is very similar but has a silvery gray back rather than a golden one. Within each species there is a great deal of genetic variation, leading some taxonomists (the scientists involved in the classification of organisms) to propose the existence of subspecies. The Eurasian golden plover population of Norway, for example, may consist of seven different subspecies according to some experts. As in the case of plants, the relatively low overall diversity of the tundra is often compensated for by the genetic variability of its inhabitants.

Many species of wildfowl (geese, duck, and swans) join the shorebirds in using the tundra as their main breeding area, and many of these also migrate considerable distances between the summer tundra and their winter quarters. The white-fronted goose (*Anser albifrons*), for example, breeds in three main populations, in Alaska and Canada, western Greenland, and northern Russia (see diagram below). In winter the North American population uses inland flyways to reach the Gulf of Mexico and Central America. The Greenland birds favor Ireland and the west coast of Britain, while the Siberian birds move to western Europe, the Mediterranean Sea, the Caspian Sea, and the east coast of Asia. The three populations are thus very different in their migratory routes and breeding grounds, so they rarely meet and interbreed. It is precisely this kind of isolation that leads to evolutionary separation and eventual speciation. Speciation is also taking place in the black brant goose (*Branta bernicla*), a small, dark goose that is circumboreal in its breeding distribution but is concentrated in northern Alaska, Canada, and Greenland. The western populations of Alaska and eastern Siberia have a dark belly and are known as *Branta bernicla nigricans.* They migrate south along the west coast of North America and spend their winters in California and Mexico. Eastern Canadian populations migrate into the Atlantic seaboard, some crossing the Greenland ice sheet at an altitude of 9,000 feet (2,800 m) and crossing into Ireland. These are considered to be the subspecies *B. b. hrota* and have a light-colored belly. The Russian population from farther west in Siberia, known as *B. b. bernicla,* moves into

Migration patterns of the white-fronted goose. This goose spends its summer months nesting in Arctic tundra in North America and Asia. In winter it migrates south to lower-latitude coastal regions. The Greenland population has a surprising migration pattern, heading east to the British Isles for the winter.

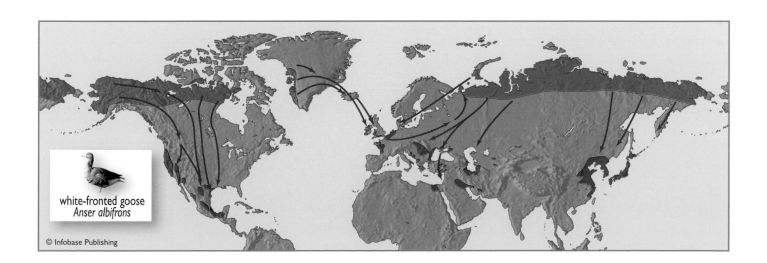

white-fronted goose
Anser albifrons

© Infobase Publishing

western Europe for the winter. Evolution here has taken a step beyond that of the white-fronted goose, and distinct subspecies are detectable.

Many ducks breed in the tundra, and some, such as the long-tailed duck (*Clangula hyemalis*), are widespread and circumboreal, whereas others, such as Steller's eider (*Polysticta stelleri*), the spectacled eider (*Somateria fischeri*), and the king eider (*S. spectabilis*) are much scarcer and restricted in their distributions. The world population of Steller's eider is only about 120,000 birds, and the spectacled eider is in even bigger trouble, with a world population of less than 6,000, most of which nest in the Yukon delta of Alaska. The king eider penetrates farther north than any of the other ducks, but, like all migrant species, it can suffer if it encounters particularly severe weather conditions on migration. The summer of 1964 was particularly cold in northern Canada, when many areas normally free from pack ice were still frozen in mid-July. As a consequence, approximately 100,000 king eider starved to death on their journey north.

The distribution and breeding ecology of some of the scarcer ducks is not well known, as is the case for the Barrow's goldeneye (*Bucephala islandica*). The western population of this duck, numbering only about 200,000 birds, breeds on inland lakes and rivers in the Rocky Mountains of Canada, but there is also an eastern population that is even smaller, consisting of about 4,500 birds, that winters on the St. Lawrence River and whose breeding range is still uncertain. Helicopter surveys have confirmed some breeding north of the treeline in northern Quebec, but some also breed on lakes in the boreal forest zone. By tagging ducks with satellite tracking equipment ornithologists have been able to confirm this more southerly breeding distribution but have also discovered that they prefer waters where fish are not present. The young ducklings feed exclusively on insects, which are much more abundant if there is no fish predation upon them. Satellite tracking has also revealed that the males have a very different agenda while the females are bringing up the family. They migrate an additional 625 miles (1,000 km) north to Baffin Island and the Hudson Bay coast, where they undergo their summer molt in isolation from all family responsibilities.

Some members of the grouse family manage to survive in the polar tundra, particularly the rock ptarmigan (*Lagopus mutus*) and the willow ptarmigan (*L. lagopus*), and their close relative the white-tailed ptarmigan (*L. leucurus*) is an alpine tundra species found in the high mountains of the west of North America, as far south as New Mexico. In Europe and Asia the rock ptarmigan is also found in the more southern mountain areas, such as the Alps and the Pyrenees, but in North America it is confined to the tundra and boreal forest regions. Unlike the terns, shorebirds, and wildfowl, these species are resi-

dent in the tundra, though they spread into more southerly regions in winter. Usually it is the females and juvenile males that move south, leaving the older males to face the northern winter. The rock ptarmigan, as its name implies, favors habitats where rocky conditions abound, especially if they form a mosaic with patches of dwarf scrub vegetation. It feeds upon the shoots of these dwarf shrubs, especially willows, birches, and crowberry, but it also eats other vegetation, including the leaves of the mountain avens in spring. Where the ranges of the rock and willow ptarmigan overlap, the rock ptarmigan tends to specialize in eating birch, and the willow ptarmigan eats willow shoots, hence its name. This is an example of niche specialization, which is a behavioral adaptation to avoid competition for limited resources. In the first few days of their lives the chicks eat mainly insects, which is a characteristic of most grouse species. The reason for this is that insect life is much richer in protein than vegetation, and this is needed by the rapidly growing chicks.

The nest of the ptarmigan is a simple scrape in the ground lined with twigs and breast feathers. The clutch size is usually six or seven eggs, and like most tundra birds, it manages to rear only one brood in the very limited season. The brown, speckled back of the ptarmigans make them very difficult to see when they are incubating, and they remain on the nest even when a predator is very close, only erupting into flight at the very last moment. Arctic foxes probably miss many nests because of the patient and daring behavior of the sitting birds. All the ptarmigans have white wings, which are not very obvious when the birds are on the ground but are striking in flight. When they land the white suddenly disappears, and they often run for a short distance over the tundra, becoming virtually invisible among the mosses and lichens. One race of the willow ptarmigan is found in the mountains of Britain, but this race, known locally as the red grouse, has brown rather than white wings and does not molt into totally white plumage in winter. This is yet another example of genetic variability in tundra birds and is advantageous in the more southern mountains, where snow cover is irregular and often incomplete.

Ptarmigans are preyed upon by arctic foxes and also by the larger predatory birds of the tundra, the gyrfalcon (*Falco rusticolus*) and the peregrine falcon (*F. peregrinus*). The gyrfalcon is the larger of the two, up to two feet (0.6 m) in length and with a wingspan of up to five feet (1.6 m). It is a heavy, powerful falcon that takes its prey in the air, combining speed with agility. The plumage of the gyrfalcon varies, with the High Arctic populations largely white and the subarctic populations gray. Those inhabiting the far north are mainly migratory and move south into the boreal zone in winter. The gyrfalcon concentrates its predatory efforts on the ptarmigans, and in the High Arctic of Canada this single prey species accounts for 95 percent of the gyrfalcon's diet. The peregrine is smaller but is still a very powerful predator, feeding upon ptarmigans, ducks, and smaller birds. It is a more widespread bird than the gyrfalcon and is found in a much greater range of habitats.

Rough-legged hawks (*Buteo lagopus*) are another important predator of the tundra. They are migratory, spending the winter in the temperate zones of North America, Europe, and Asia and moving into the tundra in the spring, nesting on the ground or on rocky ledges. They feed mainly on small mammals, especially lemmings, as does the characteristic owl of the tundra, the snowy owl (*Nyctea scandiaca*), a large, pure white owl with a length up to two feet (0.6 m) and a wingspan of 52 inches (1.3 m). Lemmings undergo very marked cycles in their populations (see "Reproduction and Dispersal in the Tundra," pages 123–129), and both rough-legged hawks and snowy owls respond to peaks in lemming populations by increasing their breeding. Usually, however, there is a gap of at least one year between the peak in the lemming population and that of the predator.

Golden eagles (*Aquila chrysaetos*) and bald eagles (*Haliaetus leucocephalus*) sometimes extend their range into the tundra, especially in Alaska and around Hudson Bay, but the Eurasian white-tailed eagle (*H. albicilla*) is more frequently found in this biome, especially around the sea cliffs of Greenland, Iceland, Norway, and the north of Russia. It is a hunter that relies upon surprise rather than speed but will also exhaust its prey by repeated attacks, especially when taking diving ducks from the surface of the sea. It feeds upon birds and fish but will also take carrion when it is available, rather like the bald eagle, to which it is closely related.

Several species of songbird, or perching bird (Order: Passeriformes), occur in the polar tundra, typical examples being the snow bunting (*Plectrophenax nivalis*) and the Lapland longspur (*Calcarius lapponicus*). Both are migratory, and the snow bunting resembles the rock ptarmigan in its combination of dark and white patches in its plumage. White underparts, white wings, and black wing tips result in a confusing series of flashes when the bird is in flight. It is a sociable species, so a small flock of snow buntings is even more disorientating as various birds take off and settle, one minute conspicuous and the next invisible, especially against a patchy background of snow and rocks. Both the snow bunting and the Lapland longspur spend their winters in the temperate zone feeding upon seeds. They move into the tundra in spring to breed, and they undoubtedly find the advantages of long day length valuable as an aid to successful breeding. Scientists have assembled information on the clutch sizes of these birds in relation to latitude. At Frobisher Bay on Baffin Island, latitude 64°N, the average clutch size for snow buntings is five and for Lapland longspurs is 4.7 eggs per nest. On Victoria Island, latitude 69°N, snow bunting average clutch size is 6.4, and Lapland longspur, 5.4. On Ellesmere Island, at latitude 80°N, the Lapland longspur is no longer able to breed, but the snow bunting produces clutches that average 6.9 eggs per nest. These data suggest that the birds are able to raise more young at higher latitudes, presumably because they have more daylight hours for feeding them during the nesting season. The young are fed entirely on insects, so this type of food must be in abundant supply even this far north. The adult Lapland longspurs and snow buntings, however, are seed eaters and carry seeds in their crops while on migration. Most of the seeds they eat are ground up and digested, but some seeds may pass through their gut and be dispersed in their droppings. When the island of Surtsey emerged from the Atlantic Ocean as a result of an undersea volcanic eruption off the coast of Iceland in 1963, it provided an ideal opportunity for ecologists to study the colonization of a freshly created tundra landscape. Some of the first land plants arrived as a result of snow buntings migrating from Europe to Greenland, some of which died of exhaustion on the new island. Their crops still contained viable seeds, and as the birds decayed the seeds germinated and established themselves on the new island. Seed-eating birds can thus be an important force in the dispersal of tundra plants.

Snow buntings have also been the focus of research into the diurnal behavior of birds during the 24-hour Arctic day. Does the pattern of waking and sleeping become completely lost when there is no environmental cue about whether it is daytime or nighttime? The answer for snow buntings is that the 24-hour pattern is maintained to a limited extent. All the snow buntings observed, whether out in the open sunshine or in the deep shade of a steep-sided valley, became somnolent at about midnight and settled for a short sleep. This is not the case with the willow ptarmigan, however. Research on the behavior of this species shows that it maintains continuous activity throughout the 24 hours, taking irregular rest breaks when needed. So different bird species evidently react to continuous daylight in different ways.

■ ALPINE TUNDRA BIRDS

Snow buntings do not occur on more southerly mountains except as winter visitors, but the mountains of Europe and Asia have a very similar bird called the snow finch (*Montifringilla nivalis*) that spends its entire life in the alpine tundra, changing its altitudinal range a little with seasons. It occurs in the Pyrenees, Alps, Caucasus, and eastward into the mountains of Tibet and China. It is very like the snow bunting in appearance, with black-tipped white wings, and it uses the same tricks to evade predators,

Golden eagles feed mainly upon rabbits, hares, grouse, and ptarmigan. They nest on inaccessible cliffs. (Harry Engels, Yellowstone Digital Slide File, Yellowstone National Park)

flashing its white patches in flight and then disappearing among rocks when it lands. It is not closely related to the snow bunting, however, and their resemblance is a good example of how evolution can come up with the same answer to a challenge on more than one occasion. This is called *convergent evolution*.

A few polar tundra birds are also found on southern mountains, such as the white-tailed ptarmigan and also the horned lark (*Eremophila alpestris*). Horned larks breed in the tundra regions of North America and Eurasia, migrating south in the winter, but there are also resident populations in North America and Asia that are nonmigratory. In North America the species has a continuous distribution through much of the continent, but in Asia there is a gap between northern and southern populations of the horned lark. The probable reason for this is the presence in Eurasia of other lark species, particularly the skylark (*Alauda arvensis*), which outcompetes the horned lark in lowland grassland. In North America there is no competitor, so there is no interruption to the geographical distribution of the horned lark. Its ability to cope with tundra conditions allows it to survive in the alpine zone of many southern mountains, such as the Sierra Nevada at Yosemite National Park in California and the Atlas Mountains of North Africa. In Yosemite it breeds at up to 12,000 feet (3,660 m) and in

the Atlas Mountains up to 11,500 feet (3,500 m). In the Himalaya Mountains of Asia it reaches 17,400 feet (5,300 m), where it encounters the snow line. The golden eagle (see photograph above) is a very widespread raptor, but is never common. It is a predator of mammals and birds in both polar and alpine tundra.

Among the game birds, ptarmigan and willow grouse are found in both polar and alpine tundra regions, but there are also some species that are restricted to the high mountains of lower latitudes. The Himalayan snowcock (*Tetraogallus tibetanus*) is found only in the high mountains of the Himalayas and Hindu Kush in southern Asia. It inhabits the bleak, rocky slopes where vegetation is sparse and can claim to be the bird most tolerant of life in very high altitudes.

Some mountain birds, such as the snowcock, are restricted to alpine tundra and do not venture into the polar areas. Vultures are among these, especially the Old World vultures, such as the griffon vulture (*Gyps fulvus*) and the lammergeyer, or bearded vulture (*Gypaetus barbatus*).

Vultures have a distinctive mode of flight depending upon upward currents of air on which they soar with motionless wings. Their flight is poor if they depend on flapping their wings, so they are restricted as a group to parts of the world where hot air creates upward thermals or where the rugged terrain of mountains leads to updrafts from valleys. The griffon vulture is a large bird of southern Europe and Asia, up to 41 inches (1.05 m) in length and with a wingspan of more than nine feet (2.8 m). It dislikes rain and snow but will soar over the mountains in fine weather looking for the carcasses of unfortunate animals that have died on the slopes. The bearded vulture is smaller and is less aggressive when competing for food at a carcass. It often has to wait until the griffon vultures have eaten their fill before it finds a place in the "front row." But it has one great advantage over the larger vultures because it can extract the marrow from the bones that are left behind. It does this by flying over rocks and scree with the bone in its beak. It then drops the bone, which shatters on the rocks below, and descends to pick out the nutritious marrow from the center of the bone. In Europe this bird is sometimes called the "bone-breaker."

It will swallow bone fragments or even whole bones, and the stomach of one bird killed in the Caucasus Mountains of Asia held the skull, legs, and horns of a chamois, the leg of a fox, and the leg of a dove. The New World equivalents of these vultures are the condors, which are among the largest of flying birds (see sidebar below).

One might not expect to find parrots in the alpine tundra, but there is one parrot that occupies this habitat in New Zealand, the kea (*Nestor notabilis*). Island habitats are famous for demonstrating how unexpected groups of animals can adapt to unlikely habitats when they are presented with no competition. The kea feeds in alpine grasslands and is relatively omnivorous, eating fruits when they are available but also being prepared to take carrion. Sheep farmers often observed groups of these parrots feeding on dead sheep, and this led them to believe that the birds were sheep slayers. It is possible that badly injured sheep or those stuck in a hole might be attacked by the parrots, but they are not likely to attack healthy animals. Their bad reputation, however, has led to severe persecution in New Zealand.

Condors

The Californian condor (*Gymnogyps californianus*) is similar in dimensions to the griffon vulture but is heavier, at 23 pounds (10.5 kg), sometimes reaching 31 pounds (14 kg). The Andean condor (*Vultur gryphus*) of South America is usually lighter but has a greater wingspan, sometimes reaching 10.5 feet (3.2 m). Fossils of an extinct condor, *Teratornis merriami* from the tar pits of La Brea, California, reveal that this ancestor of modern condors had a wingspan of 14 feet (4.3 m) and weighed about 50 pounds (23 kg). Despite their great size, condors are carrion feeders rather than predators, though the Andean condor does take eggs from seabird colonies. The Californian condor has a remarkable tongue with a deep trough running down the center and rows of backward-pointing spines. The bird is able to loosen soft flesh by working the tongue backward and forward, and the spines ensure that the flesh moves into its throat. A condor needs to eat about two pounds (1 kg) of meat each day, but in practice it will take as much as it can obtain at a carcass and then may starve for several days. The main wing feathers of the condor are highly modified to provide maximum lift for these heavy birds, especially after a meal. The primary feathers are long, curved, and stiff, serrated on both sides so that air currents can move through them and create eddies that result in the upward lift. Although condors are remarkably good at soaring, they have a very wide turning circle, so they are far from agile in the air.

Condors are gregarious birds, like many vultures, and flocks of up to 60 Andean condors are often seen together. They mate for life and will only take a new mate if the original one dies. They are unusual among birds because their sex can be determined by eye color: Males have gray eyes and females have red. They nest on remote cliff ledges and lay just a single large egg, about 10 ounces (250 grams) in weight, and they need to incubate it for more than 50 days before it hatches. The young bird remains at the nest five months before it is able to fly, and it is still dependent on the parents for food for another year before it is able to fend for itself. This means that a pair of condors can lay their single egg only every two years, leading to a very slow rate of population recruitment and leaving the species extremely vulnerable in the event of any catastrophe affecting the whole population. No doubt this, coupled with human persecution, has contributed to the very fragile state of the Californian condor and its current reliance on captive breeding and subsequent release for survival in the wild.

■ BIRDS OF
THE SOUTHERN TUNDRA

Mountain tundra has thus seen the evolution of some remarkable birds strangely equipped to cope with the rigors of this extreme environment, but the polar tundra of the Southern Hemisphere has some even more remarkable bird life, ranging from the flightless penguins to those masters of the aerial world, the albatrosses. There are 17 species of penguins in the world, all flightless, but not all are tundra inhabitants. The temperate Cape of Good Hope in South Africa has a penguin population, and one species is tropical, living on the Galapagos Islands in the Pacific Ocean. But most of the penguins are found in Antarctica and the subantarctic islands of the Southern Ocean. The smallest of the penguins is the appropriately named little penguin (*Eudyptula minor*), which is just 16 inches (40 cm) tall and weighs 2.25 pounds (1 kg). At the other end of the scale, the emperor penguin (*Aptenodytes forsteri*) is 45 inches (115 cm) tall and weighs 84 pounds (38 kg). In all species the males are generally larger than the females and have stronger flippers.

One feature that all penguins have in common, apart from their flightlessness, is their dark back and white front. The dark, often black, back does not offer any camouflage when they are at their nesting colonies on the ice, but penguins spend much of their lives at sea when the dark upperside makes them less visible when floating and the light underside makes them inconspicuous from below. The same basic pattern of coloration is found among the murres and other auks. Flight is of limited value for a fish-eating seabird that spends a great deal of time swimming, but one advantage of losing the ability to fly is that weight is no longer a problem. Penguins are able to accumulate much greater proportions of fat than other birds, providing them with an insulating layer to keep out the cold. They can remain in water close to freezing point almost indefinitely without suffering hypothermia. Indeed, they are more likely to suffer from overheating when standing in the sunshine because they have limited powers of heat dissipation. The feathers also contribute to the penguin's insulation, but like all birds they need to replace their feathers periodically because of the wear and tear upon them. The feathers are lost during a complete molt in the summer when the birds are on land at the close of the breeding season. At this time they are vulnerable to the cold and cannot return to the sea until their new feathers are fully grown, renewing their thermal insulation. Penguin feathers are very short and broad, and they interlock more effectively than most feathers, giving a tighter cover and better insulation.

Breeding takes place in colonies, often of 5,000 or more birds, which has the advantage of communal protection from predators such as gulls and jaegers. The single egg is tucked into a brood pouch, which is a layer of fat and skin above the feet that can be wrapped over the top of the egg to keep it warm. There is no need for a nest. Indeed, those species that collect on the ice sheet to breed would find little benefit in placing the sensitive egg upon the frozen floor. Emperor penguins stand upon the bare coastal flats of Antarctica for nine weeks in constant darkness and temperatures of –76°F (–60°C).

When they first hatch the chicks lack sufficient fat to keep warm, so they stand on the parent's (usually the male's) feet and shelter beneath the fatty paunch of the adult, but they gain weight very rapidly. In the case of the king penguin (*Aptenodytes patagonicus*), the chicks reach their parents' weight in just two months, but winter then arrives and prey becomes scarce. So they are fed less frequently, and they lose weight. It takes more than a year for the young king penguins to gain their final coat of feathers and take to the ocean to feed themselves, so this species is able to breed only once every two years. The emperor penguin breeds every year because the young hatch in the middle of winter, so conditions are improving as the young bird becomes more demanding. It is mature enough to look after itself by the time the next winter arrives. Only about 20 percent reach this stage. Subsequently, their chances of survival are good, and they can expect to live for 20 years. At this stage in their lives the young penguins head out to sea, and until recently their destination was unknown. Using radio transmitters scientists have now discovered that they move north into the Atlantic and then eastward following the course of the currents in the ice-free waters. Within a few weeks the tagged chicks had traveled almost 2,000 miles (3,000 km).

While incubating the egg and caring for newly hatched chick, the male emperor penguin has to fast for between four and six months and loses about half of its body weight. The hungry male is eventually relieved by the female returning from the sea to take over the care of the youngster. King penguins breed on the subantarctic islands rather than on the Antarctic continent, which is favored by emperor penguins. They also share the responsibilities of parenthood more equitably between the male and the female, both taking turns at incubating and food hunting. Their very slow breeding routine, however, leaves them open to catastrophic declines, from which they are slow to recover. For many years king penguins were harvested for their oil by hunters, and the population of this species declined drastically. They are now protected, and their numbers are increasing once more, with a population growth rate of more than 5 percent per year.

Different species of penguins have their own peculiarities. Chinstrap penguins (*Pygoscelis antarctica*), for example, named thus because of the dark line running across their throats (see p. 114), are extremely noisy in their breeding colonies and keep up a constant cacophony of calls. They also have the engaging habit of collecting attractive stones while on their way back from the sea and give these as presents to their mates who are guarding the egg or young. Gentoo penguins (*P. papua*) build nests out of stones and driftwood from the tideline. They nest on rocks and shingle rather than ice, so they are able to place their egg upon the ground. Rockhopper penguins (*Eudyptes chrysocome*), as their name implies, jump from rock to rock with both feet, which is a habit peculiar to this species.

Some penguin species show considerable genetic variation, especially the island species, such as the rockhopper, which has a number of different races. Some species are relatively common, including the macaroni penguin (*E. chrysolophus*), with a population of about 24 million birds, while others are scarce and vulnerable, such as the yellow-eyed penguin (*Megadyptes antipodes*), found only on the South Island of New Zealand.

If penguins are remarkable for their inability to take to the air, the albatrosses are famed for their unwillingness to leave the air. The wandering albatross (*Diomedea exulans*) has the largest wingspan of any bird, measuring 11.5 feet (3.5 m), and it is just one of 24 albatross species that nest on the islands of the Southern Ocean. The wings are long and very narrow, ideal for gliding over the ocean using the updraft from waves to maintain their flight with a minimum of effort. Using this energy-saving mechanism, they are able to travel considerable distances, estimated at 5,000 miles (8,000 km) in a week. They feed upon fish and hunt the Southern Ocean for food to bring back to the young at their nests. Like penguins, they breed in colonies and lay just one egg. When the egg is laid the female leaves for the open ocean, and the male is left to incubate. Her trip may take her more than 9,000 miles (15,000 km) before she eventually returns to assist her mate, allowing him to go fishing. After it is fledged, the young albatross does not breed for 10 or 11 years, so the rate of population replenishment in this bird is extremely low. This makes the albatrosses very vulnerable, and there is currently great concern about the frequency of albatrosses drowning when they take fish bait from the lines of boats and become entangled with the gear. The wandering albatross has a world population of only about 100,000, and with such a slow rate of reproduction, any increase in mortality could spell extinction. The birds of the Antarctic and subantarctic tundra are thus marine feeders that use the land only for breeding, which contrasts with the Arctic tundra and alpine tundra, where many birds are entirely terrestrial.

■ POLAR TUNDRA MAMMALS

As in the case of birds, the large mammals of the Antarctic and subantarctic tundra are entirely marine and use the land only for breeding. Spending most of their lives at sea, mammals such as seals pull themselves onto the tundra shores for the purpose of giving birth to their young.

There are five species of seals in the Antarctic, including the southern elephant seal (*Mirounga leonine*). Males of this species can grow to 15 feet (4.6 m) in length and have large inflatable noses that give the species its name. Most seals are fish eaters, but the leopard seal (*Hydrurga leptonyx*) also hunts penguins and seal pups in the ocean. Fur seals breed outside the tundra zone in the Southern Hemisphere, but northern fur seals (*Callorhinus ursinus*) extend into the Bering Sea in the North Pacific region, where they are joined by ribbon seals (*Phoca fasciata*), which are a truly ice-loving species. In the North Atlantic the harp seal (*P. groenlandica*) is one of the most characteristic seal species of the pack ice, and the females give birth to their pups during the winter. The walrus (*Odobenus rosmarus*) is confined to the Arctic and is a massive animal, the males growing to 13 feet (4 m) in length and weighing more than 4,400 pounds (2,000 kg). It is distinguished from the seals by its tusks, which are used for bottom feeding in the ocean, where they assist the walrus in digging for clams.

On the whole these marine mammals contribute little to the tundra ecosystem beyond the local deposition of feces in their breeding colonies. The exception is found in the Arctic, where seals are the main prey of the tundra's largest predator, the polar bear (*Ursus maritimus*). It grows up to nine feet (2.8 m) in length, can weigh 1,600 pounds (720 kg), and is unusual among bears in being totally carnivorous. It usually catches seals when they are basking on the ice or when they come up for air at their breathing holes. The polar bear has very thick fur that protects it from cold (see sidebar "Hair," page 105). It can spend many hours swimming in the icy polar waters without suffering any harm, but it is cumbersome in the water and no match for seals in their native element. Polar bear fur is so efficient at heat retention that ice can form on the outer hair, and so little heat is lost from the bear's body that it often fails to melt. Even the soles of the feet are covered with hair, preventing frost bite as it pads across the ice. The only predators of polar bears are humans, and the Inuit people of the Arctic tundra hunt and eat them but do not consume the liver because it is so rich in vitamin A that it can prove toxic, causing a condition called hypervitaminosis A. Affected humans feel drowsy, develop headaches, vomit, and then suffer from peeling of the skin. The livers of walruses and seals are also extremely rich in vitamin A and are not eaten by hunters.

Polar bears spend most of their lives close to the sea or even out on the floating ice. There are risks involved in living on the ice, for it can break up into drifting sections and move away from land, and some polar bears have been recorded on ice floes as far as 200 miles (320 km) from land. But their powerful swimming ability enables them to move from one raft to another and ensures their survival. They seem to be incessantly on the move, either in search of prey or, in the case of males in the spring, in search of a mate. It is at this time of year that the wandering males are in their most irritable and dangerous moods and may attack people.

The female polar bear hibernates through the long dark winter. It burrows beneath the snow, which insulates it from the worst of the cold conditions, and there it lives off its fat until the arrival of spring. Females that have become pregnant as a result of mating in the spring give birth to the tiny, rat-sized cubs inside the winter den. Although the polar bear is among the largest of the four-legged carnivores of the world, its babies are among the smallest in relation to the mother's size. Most commonly twins are born in December or January. Here they suckle on fat-rich milk until they emerge in the spring, and the female at this stage is in a light state of dormancy and wakes easily if danger threatens. When they emerge the young are weaned onto a diet of meat, usually arctic hare rather than seal in their first year. The cubs stay with the mother for the whole of the following summer, so that she mates only every other year. Males as

well as females with growing cubs tend not to build snow dens but remain active through the winter.

While polar bears stay in place through the winter and endure the hardships, another large mammal of the tundra, the caribou, escapes the worst of winter conditions by moving south to areas of better food supply. The caribou, or reindeer as it is called in Europe and Asia (*Rangifer tarandus*), is the only deer that survives north of the Arctic treeline throughout its life. Although there is just one species of caribou, it is widely distributed throughout the polar tundra of the north and is divided into several subspecies that have different ways of life. In Siberia, for example, and also in western North America, there are subspecies that spend their lives in the forest rather than out on the tundra, but the most common forms of this animal graze on the open tundra vegetation. The North American race is generally larger than the Eurasian reindeer, a bull often growing to a height of more than four feet (1.2 m) at the shoulder and weighing up to 600 pounds (270 kg). The cows are smaller. Once again, this is an example of genetic diversity in a tun-

The range of caribou in the tundra of North America. The species is distributed in a number of herds, each of which has its own calving grounds that it occupies in the spring, migrating south in the winter.

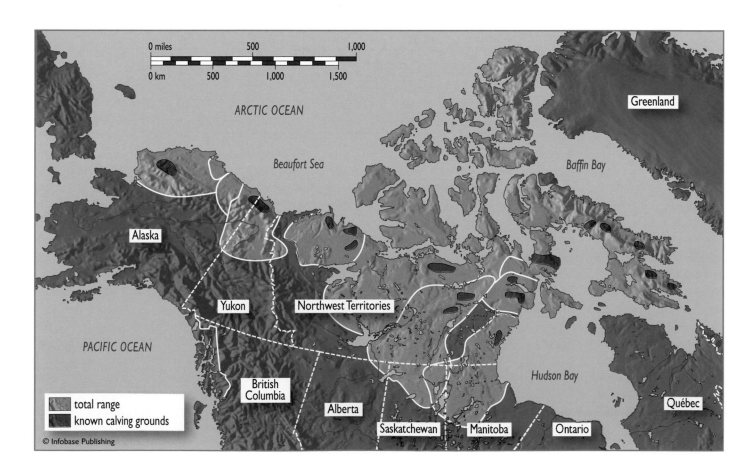

ARCTIC OCEAN

Greenland

Beaufort Sea

Baffin Bay

Alaska

Yukon   Northwest Territories

PACIFIC OCEAN

British Columbia

Hudson Bay

Alberta

Québec

Saskatchewan   Manitoba   Ontario

total range
known calving grounds

© Infobase Publishing

Caribou have broad hooves that enable them to run over soft tundra wetlands without sinking into the mud.   *(NOAA)*

dra organism. This is partly due to the isolation of different populations, and North America's caribou separate into herds, each of which has its own preferred calving grounds, as shown in the diagram on page 166.

Like the polar bear, caribou are covered in dense hair that is hollow and provides extra thermal insulation. These air-filled hairs also enable the caribou to float very efficiently, and they swim strongly with about a third of their body above the water. In summer the caribou are brown or gray in color, but they become paler, almost white, in winter, especially on some of the Arctic islands. The feet have spongy footpads that effectively spread the weight of the animal and allow it to move over areas of soft, peaty soils without sinking into the mire, as shown in the photograph above. In winter the pads harden and develop bunches of hair, which serve both to insulate the feet from frost and to give additional grip on the slippery ice and snow. With these adaptations the caribou can run rapidly, achieving speeds of about 50 mph (80 km/h), and can escape such predators as wolves in this way. But a caribou is a sprinter and cannot keep up such speed for very long, so it may fall prey to wolves if these animals persist in their pursuit until it weakens. Migrating herds of caribou are often accompanied by wolf packs that pick off the old, the fragile, and the young as their victims.

Both male and female caribou have antlers, but those of the male are usually larger. In the North American race the antlers grow up to five feet (1.5 m) in length. Males fight one another with their antlers during the rutting (breeding) season in the fall, as shown in the picture on the right, and ant-

lers also provide a means of defense against predators, such as wolves and wolverines. After the rut (by about December) the males shed their antlers, but the females retain theirs through the winter, shedding them in April when the calves are born. The birth of calves usually takes place within specific calving areas, where the females assemble in spring, as shown in the map on page 166. During winter the females are able to assume a degree of dominance over the males because they have retained their antlers, so they can gain better access to the richer vegetation when they are pregnant and need additional nutrients.

Caribou are very particular about their food. Although they will eat the young leaves of willow and birch, they prefer the new growth of grasses and sedges. They also eat lichens, especially those of the genus *Cladonia*, the so-called reindeer mosses. These small, slow-growing organisms are dry and crisp and do not appear very attractive as food, but they are actually quite nutritious. Their very slow growth, about one inch (2.4 cm) every five years, means that a grazing herd of caribou can remove many years of growth in a very short time. Consequently, the caribou have to keep moving over the tundra to seek out new resources while the grazed areas slowly recover (see "Grazing in the Tundra," pages 129–130). The outcome is that the tundra can support only low densities of such grazers; a density of one caribou per 150 acres (60 ha) is about as high as can be maintained.

A large proportion of the world's caribou live in Russia, where almost half of the 3 million or so animals

During the rut male caribou enter into fierce contests, using their antlers to test their strength against rival bulls. The strongest gain the right to breed.   *(Eighth World Wilderness Congress)*

are semidomesticated. Pastoral peoples follow the herds on their migrations, protect them from predators, especially in the calving season, and harvest them as a source of milk, meat, and skins. It is likely that caribou were once very much more abundant than they are now, and one estimate indicates that there were once more than 3 million caribou between Hudson Bay and the Mackenzie River. Human predation has undoubtedly contributed to their decline, but they also undergo strong population fluctuations. Many researchers have investigated what causes caribou populations to rise and fall every 60 years or so. Natural and human predation, together with climate and food supply, influence population levels and may lie at the root of the regular cycles.

The musk ox (*Ovibos moschatus*) is another large grazer of the tundra, and this animal ranks as one of the most highly adapted of Arctic mammals. It is totally confined to the tundra, particularly the polar desert regions. Hunting has greatly reduced populations of musk oxen, driving the species to extinction in Alaska by 1860, and these animals have also been eliminated from the Russian Arctic apart from one island where they has been reintroduced from North America. Musk oxen are now found largely in the Canadian Arctic and Greenland. Actually more closely related to sheep and goats than to cattle, they are stocky animals, standing about five feet (1.5 m) in height and weighing up to 670 pounds (300 kg). The musk ox grows a coat of dense shaggy hair with an undercoat of very soft hair that some Native American tribes use in the construction of cloth. It bears massive horns that curve downward on either side of the head. The musk ox owes its name to its strong smell, which is not, in fact, due to musk but to its fetid urine. Like caribou, musk oxen are herbivores. They prefer the woody tissues of dwarf shrubs, although they also eat grasses and sedges during the spring growth.

Musk oxen are not migratory but wander around in small groups of about 20 to as many as 100 animals, consisting of both males and females. When they are threatened by predators, they form a defensive circle with their heads and formidable horns pointing outward, keeping any young animals safely within the circle. It would take a brave or foolhardy predator to attempt to penetrate the wall of horns that meets them.

Of all the animals of the Arctic tundra, few have generated more interest or legends than the lemming. This small rodent has a reputation for mass suicide; folk tradition asserts that lemmings leap from cliffs in hundreds during episodes of population explosion. There are several different species of lemmings, all of them small rodents closely related to voles that survive under very extreme conditions of cold. The one that has generated most of the fabulous stories of suicide is the Norwegian lemming (*Lemmus lemmus*), which occupies the mountainous tundra of Scandinavia (see photograph on page 101). The populations of the Norwegian lemming undergo considerable fluctuations, and there are years in which large numbers of the animals migrate in massive numbers, usually down the valleys from their mountain homes. When they reach a water barrier, Norwegian lemmings are often undecided about what to do because although they swim well, they are reluctant to take to water unless they can clearly see the other side. The pressure of the crowd at such times forces many individuals off the edge of cliffs or riverbanks and into the water, where they drown, but this cannot be regarded as mass suicide, as was once believed. The discovery of large numbers of drowned lemmings washed up on the shores of fjords is probably what inspired the mass suicide myth.

Exactly what causes the population explosions among lemmings is still not entirely clear (see "Productivity and Population Cycles in the Tundra," pages 101–102). There is an approximate three- to four-year cycle in the population fluctuations. When the population is at its peak size, there may be as many as 160 lemmings on each acre of tundra (400 ha^{-1}). The buildup in population probably results from an abundance of high-quality food. During low lemming years the grazing pressure is much diminished, and vegetation grows more rapidly. The growing vegetation then becomes richer in certain essential elements that are in short supply, such as phosphorus. The outcome is a lemming plague that consumes most of the available resources, so the population crashes once again as quality food becomes scarce. During the stage when food resources are beginning to run low, the lemmings migrate outward from their centers of population, giving rise to the myths associated with them. In peak population years the lemming is a major source of food to a wide range of predators, including snowy owls, arctic foxes, rough-legged hawks, and parasitic and long-tailed jaegers (*Stercorarius longicaudus*). These predators also undergo population cycles that match the lemming cycles, illustrating how dependent they are upon this source of food. Snowy owls, for example, may manage to raise 10 or more young in a good lemming year but may fail to rear any when lemming populations are low. It is unlikely that the intensity of predation actually causes the crash in lemming populations, however. Even heavy predation is likely to account for only about 10 percent of the lemming population, which is hardly enough to cause a population crash; food limitation is much more likely to be the cause. A drop in the availability of food affects the rate of reproduction, which then has a rapid impact on the overall population of the lemmings.

Like other small mammals, lemmings are capable of very rapid breeding. Unlike many of the other small rodents of the tundra (such as voles), however, they have the remarkable ability to breed all year round. In one study of brown lemmings (*Lemmus sibiricus*), 33 percent of females were found to be pregnant in the December to April period and about 80 percent in July. The Arctic lemming (*Dicrostonyx torquatus*) has even higher rates of breeding, with 40 percent of females pregnant in winter and 90 percent in July.

In the Arctic summer, of course, there is no opportunity for nocturnal activity, which in temperate animals is often a behavior pattern adopted to avoid the bird predators that mainly hunt by daylight. As the summer wears on, however, the lemmings tend to confine their activity to the hours of darkness to escape the attention of predatory birds and foxes. Lemmings are herbivores, and the different species have varying food preferences, which is why they are able to coexist in the same area. The arctic lemming, for example, prefers broad-leaved herbs and dwarf willows, and it eats very little in the way of grasses, sedges, mosses, and lichens. The brown lemming, on the other hand, concentrates on these latter plants and takes very little of the broad-leaved and woody plants. In winter lemmings burrow beneath the snow, and the breeding females construct well-insulated nests of sedge stalks to protect themselves against the penetrating cold.

The arctic hare (*Lepus arcticus*) is a larger tundra grazer and has the most northerly distribution of any of the world's hares and rabbits (lagomorphs). It occurs in the tundra regions between Labrador and the Mackenzie River, being replaced in Alaska by the Alaskan, or mountain, hare (*L. othus*). The arctic hare is one of the largest of the hares, weighing around 12 pounds (5.5 kg). It is quite fussy about its food, concentrating upon the arctic willow (*Salix arctica*), which means that its distribution is limited by this particular food species. It is quite unusual in such a harsh environment for a large herbivore to be so strictly limited in its food preferences, but it seems to thrive despite this specialized diet. The arctic hare remains active through the winter and therefore has to cope with extreme cold in the months of darkness, often with monthly average temperatures of –35°F (–38°C). Unlike the lemming, which burrows below the snow, it has to survive without any insulation apart from its thick fur. Its breeding, however, is restricted to the summer, and it produces a litter of five or six young in a scraped hollow, often in the shelter of a rock. The female arctic hare is a more devoted mother than is the case with most lagomorphs. She remains with the young for as long as two or three weeks, then they form small groups, or crèches, and the mothers return regularly to nurse their own offspring. After about nine weeks they are left to fend for themselves. Adult arctic hares remain quite solitary animals during the summer, but in the fall they form large herds. Flocks numbering into the hundreds have been recorded on Ellesmere Island, and this may offer them some protection against such predators as the arctic foxes, which find it more difficult to stalk a grazing flock without being detected. The hare herds can cause intensive grazing pressures on the tundra vegetation and may outcompete some of their fellow grazers, such as musk oxen and caribou, especially in years of high populations. Like lemmings, they exhibit a cyclic pattern of population variation, with peak populations roughly every nine years.

The Alaskan hare is closely related to the arctic hare and is found throughout northern Asia and as far west as Europe. It is the only hare found in Ireland. The Alaskan hare is unusual among tundra mammals in being found also in the more southerly mountains of Europe, including the Alps, so it is a truly arctic-alpine species. It is sometimes called the blue hare because its summer coat is a blue-gray color, but in winter it molts and produces a white coat with just the tips of the ears remaining black. It has a wider range of food preferences than the arctic hare, feeding on grasses and heathers. The change in coat color is of obvious advantage in habitats that are dark in summer and white in winter, but the timing of the molt is crucial in determining how effectively this camouflage defends the animal against predators. A combination of day length and temperature determines the onset of the molt, but if the timing is wrong because of early snow in fall, for example, then the outcome can be fatal. White hares on the dark surface of the snowless tundra or gray hares on the surface of the snow are easily spotted by the foxes and eagles that prey upon them.

One group of mammals that may not immediately spring to mind in the context of the tundra is bats (Vespertilionidae). It is certainly true that bats are scarcer and less diverse in the tundra regions than they are in the temperate and tropical regions, but they do occur as seasonal migrants, taking advantage of the great abundance of insects found in the Arctic summer. Bats are largely nocturnal insectivores, flying and hunting at night and thus being able to avoid the effects of daytime predation by such birds as hawks and falcons. By confining their activity to dusk and nighttime, they also avoid direct competition with insect-eating birds, such as swallows and swifts. In the Arctic, of course, bats face the problem of the absence of any real night during the summer. Studies on bat activity times in northern Norway have shown that they maintain a period of hunting that corresponds with their perceived "night," generally between 10 P.M. and 2 A.M. During that period the activity of bird insectivores is much lower, so the bats and the birds seem to divide the available time between them despite the absence of darkness as a cue. This division is not ideal for bats, however. Insect activity is lower at "night," so the bats have the poorest slot for their insect hunting and yet are still exposed to attack by bird predators. It is not surprising that bats are scarce in the tundra.

■ ALPINE TUNDRA MAMMALS

One of the most typical of mountain mammals is the marmot. There are, in fact, many species of marmot, and they are widely distributed around the mountains of the world, including the Rockies, the Alps, and Siberia. They are not large mammals, generally about two feet (0.6 m) in length, although the hoary marmot (*Marmota caligata*) of western Canada can grow to almost three feet (0.9 m). What they lack in length, they make up in girth, being rotund animals with a stocky, rounded build. Marmots are closely related to squirrels and, like them, are herbivorous in diet, spending the summer grazing on the herbage of the alpine tundra above the treeline. They enjoy well-drained, sloping grassland, often with rocky outcrops and scree, where they sit and watch the skies for predatory eagles and give warnings to their grazing colleagues by emitting loud and penetrating whistles when they see danger threatening.

Marmots are largely social and cooperative mammals, as shown in the photograph above. Living in extended family groups, they defend their local territory, chasing off individ-

A pair of yellow-bellied marmots (*Marmota flaviventris*). These alpine tundra mammals spend most of the year hibernating and need to consume a large volume of vegetation in the brief summer to build up fat for the long winter. *(Carolyn McKendry)*

uals that seek to graze on their territories. Males also wrestle with one another during the breeding season, standing on their back legs and pushing one another with their forelegs. They live in burrows or rock crevices beneath the ground, and they have short, strong limbs for burrowing in the rocky soils. Here they hibernate through the long cold winter. Once the spring arrives and the melting snow reveals green plants, the marmots emerge, bask in the sun, and clean themselves. They are active in the daytime and are relatively unafraid of people, so they are easy to observe in their activities. One of their first activities on emerging in the spring is to clear out their bedding and replace it with clean materials. When a marmot tries to carry too much new bedding material, it may end up rolling on its back holding a mass of bedding with all four legs. When this happens its companions drag the marmot back by the tail into its burrow. This behavior was first recorded by the Roman naturalist Pliny (23 to 79 C.E.) from his observations in the European Alps, and, like some of Pliny's other tall tales, it was once regarded as suspect. But scientists have observed this taking place in recent times, so Pliny's unlikely sounding story has been confirmed. When their winter bedding has been replaced, marmots then turn their minds to breeding, and the first young are born within about six weeks of mating. The family stays together for several years before the young are sufficiently mature to set off on their own.

In the short season of the high mountain tundra, marmots must spend much of their time feeding, especially as the winter approaches. Marmots do not store food in their burrows but build up large reserves of fat to keep them alive through the hibernation. This together with their dense fur also serves as an insulating layer to retain some body warmth, and human hunters prize both the fur and the fat. The black-capped marmot (*Marmota camtschatica*) of eastern Russia is one of the most proficient sleepers in the animal kingdom. These animals can spend up to nine months of every year asleep. By mid-September they are covered with a thick layer of fat and weigh about 11 pounds (5 kg). They descend into their burrows along with the other members of their families and curl into tight balls with their forepaws covering the sides of their heads. As the external temperature drops, so does the body temperature of the marmot, often becoming as low as 40°F (4°C) and sometimes even below freezing. At

this stage it is very difficult to detect any sign of life in the animal. Its heart beats just once every two or three minutes, and it breathes only once every 10 minutes. Even this very low level of body activity, however, does generate some waste materials, so the marmot warms up and wakes about every three weeks so that it can urinate. By the time it finally wakes in May, it has lost half its body weight, so it is easy to appreciate their eagerness to emerge and start eating.

On the whole the alpine tundra habitats of the world's mountains contain a different set of mammals from those of the polar tundra. Very few mammals (one of these is the blue hare) can be found in both habitats. The marmots, as we have already seen, are a group of mammals that are essentially alpine tundra rather than polar tundra in distribution. The same can be said of the sheep and goats; many of these are mammals of the mountains.

The American mountain goat (*Oreamnos americanus*) is not actually a true goat but belongs somewhere between goats and the antelopes. It occurs above the treeline in the mountains of Alaska, British Columbia, and the Rocky Mountains south into Colorado and South Dakota. Like many of its relatives, its hoof is adapted for gripping, hav-

Bighorn sheep are important grazers in many areas of alpine tundra in North America. This flock resides in Yellowstone National Park.   *(Yellowstone Digital Slide File, Yellowstone National Park)*

ing a hard outer rim and a softer inner part that can grip onto slippery surfaces. It lives on the high, steep, and inaccessible cliffs of the mountains, producing young in the spring that are able to climb on the precipices virtually from birth. The kids may fall prey to golden eagles, but their greatest danger is death by misadventure. A slip on a mountain ledge or a loose rock can easily lead to a fatal fall. The chamois (*Rupicapra rupicapra*) of the European and western Asian mountains is a close relative of the American mountain goat and has many of the same adaptations. Like the mountain goat, it is polygamous, and the males compete for harems of females in the November rutting season. The chamois has been extensively hunted in Europe but is now widely protected and is becoming a common animal of the Alps and Pyrenees. It can live up to 22 years and is capable of survival at very high altitudes, the highest record being 15,430 feet (4,750 m) on Mont Blanc, Europe's highest mountain.

Another agile, rock-climbing mammal of the American mountains is the bighorn sheep (*Ovis canadensis*), shown in the photograph on page 171. It occupies the more remote areas of the Rockies and the Sierra Nevadas, avoiding contact with human beings. It lives in groups consisting of up to 15 animals, mostly females and young in summer, and congregating into herds of up to 100 for the winter. The males join the female groups during the rut, when battles

for access to females reach epic proportions. Competing males charge one another head-on, and the crash of their colliding horns can be heard a mile away. The heads of the males have a very dense cover of bone at the front so a direct clash rarely causes serious injury. These battles may last many hours until the fitter male wins the day and has the privilege of passing on his genes to the next generation. The lambs are susceptible to predation, particularly from golden eagles and mountain lions. Wolves, coyotes, bears, bobcats, and lynx can also pose a threat to them if they wander from their precipitous habitats into lower parts of valleys. This animal tends to migrate from the higher regions that it occupies in the summer, feeding on grasses and sedges, to the lower parts of the mountain in winter, where woody materials, including willow, form its major food. Like the chamois, it has been intensively hunted in the past, but such activities are now carefully controlled to ensure its long-term survival. Trophy hunters concen-

A wolf running through the snow. Following centuries of persecution, conservationists favor programs of reintroduction of this predator into some mountain areas of North America and Europe, but the consequences for pastoral farmers is putting a brake on such policies. *(Barry O'Neill, Yellowstone Digital Slide File, Yellowstone National Park)*

The snow leopard (*Uncia uncia*) is the top predator of the alpine tundra of the Himalaya and Hindu Kush Mountains in Asia. It is a nocturnal hunter and feeds on wild sheep and other mountain mammals. *(Kenneth Fink, Yellowstone Digital Slide File, Yellowstone National Park)*

trate their attentions on very large males, claiming that the removal of males from a population has less overall impact than killing females, but geneticists have recently shown that by killing the very best males, the genetic quality of the entire population eventually becomes impoverished.

There are several species of mountain sheep in Asia, but on the whole these are animals of rolling hills rather than steep cliffs and rocky crags. The nearest equivalent to the bighorn is the snow sheep (*Ovis nivicola*) of eastern Asian mountains, which extends into the northern highlands beyond the Arctic Circle. In most of the Eurasian mountains, however, the ecological equivalent to the bighorn sheep is a type of goat, the ibex (*Capra hircus*). This is a large goat, up to three feet (0.9 m) in height, and is notable for its conspicuous, backwardly curved horns, often around 30 inches (7.5 m) in length. The horns are black and strongly ridged, and, as with bighorn sheep, they are used in battles between males during the rutting season. Human hunters almost drove the Alpine populations of the ibex to extinction, partly for the trophies supplied by its enormous horns but also for the so-called bezoar stone. The animal's stomach contains this "stone," which is a hard concretion of undigested material. People believed that the bezoar stone possessed great potential for healing, including a cure for

various types of poisoning, and slaughtered the animals to obtain it. Fortunately, when the ibex population had been reduced to just a small herd in northern Italy, authorities imposed protection, and reintroduction over its former range is now proceeding.

The alpine grazers provide a source of food for a number of predators. Wolves, as shown in the photograph on page 172, are highly adaptable predators found in all habitats from subtropical deserts to alpine and polar snows. Hunting in packs, they are capable of overpowering animals far larger than themselves. Other predators hunt singly, including the snow leopard, which is found in the high mountains of southern Asia. Like all leopards, this animal is mainly a nocturnal hunter, stalking and killing wild sheep and other grazers of the high mountains. With eagles they are the top predators of the alpine tundra.

■ CONCLUSIONS

Biodiversity encompasses the range of different species in an area together with the genetic variation within each species and the variety of habitats available. The tundra is poor in species compared to other habitats, probably because of its relatively low annual productivity. On a global scale biodiversity often relates closely to productivity of different biomes. Although low in species diversity, however, the tundra contains many species found in no other biome, so the region contributes substantially to global biodiversity. The species contained in the tundra also exhibit a high degree of internal genetic variability, and many species of plants, insects, birds, and mammals have a variety of genetic races contained within them.

Bacteria and fungi are present in abundance in tundra soils. They disperse very efficiently, having small, airborne spores, and they are responsible for the decomposition of organic matter. Some, such as the cyanobacteria, are photosynthetic and also fix atmospheric nitrogen, thereby contribute to nitrogen cycling in the tundra. Lichens are combinations of fungi with unicelluar algae or cyanobacteria, and they are able to cope with drought as well as intense cold, so they form an important component of tundra ecosystems.

Mosses and liverworts (bryophytes) are present in most tundra habitats, including patches of late snow, where few other plants can survive. Although apparently fragile in structure, bryophytes are resilient and, like lichens, cope well in situations of extreme cold and drought. In tundra wetlands the bog mosses, *Sphagnum* species, play an extremely important part, covering large areas of the tundra landscape.

The soil also contains a wide range of microscopic and larger protists and animals, ranging from amoebas to earthworms. These are active mainly in summer and occupy the thin zone between the surface and the underlying permafrost. Some of these organisms are detritivores, feeding on the dead remains of plants and animals falling upon the soil surface, and others are carnivores, preying upon the populations of protists and small animals that make up the detritivores' community.

Vascular plants are very scarce on the continent of Antarctica but are frequent in the Arctic and in alpine tundra. Many have a cushion (chamaephyte) form, such as the very successful saxifrages, which are found in both polar and alpine tundra. Grasses and sedges avoid the rigors of winter by dying back to ground level, while dwarf shrubs overwinter close to the ground surface. Pollinators are relatively scarce, so there is much competition for the attentions of insect visitors in summer. As a result many tundra plants have large flowers, frequently with parabolic shapes and solar tracking movements, which focus the warm sunlight on the center of the flower and encourage visits from insects.

Tundra arthropods include some detritivores, such as springtails and mites, and also large numbers of blood-feeding insects, such as mosquitoes and midges, the females of which need a blood meal before laying their eggs in the tundra wetlands. Beetles are present, but the species tend to be small and so prove an exception to Bergmann's rule. Butterflies are abundant for the short summer season, especially in alpine tundra. Some species are migratory and others, especially in alpine habitats, have lost the power of flight, which avoids the problem of being blown away from mountain peaks. Like many insects, butterflies together with spiders bask in sheltered spots to enhance their energy supply.

There are very few amphibians that extend their range into the tundra, but birds are plentiful. Seabirds use the polar tundra coasts for breeding, including loons, gulls, terns, and jaegers. The loons, gulls, and terns live by fishing, but the jaegers are pirates, harassing other birds and stealing the food they have gathered. Some seabirds, such as the arctic tern, migrate over great distances in the course of the year, traveling to the Southern Ocean and back. Shorebirds, including waders and plovers, also use the tundra for breeding, feeding in both freshwater and coastal wetlands and using the long days for food gathering to provide for their young. Wildfowl nest in large numbers in the polar tundra, including swans, geese, and ducks. Wildfowl, like most seabirds and shorebirds, are migratory, many tundra breeders wintering in the Gulf of Mexico or Central America. Ptarmigans are resident in the tundra, and they change their plumage according to the season to maintain their camouflage in dark shrub cover of summer and the snow cover of winter. They are preyed upon by gyr falcons, which are the largest and most powerful of the tundra falcons. The rough-legged hawk and the snowy owl are mainly predators of small mammals, such as lemmings. Small perching birds also inhabit the tundra in summer, including snow buntings and Lapland longspurs, which are seed-eaters but which feed their young on insects.

Alpine tundra is not in general attractive to seabirds, shorebirds, or wildfowl, but small birds, including snow finches are found there together with large predatory birds. Vultures are present in many Eurasian mountain regions, feeding on the carrion of large grazing mammals. The lammergeyer extracts marrow from bones by dropping them on rocks. In the New World condors are the equivalent of the Old World montane vultures and are the largest of all flying birds.

The Antarctic tundra is particularly important for its populations of penguins and albatrosses. Penguins are flightless, spend most of their lives at sea, and live on a diet of fish or plankton, but they come to the tundra to breed. Large colonies are found both on the mainland of Antarctica and on the subantarctic islands. They lay only one egg, so population growth is slow and the birds are vulnerable to catastrophe. The same is true for the albatrosses, which also use the tundra of the subantarctic islands for nesting only. Unlike penguins, they are excellent fliers and can travel thousands of miles over the oceans in a week on the wing.

Some mammals of the tundra are essentially marine creatures and, like many seabirds, use the tundra only for breeding, among which are the seals and the walruses. These are the main prey of the largest tundra predator, the polar bear, which is entirely carnivorous and hunts seals on the ice. It is a permanent resident of the tundra, unlike caribou, some of which migrate south in winter away from their tundra breeding sites. These large, communal herbivores consume a great deal of vegetation, including lichens, and need to move around to avoid overexploitation of the low productivity of the tundra. Musk oxen are heavier animals and are residents of the tundra, living in smaller herds.

Small mammals abound in the tundra, including hares, voles, and lemmings. The latter undergo population cycles related to the availability of food supplies. They live above ground but below the dwarf shrub vegetation and the cover of snow during the winter. Bats migrate into the Arctic tundra to hunt insects in the summer, but in the absence of darkness at night they have to hunt in the light. They are mainly active in the time slot when they would normally expect darkness, however, while insectivorous birds, together with the insects themselves, are more active in what they perceive as daytime. Bats therefore hunt at a time of day when insects are less abundant but predatory birds are a threat. Relatively few bat species are found in the tundra.

Mountain mammals include the marmot, which lives in burrows and hibernates during the winter. It may spend more than half the year in a state of dormancy and low metabolic activity. Wild sheep and goats are found in many mountain regions and are adapted to a precarious life of grazing on steep and dangerous cliffs and mountain ledges.

Tundra organisms show a wide range of adaptations to an extreme environment, and many of the permanent residents are so specialized that they are unable to live in any other biome. Many plants are found in both polar and alpine tundra and are called arctic-alpines. A few birds are also found in both types of tundra habitat, but very few mammals have this type of distribution. Mountain mammals tend to be restricted to alpine habitats and are not found in the polar tundra regions.

7

Geological and Biological History of the Tundra

The tundra is a biome that develops only in conditions of extreme climatic coldness, so cold that trees cannot survive and all vegetation lies close to the surface of the ground. Geologists now assert that the Earth has very recently emerged from an ice age, so the presence of tundra on the planet is not surprising, but how long has this biome been present? Has the Earth always supported lands of ice and snow at its poles and on the high mountaintops? The record of the rocks suggests that this is far from the truth and that the presence of cold climatic zones upon the Earth is very much the exception rather than the rule. In order to understand the tundra, it is necessary to look back on the history of the Earth to detect when conditions were right for the development of this bleak biome.

■ GLACIAL HISTORY OF THE EARTH

The word *history* is often reserved for the documented record of human life that is available only for those recent millennia in which writing has been available and written records created. Before that time, sometimes called prehistory, any investigations of events or conditions must be based upon indirect evidence, such as archaeological remains or changes in the fossil contents of sediments and rocks. It is precisely this type of approach that must be applied to the study of ancient climates and the past ice ages of the Earth.

When the climate of the Earth is cold the polar regions and the higher parts of mountains accumulate permanent snow that packs down to ice over the course of years (see "Ice Caps and Glaciers," pages 40–44). Ice advance and retreat leaves behind records of its movement, such as tills, erratics, and features associated with glacial erosion (see

"Glacial Debris and Its Deposition," pages 44–49). These clues to past glacial activities can remain in a kind of "fossil" state for many millions of years, and geologists have used this type of evidence to reconstruct the climatic history of the Earth and to detect those times when the planet entered an ice age.

When the Earth condensed from a rotating cloud of interstellar gas and dust about 4.6 billion years ago, it was exceedingly hot. The heat was generated by the kinetic energy of the colliding particles as they condensed and also by the radioactive decay of such elements as potassium, uranium, and thorium in the dust cloud. This radioactive heat production continues to the present day and contributes significantly to the current warmth of the Earth; it is not just solar radiation that keeps the planet warm. As the surface of the Earth cooled, rocks were formed, and the first primitive forms of life came into existence. The oldest rocks on Earth date back about 3.8 billion years, and some of these very ancient rocks are currently found within the tundra regions of the world (see "Polar Rocks," pages 35–36). By about 2.3 billion years ago, still long before any trace of organic life above the organizational level of bacteria, there are the first traces of glaciation, indicating that the cooling of the Earth had proceeded so far that permanent ice was present on the planet (see diagram on page 177). Several other ice ages occurred late in the Proterozoic era, prior to 542 million years ago, but separating these from one another and dating them precisely is difficult. Such cold times did not give rise to tundra in the mod-

(opposite page) Changes in the mean global temperature of the Earth over the course of geological history. Occasions when ice caps have been present on Earth are relatively few. The recent Ice Age, therefore, is an unusual state for the world's climate.

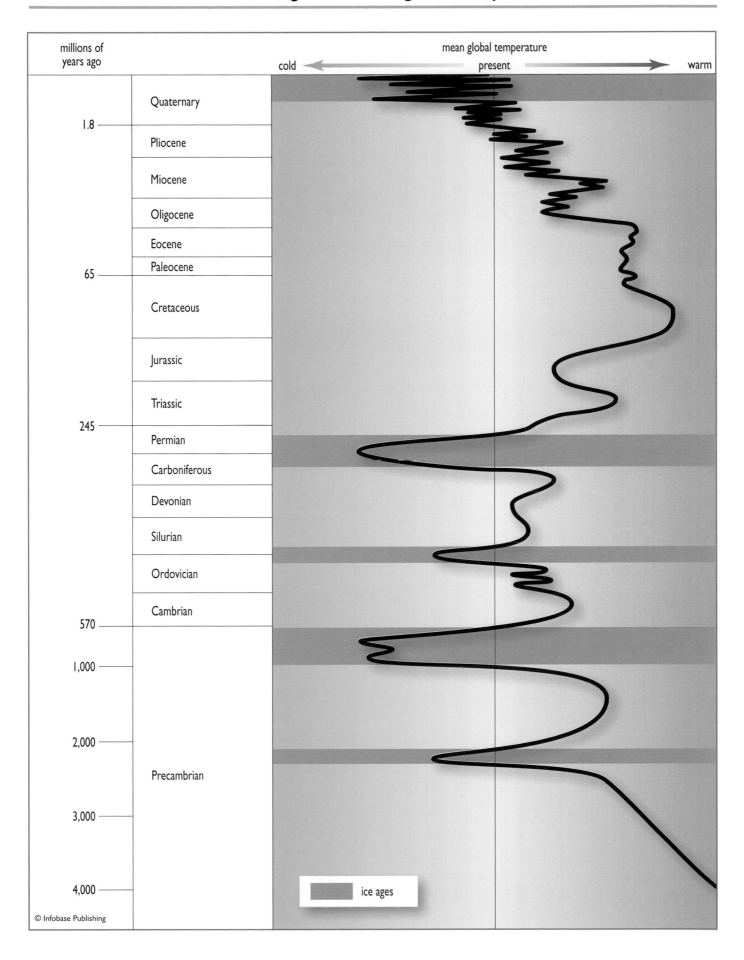

ern sense because there were no plants or animals present on Earth.

The next cold period in the history of the Earth arrived in late Ordovician times, about 450 million years ago. Land plants had first appeared in the rocks by this time, but there is not sufficient evidence available to trace any possible impact of glaciation upon the plant cover because most of the plant fossils from this period are located in the tropical latitudes. The next time of major glaciation in the Earth's history occurred at the end of the Carboniferous period. This was a time when the carbon dioxide levels in the Earth's atmosphere were rapidly falling, probably as a result of the deposition of large amounts of lime in the shallow oceans and the accumulation of the organic matter that would one day become coal in tropical swamps. By the end of the Carboniferous period, about 300 million years ago, atmospheric carbon dioxide reached an all-time low, and the atmosphere had lost much of its greenhouse properties (see sidebar "Carbon Cycle and the Greenhouse Effect," page 220). The result was the onset of glaciation in the colder parts of the world. At this time the major continents of the Southern Hemisphere were fused into one supercontinent called Pangaea. The southern tips of what are now South America, South Africa, and Antarctica lay together and formed the center of the late Carboniferous glaciation. The evidence for this glaciation together with the fossils of a seed-forming plant called *Glossopteris* that spread at the end of this glaciation have provided proof that these continents were once joined together in a single landmass. Some fossil localities in northern Russia indicate that trees were replaced by herbaceous growth forms in these high latitudes, perhaps indicating the first true tundra vegetation.

Global temperature rose without major interruption through the subsequent course of Earth history until relatively recently in geological terms. The Cretaceous period (146 to 65 million years ago) was generally a period of global warmth, but there were episodes when the climate cooled. Even more important from the point of view of possible tundra development, the seasonal range of temperature during the cooler episodes was similar to that of the present day. It is possible, therefore, that there was some high-latitude glaciation during late Cretaceous times. Fossils of coniferous trees with an understorey of flowering plants are found at 85°N in mid-Cretaceous times, so if tundra was present it must have been very local. By the end of the Cretaceous the Arctic was rich in herbs, suggesting the spread of tundra, but in Antarctica a cool temperate rain forest was present.

The last 65 million years has seen a gradual decline in global temperature. Broad-leaved evergreen trees extended into the Arctic in the early stages of this Cenozoic era, but they were gradually replaced by conifers. Forests persisted on the Antarctic Peninsula at this time, but by 35 million years ago (Oligocene epoch) cooling became rapid, and these Antarctic forests showed a sudden decrease in leaf size, often an indication of colder conditions. This was rapidly followed by Antarctic glaciation and a decline in the frequency of fossil evidence. There were times when trees reestablished themselves on the Antarctic mainland, particularly during glacial recessions between 5 and 2 million years ago.

In the Arctic fossils from about 2 million years ago can be matched with some modern plant species. Larch (*Larix* species), black spruce (*Picea mariana*), dwarf birch (*Betula nana*), and *Dryas* species are often present, and boreal forest at that time extended to the north of Greenland (82°N). Spruce, birch, and pine occupied the north of Alaska, so clearly conditions were less severe than at present. But over the last 2 million years the forests withdrew from these areas, and the tundra biome came into its own. The latest ice age had truly begun.

■ THE PLEISTOCENE GLACIATIONS

It is difficult to place a precise date on when the current Ice Age began. The climate of the Earth has been on an overall downward track for 65 million years, accelerating over the last 35 million years, and reaching a sufficiently cold state in the last 10 million years to ensure that ice was present somewhere on the planet. Dating the climatic declines is still problematic, but evidence suggests that there was a further downturn in the climate at 2.4 million years ago and again about 1.75 million years ago, the date taken as the beginning of the Pleistocene epoch. The fact that the Earth is currently in the course of one of its very rare cold episodes may come as a surprise given the present concerns with global warming, but the relative warmth of the present day has developed only over the last 10,000 years, a mere blink of the eye in geological terms. Over the past million years the Earth's climate has swung backward and forward like a pendulum, so there is no reason to assume that these swings will cease. The possibility of a return to another time of glacial advance remains.

The glaciations of the Pleistocene era have been the focus of a great deal of geological research in the last 100 years or so, since the recognition and general acceptance of the past occurrence of glacial episodes. Before the concept of ice ages could be contemplated, a great leap in conceptual thinking was required on the part of geologists, and this occurred with the development of the principle of *uniformitarianism* (see sidebar on page 179).

The idea that all past events could be explained by the careful observation of current processes was taken up by Charles Lyell (1797–1875), a British geologist whose work was to have a strong impact on his close friend Charles

Uniformitarianism

In the 18th century the sole basis for understanding the history of the Earth was the account given in the book of Genesis in the Bible. Contained there is the story of a great Flood at the time of Noah that covered the entire Earth and destroyed all former life except that which escaped on the ark. All geological features, such as the fossils of extinct animals, the deposition of large volumes of sediments, the erosion of past shorelines, and the scouring of mountain valleys, were accounted for by the Flood and its subsequent retreat. A Scottish scientist, James Hutton (1726–97), was unhappy with this explanation and began to apply principles of scientific analysis to the understanding of landscape. He proposed the simple dictum that events in the past were caused by essentially the same kind of processes and at the same speed as those in the present. The history of the Earth, he suggested, has not been determined by a series of unrepeatable events of a catastrophic nature but by the gradual processes seen at work in the modern world. Streams and rivers can erode mountain valleys, but they operate relatively slowly, so they will take a long while to do so. The rates of erosion under current conditions can be calculated, so it is possible to work out how long it has taken for a particular valley to form. The idea that the present is the key to the past and that geological mechanisms have not substantially changed in the course of time is called *uniformitarianism.*

The idea, when first proposed, attracted little support mainly because the slow speed of the processes involved implied that the Earth would have to be much older than the then estimate of 6,000 years in order to account for landscape conditions. If the Earth were indeed as young as it was believed to be (on the basis of biblical records of genealogy), then only *catastrophism* of the type described in the Flood account could explain the current state of the Earth.

tance transport of such heavy objects. He finally returned to Noah's Flood and suggested that erratic rocks could have been caught up in floating icebergs on the floodwater and transported great distances in this way. In many respects he was on the right track; ice was involved, but not floating on the floodwaters.

Perhaps it was inevitable that the mysteries of glaciation should be solved by a Swiss geologist, Louis Agassiz (1807–73), who grew up in a land of glaciers, tills, U-shaped valleys, hanging valleys, moraines, and ice-scratched rocks (see "Glacial Debris and Its Deposition," pages 44–49). In 1837 he presented a research paper to the Swiss Society of Natural Sciences at Neuchâtel in which he came to the logical but extremely novel conclusion that the glaciers he knew so well had been much more extensive in times past than they were in his day. He showed how erratic rocks could have been moved by glacial ice and provided a means of mapping the course of glacial movements, and he rightly explained the presence of scratches on rocks as evidence of past grinding by ice. He used the term *Eiszeit*, the German equivalent of "ice age," and proposed that colder conditions had prevailed in the recent past leading to a greater extent of ice cover in the European Alps. The idea was regarded as entirely eccentric by most geologists at the time, but work in northern Germany seemed to support the ice age theory. One German geologist claimed that the Arctic ice had once extended as far south as northern Germany. He was ridiculed for such a bizarre suggestion, but Agassiz then raised the claim to a higher level by proposing that ice sheets had not only covered much of northern Europe but also northwestern Asia and the northern regions of North America. American geologists were more ready to consider such claims than the Europeans, and they responded by inviting Agassiz to take up Harvard University's chair in geology in 1846. He accepted and spent much of the rest of his life surveying glaciated terrain in North America and extending his work on past glacial advances.

The subject remained one of intense debate and controversy until well into the 1860s, after which many geologists became increasingly convinced that the Earth had recently undergone a very cold climatic episode causing ice sheets to extend. Questions of detail then arose. Was the Ice Age a single cold event, or had there been a number of glacial advances and retreats? Agassiz assumed that there had been just one ice advance. His work in North America provided little evidence of any alternative explanation, but the Europeans began to develop more complex schemes. The last glacial advance in Europe did not extend as far as some previous ice ages, so there were regions that showed the effects of glaciation that preceded the final glaciation. In England Joseph Trimmer examined a series of tills, or boulder clays, that varied in their physical and chemical constitution and were evidently derived from

Darwin (1809–82). Like many geologists in the 19th century, he was concerned with the problem of how some rocks (erratics) were found far from their source areas. They were often massive, difficult to move, and many hundreds of miles away from any matching bedrock. Even using the principle of uniformitarianism, it was difficult to come up with a mechanism that would account for such long-dis-

different source areas. When he studied the orientation of the stones embedded in the clays, he found that the different tills displayed different orientations, suggesting that the ice responsible for depositing them had been traveling in different directions. There must have been several ice advances, each of which spread in its own characteristic manner. This implied that there must have been warmer episodes of glacial retreat between each glacial advance, and the Scottish geologist Archibald Geikie proved this by finding fossils indicating warmth interleaved with the glacial tills. A history of alternating glacials and interglacials was thus confirmed.

Meanwhile, in the Swiss Alps, where Louis Agassiz had first found his inspiration, evidence for glacial advances and retreats was also being collected. The terminal moraines of glaciers provided evidence of the maximum extent of a particular glaciation, and these, like the English tills, indicated several separate advances. In 1909 the work from the Alps was formally published, proposing that there had been four main ice advances. Geologists named these four glacial episodes after the rivers that marked their maximum extents, Günz, Mindel, Riss, and Würm, all of which are tributaries of central Europe's Danube River. Between these cold episodes, or *glacials,* there were periods of warmth, called *interglacials,* in which temperate conditions and vegetation replaced the tundra of the glacials. The Agassiz model of a single glacial advance was thus replaced by a model in which four advances were demonstrated, and this idea dominated geological thinking for the next 50 years.

There is a danger in science that any new and well-demonstrated proposal becomes accepted to such a degree that no variation is acceptable even as a working hypothesis, and this was true of the fourfold glaciation theory. All new evidence from different parts of the world was fitted into the four-glaciation model. At that time it was difficult to determine the precise date of any newly discovered till or interglacial deposit, so the identification depended on the location of the site in a stratigraphical sequence. Traditionally geologists use fossils for this purpose, matching different deposits by the fossils they contain, but this can lead to a circular argument. For example, if fossils A, B, and C all occur together in interglacial X, then wherever the three are found together, this must be interglacial X. One can then propose that organisms A, B, and C had widespread distributions and were found in a regular community during interglacial X. But this is a circular argument because the organisms have been used to define the time period in the first case. What was needed was a secure system of absolute dating so that different deposits could be matched without reference to indicator fossils, and this has been very difficult to achieve.

The technique of radiocarbon dating was first developed in the 1950s (see sidebar), but although this method has proved very valuable in archaeology and in very recent geological studies, it has limited value when applied to materials that are older than about 40,000 years. Also, it can only be used for materials containing organic carbon.

There are alternative dating methods based on radioactive decay, such as the decay of potassium 40 to argon 40, but

Radiocarbon Dating

The Earth's outer atmosphere is constantly bombarded by high-energy cosmic rays from the Sun. These cosmic rays consist of neutrons, which are streaming through space from the Sun and collide with atoms in the atmosphere, resulting in nuclear changes. Nitrogen is the most common element in the Earth's atmosphere, and its nucleus consists of seven protons and seven electrons, giving it an atomic mass of 14. When neutrons strike the nitrogen nucleus, one of the protons is replaced by a neutron, resulting in a nucleus with eight neutrons and six protons. This is no longer a nitrogen atom but a radioactive isotope of carbon, called carbon 14 (^{14}C). It is an unstable atom and decays, emitting β radiation and reverting to a nitrogen atom, but this decay is quite slow. The half-life of a radioactive substance is the time taken for 50 percent of the radioactive atoms to decay, and in the case of carbon 14 this is 5,730 years. This may seem like a slow rate of decay, but in geological terms it is fast. Within 40,000 years less than 1 percent of the original radiocarbon remains. Physicists determine the age of an object by counting the number of ^{14}C atoms in relation to the "normal" ^{12}C atoms in a piece of organic matter. Assuming that the ratio of these atoms in the atmosphere is constant, it is possible to determine how long the material has been losing its radioactivity. But once the level of ^{14}C has fallen below 1 percent of its original abundance, it becomes difficult to measure with accuracy, so the technique is of limited value for materials older than 40,000 years. There is also a problem with the assumption that the original ratio of the carbon isotopes is constant. In fact, this varies with the output of cosmic rays by the Sun, but this difficulty can be overcome by analyzing materials of known age and fitting a calibration curve to the radiocarbon timescale. In this way it is now possible to apply a correction to radiocarbon dates to bring them into line with true dates.

this is applicable mainly to volcanic rocks in which there is sufficient potassium to permit the analysis. This method is applicable only for materials with an age older than 100,000 years, but it has been useful in the study of early stages in the recent ice ages.

One of the most important developments in techniques for the study of the glacial-interglacial fluctuations of recent times, however, has been the analysis of oxygen isotopes in marine sediments and in the cores of ice sheets themselves. The technique is based on the fact that ratio of the isotopes of oxygen present in these materials varies with the temperature of the Earth at the time when they were formed (see sidebar). Successive layers of undersea sediments or of ice from the ice sheets in Greenland and Antarctica display an alternation of cold and warm episodes that reflects the climatic instability of the Earth over the last 2 million years.

The pattern of climatic change deduced from oxygen isotope studies of marine sediments and ice cores is now well established. Fluctuations between warm and cold conditions have been frequent and extreme over the past 2 million years. Between 2 million and 1 million years ago the fluctuations were relatively weak; the amplitude of the temperature changes was therefore small. But over time the intensity of the warm and cold alternations has increased, and the last million years has seen much more extreme fluctuations. There is a pattern to each warm and cold stage that seems to repeat itself. The coldest part of each cold stage comes late, immediately before a rapid warming, whereas the warmest part of a warm stage comes very early. There is then a progressive cooling, sometimes with sudden downward steps, into the next episode of maximum cold. This pattern means that the times of warmth in the last million years, when added together, occupy far less time than the periods of relative cold. Climatic warmth, it seems, has become something of a scarce event when viewed at this scale.

The current warm stage, the last 11,500 radiocarbon years, which geologists call the Holocene to distinguish it from the Pleistocene epoch, is clearly part of an ongoing pattern of fluctuation. There is no reason to believe that we have finally emerged from the Ice Age and that the warmth of the current stage will be indefinitely prolonged. If the established pattern continues, the peak of the current interglacial has passed, having been attained in the early part of the Holocene, perhaps about 10,000 to 8,000 years ago, and the temperature curve should be on its descent into the next glacial. All of the climatic evidence supported this position until 150 years ago, when global climate began to rise again (see "The Threat of Climate Change," pages 222–224).

When oceanographers produced the first oxygen isotope information in the 1950s, terrestrial geologists immediately

Oxygen Isotopes

The abundant element oxygen exists in a number of isotopic forms in nature, two of the most stable forms being ^{16}O and ^{18}O. Of these, ^{16}O is by far the more common, making up approximately 99.8 percent of the Earth's oxygen, and is also lighter in weight than the heavy isotope ^{18}O. Water containing the heavy isotope is slower to evaporate, so these molecules tend to be left behind, while the lighter molecules escape from the water masses of the oceans and precipitate over land or over the ice sheets. When the precipitation falls as snow over glaciers and ice sheets, the lighter form of oxygen is thus more strongly represented in the ice than in the sea, where the heavier form begins to concentrate. If more of the world's water becomes locked up in ice sheets, as occurs during an ice age or glacial stage, then the oceans become increasingly rich in water containing the ^{18}O isotope. In warm times, on the other hand, the heavy isotope is diluted in the oceans by the lighter isotope because less of this light form is trapped within the ice masses of the world. Also, in warm times water with the heavy isotope is more likely to evaporate, so its contribution to any ice present on Earth will be proportionally greater than under cold conditions. The ratio of ^{18}O to ^{16}O in marine sediments and ice cores thus provides a clue to the temperature of the Earth at that time, but the results need to be considered carefully. A high ratio of ^{18}O to ^{16}O in marine sediments means colder conditions, whereas a high ratio in an ice core means warmer conditions. Oceanographers and glacial geologists have now taken cores from different parts of the world's oceans and from the two major ice sheets, and the pattern of warm and cold fluctuations over the past 2 million years has been well documented using the oxygen isotope ratio technique.

began to search for the four great Alpine glaciations that had dominated their thinking over the preceding 50 years. But the oxygen isotope record showed that the actual history of glaciation was much more complicated than had been expected. There were at least a dozen major cold episodes in the last million years and many more minor fluctuations (called *stadials* and *interstadials* rather than glacials and interglacials) that could have caused local glacial advances in some areas, so the concept of a simple fourfold glacial his-

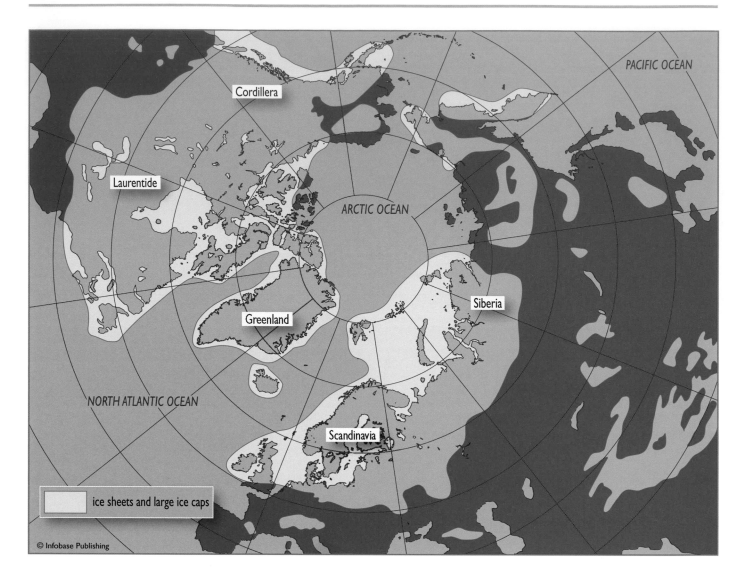

tory is no longer tenable. Geologists now describe and name their glacial and interglacial events independently and gradually try to date and match them as evidence becomes available. Only the most recent of glacial advances and retreats can be correlated with any degree of confidence.

The latest glacial advance reached its maximum extent between 22,000 and 20,000 years ago and was widespread in the Northern Hemisphere, as shown in the diagram above. Almost all of Canada and Alaska were covered, apart from certain parts of the northwest, and the ice sheet extended well south of the Great Lakes region. The Greenland ice sheet linked with the North American Laurentide ice sheet through Baffin Island but was probably separate from the small ice sheet covering Iceland. In western Eurasia a single ice sheet covered much of western Siberia and Scandinavia, extending west over much of the British Isles and the northern parts of continental Europe. Other separate ice sheets developed in the mountain regions of eastern Russia, the Tibetan plateau, parts of the Rocky Mountains, the Alps

The extent of ice cover in the Northern Hemisphere at the height of the last glaciation, about 20,000 years ago

and Pyrenees of Europe, and the Caucasus, and extensive glaciers developed on many low-latitude and Southern Hemisphere mountains. This is one glacial event that can be correlated with some certainty in different parts of the world. It is known as the Wisconsinian glaciation in North America and corresponds to the Devensian of the British Isles, the Weichsellian of northern Europe, and the Würm of the Swiss Alps.

The oxygen isotope technique for climatic reconstruction is based on the fact that more of the Earth's water is in the form of ice during cold periods. One important consequence of this change in the balance of the Earth's hydrology is that during glacials sea levels fall (see sidebar on page 183). Lower sea levels, in turn, affect the movement of ocean currents and provide new areas of land that can be occupied by plants and animals. Where these newly exposed lands were close to the ice sheets, they were occupied by tundra vegetation, so the tundra biome not only extended to lower latitudes than at the present day, it was also present in areas now covered by the sea.

Global sea levels continue to rise at the present day in response to increasing temperature. Over the past 100 years the world's sea level has risen by approximately three inches (8 cm), and present indications are that it will continue to rise at this rate. This could prove a threat to the low-lying coastal areas of the world, including the polar tundra regions.

■ CAUSES OF GLACIATION

Geologists constantly debate the possible causes of glaciation, and they have come up with many theories. They usually begin their analysis by looking for repeating patterns, possible periodicity in the occurrence of glacial events. Looking back at the last 600 million years of Earth history, there is some evidence for regularity in ice ages (see the diagram on page 177). Ice ages occurred about 600 million years ago, 450 million years ago, 300 million years ago, and at the present time. There is some indication of a cycle with a periodicity of about 150 million years, apart from the gap

Glaciation and Sea Level Change

During glacials and interglacials the volumes of the ice caps increased and decreased, respectively. When the volume of water locked up as ice increases, the volume of water in the oceans must decrease, because the amount of water present on the Earth's surface is finite. Consequently, during glacials the world's sea level fell, exposing very considerable areas of land, which could be invaded by plants and animals. The area occupied by ice at the height of the last glaciation can be mapped with considerable accuracy because of field surveys of glacial features on landscapes (see map on page 182). The depth of the ice sheets was considerable, often exceeding a mile, so the total volume of water contained in the ice can be calculated. Allowance must also be made for the lower temperature of the oceans, which would have led to a contraction in water volume. When this is deducted from the volume of water in the oceans, it is possible to calculate that the sea level during the Wisconsinian glacial was about 330 feet (100 m) below the current level. This is confirmed by various other kinds of evidence, such as the growth of corals on tropical coasts, many of which were exposed and eroded during the glacials of high latitudes. Changes in sea level caused by alterations in the volumes of ice and seawater on the planet are called *eustatic* changes.

Studies of sea levels, however, are usually conducted in relation to land surfaces. Geologists record past shorelines using the evidence of cliff lines that are now either above or below the current shoreline or beaches that are higher or lower than the present. Sunken shorelines often take the form of fiords, or drowned valleys. But what if the land surface itself has risen or fallen in the past? This is precisely what has occurred, and it greatly complicates

the study of past sea levels. In the British Isles, for example, the land surface of Scotland is rising at a rate of 0.8 inches (2 cm) per century, while in the southeast of England the land surface is sinking by 0.4 inches (1 cm) per century. The reason for this is that the land surface of the island was distorted by the uneven load of ice during the last glaciation, and it is still recovering from that effect. The center of glaciation was in the north, where an ice cap formed over Scotland and linked up with the North Sea ice from Scandinavia. The southern part of Britain remained ice free. Now that all the ice has melted, Scotland has been relieved of a massive load that had depressed the Earth's crust in that region, so it is rising gradually as it recovers, resulting in the British Isles slowly tilting from the northwest toward the southeast. The apparent changes in sea levels caused by crustal warping of this kind are termed *isostatic* changes. Both eustatic and isostatic effects must be taken into account in the study of past sea levels and their relation to climate.

In the region of what is now Hudson Bay, these two processes have resulted in successive sea level changes. Following the final retreat of the ice sheet at the end of the last glaciation, the first effect on sea level was eustatic, resulting in a rise in sea level greatly enlarging the area covered by water. A water body called the Tyrrell Sea, named for the explorer Joseph Burr Tyrrell (1858–1957), was much larger than the current Hudson Bay, extending south halfway to the Great Lakes. The loss of such a weight of ice, however, led to an isostatic rebound of the Earth's crust in the region, raising the level of land in relation to the sea. As a result, the Tyrrell Sea retreated, and the coastline of the current Hudson Bay came into being.

in the sequence 150 million years ago. Why was there no Jurassic ice age if this is a genuine pattern? As yet no one has been able to explain this apparent periodicity, but the one feature of the Jurassic that could explain the lack of an ice age at that time is that there was no continental landmass in a polar position. The formation of a polar ice sheet seems to depend upon this condition. Beyond that, there is still speculation about possible astronomical cycles, such as periodic variations in the energy output of the Sun or belts of asteroids that reflect some of that energy and cause a cooling of the Earth, but there is currently no convincing explanation for the Earth entering into ice ages on this long time scale.

When one of these cold periods strikes, however, the cyclic events of glacial and interglacial episodes are more

(this page and next) Certain astronomical events that affect the climate of the world take place in a cyclic form, known as Milankovitch cycles after their discoverer. (A) The Earth's orbit around the Sun varies between circular and elliptical, and a cycle takes 96,000 years to complete. (B) The tilt angle of the Earth on its axis varies by 3° over a cycle of 42,000 years. (C) The Earth wobbles on its axis, which affects the season at which the Earth approaches nearest to the Sun on its elliptical orbit. This variation operates in a cycle of 21,000 years. (D) The changes in temperature over the past 2 million years, as deduced from isotopic studies of ocean sediments. During the last million years Earth's temperature has fluctuated more strongly than was the case earlier, resulting in a sequence of glacial and interglacial events. These have developed a cycle of approximately 100,000-year intervals, which fits quite well with the pattern resulting from superimposing the three Milankovitch astronomical cycles.

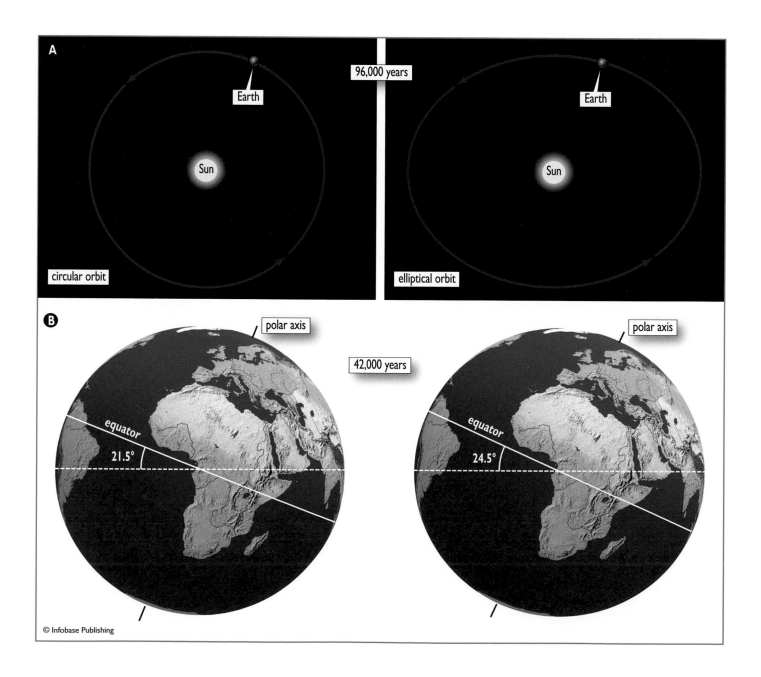

amenable to geological and astronomical research. For example, during the last million years the peaks in glacial advances are fairly regularly arranged, with an interval of about 100,000 years. This raises the question why this periodicity should have occurred during the Pleistocene. What factors could account for such apparent regularity?

A geophysicist from Yugoslavia named Milutin Milankovitch (1879–1958) came up with a novel suggestion that he first published in 1920. He claimed that the pattern of glacials and interglacials corresponded with solar cycles. The orbit of the Earth around the Sun is sometimes circular, but not always (see diagram on page 184). At times it is more elliptical, and this results in a greater difference between the seasons experienced on Earth, because the annual distance between the Earth and the Sun varies. The cycle of change in

orbit from circular to elliptical to circular again takes 94,000 years to complete, which is remarkably close to the 100,000-year cycle of the glacials. But there are additional complications because the tilt of the Earth on its axis also varies in a cyclic manner. The Earth is not precisely upright on its axis but is tilted at an angle of more than 20° from the vertical. The precise angle varies from 21.5° to 24.5° in a cycle that takes 42,000 years to complete, and the seasonal variation in climatic conditions will be greater when the angle is at its maximum. Finally, the Earth also undergoes a wobble on its axis, which again affects seasonality and has a cycle length of 21,000 years. Put these three cycles together, Milankovitch claimed, and the periodicity of the glacial and interglacial cycles can be accounted for. It took some time for the scientific community to be convinced by these proposals, but

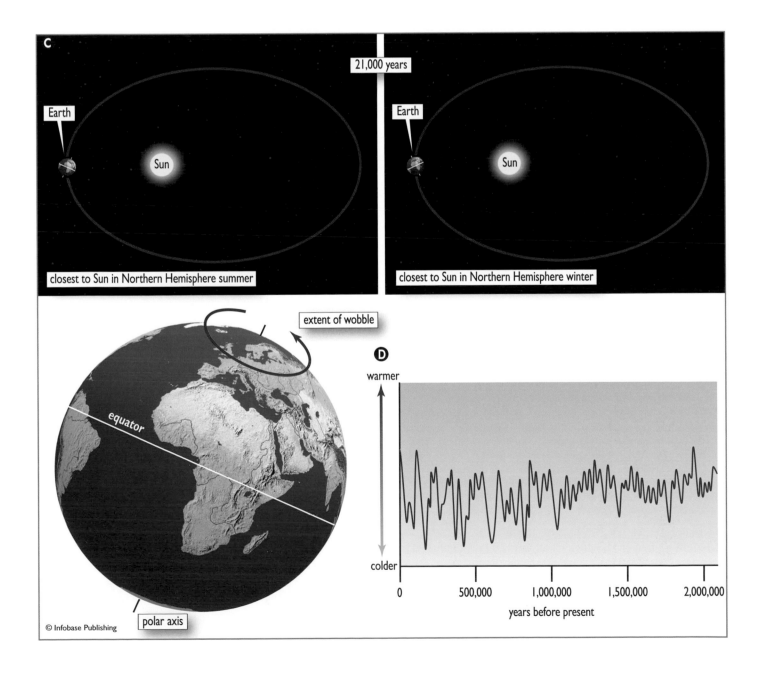

they now form the basis of all modern accounts of climatic variations in the recent past. Armed with the data-handling power of modern computers, it is possible to model changes very precisely by combining the three cyclic patterns of the Earth's orbit and its wobbles, and the pattern that emerges correlates extremely closely with the pattern of climate changes documented in the oxygen isotope curves from marine sediments and ice cores.

There are additional variables involved in climatic cycles, however, some regular, such as the 11-year sunspot cycle that affects the energy output from the Sun, and some unpredictable, such as the creation of atmospheric dust clouds by massive volcanic eruptions. The eruption of Mount Toba on Sumatra 73,500 years ago is associated with a downturn in global temperature, which was already cooling as the last great glaciation advanced. There are also modern examples of volcanic eruptions affecting climate, as in the case of the eruption of El Chichon in the Yucatán Peninsula of Mexico in 1982. This eruption forced an estimated 16 million tons (16 million tonnes) of dust into the atmosphere, where the veil of dust prevented much solar radiation from reaching the Earth's surface. In 1783 the temperature of the Northern Hemisphere dropped by 3°F (1.5°C) following volcanic activity. Carbon dioxide and methane levels in the atmosphere vary in a cyclic way, and the effect of these and other greenhouse gases on climate is now well known (see sidebar "Carbon Cycle and the Greenhouse Effect," page 220). But it is unclear at present whether the composition of the atmosphere is a cause or an effect of cyclic climate changes. Oceanic circulation patterns change during the glacial and interglacial cycles, and these are an important way of redistributing tropical warmth around the planet. So any change is likely to have an impact on global climate (see "The End of the Ice Age," pages 190–192). The Milankovitch cycles, therefore, provide the basic framework for understanding the pattern of climate change during an ice age, but many other factors can dislocate the pattern and introduce unexpected changes.

The ice of glaciers proves a very valuable source of information concerning past volcanic activity and past atmospheric composition. The dust and ash thrown into the atmosphere during volcanic eruptions spreads widely in the upper atmosphere, which is why it has far-reaching climatic effects, but in time the dust settles and falls upon land surfaces, including ice sheets, lakes, and peat bogs, where it becomes permanently incorporated into the accumulating materials. In ice sheets and glaciers the layers of ash can often be detected by eye when viewed in a coring taken through the deposit. The ash is known as tephra, and microscopic examination reveals small glasslike particles forming distorted shapes and angular fragments. Such a layer indicates that an eruption has occurred at the time when this ice sample was accumulating, but geologists usually want to know where the eruption took place. This is difficult to discover because of the long distances tephra can travel in the air. Tephra layers in the peat bogs of Scotland and Ireland have often been derived from eruptions in Iceland 800 miles (1,300 km) away. Fortunately, different volcanoes leave their own distinctive chemical signature in the tephra they produce, so chemical analysis of a tephra layer can usually identify the source volcano. Indeed, each eruption has distinctive chemical features, so it is often possible to determine precisely which eruption a tephra layer represents, and this in turn provides a means of dating the ice or peat horizon. In an ice core, therefore, scientists can trace the immediate impact of a volcanic eruption by analyzing the oxygen isotope ratios above and below a tephra layer.

Ice sheets are also a source of information about past atmospheric composition. When snow accumulates on the surface of an ice sheet or glacier and gradually becomes compacted to form ice (see "Snow and Ice," pages 37–38), the air trapped between the snowflakes is also compacted into bubbles that remain enclosed in the developing ice. These air bubbles form small pockets of fossil atmosphere, retaining a record of the atmospheric composition at the time they were first captured by the ice. Geophysicists have taken samples from different layers in ice cores from Greenland and the Antarctic, crushed them, and extracted samples of fossil atmosphere. When they analyzed these samples they found that the atmospheric composition had varied in the past. Carbon dioxide is low during the glacial stages, often falling to about 180 parts per million (ppm) by volume (0.018 percent) and rises rapidly to more than 300 ppm (0.03 percent) during interglacials. Methane is also low during the glacials, with a concentration of only 300 parts per billion (ppb) by volume (0.00003 percent), rising to 700 ppb during interglacials. Researchers in the Antarctic continue their search for deeper and older ice cores, and they hope eventually to be able to reconstruct atmospheric conditions going back 1.5 million years on the basis of ice core analysis.

The ice sheets are thus a very valuable scientific resource. They are an archive of past conditions, and their study will undoubtedly help scientists to study the interrelationships between the atmosphere, climate, and vegetation. Work on glaciers has not developed as fast as research on the world's two major ice sheets, but the use of glaciers, particularly in tropical mountains, should provide a great deal of additional information about the Earth's recent climatic history.

■ STRATIGRAPHY AND RECENT HISTORY OF TUNDRA

The reconstruction of past history on a geological timescale depends upon materials accumulating in layers. The study of layered sediments is called *stratigraphy,* and this is the major source of detailed information about tundra conditions in the past. There are several types of situations in which layered sediments can accumulate. Two of these have been mentioned already, the ocean bed and the ice of glaciers.

In the oceans there is a constant rain of detritus that sinks through the water and accumulates on the ocean bed, gradually building up over the course of time. Close to land these sediments may consist largely of eroded materials brought to the ocean by rivers, including suspended mineral materials from rocks and soils. But far from land this component of the sediments is less abundant, and most of the sedimenting detritus consists of the dead remains of the organisms that occupy the ocean waters, especially the planktonic organisms that are microscopic in size but are present in such large numbers that their bodies supply a steady and considerable source of deposition. Some of these planktonic organisms have shells or cases that do not easily decompose, so they contribute more to the sediments than organisms made entirely of organic matter, which is a source of food for marine bacteria. The foraminifera (phylum: Granuloreticulosa) and the coccolithophorids (phylum: Haptomonada) are groups of protists that form shells out of calcium carbonate, and these shells survive when their organic contents have decomposed, creating a lime deposit on the ocean floor. Calcium carbonate ($CaCO_3$) contains oxygen, and this is the source of oxygen that is used in the oxygen isotope studies of marine sediments (see sidebar "Oxygen Isotopes," page 181). Diatoms (phylum: Bacillariophyta) have silica shells, so they also accumulate in the ooze beneath the oceans.

The shells of foraminifera and diatoms can be identified under a microscope, so it is possible to gain a picture of the planktonic assemblages of former times by counting the numbers of the different types of microfossils in the sediments. Repeating the exercise in successive layers provides a means of reconstructing past changes in the plankton communities, and these changes are often closely related to the temperature changes indicated by the oxygen isotope studies.

Stratigraphic studies of this sort are not confined to the ocean bed but can also be undertaken using the sediments of freshwater lakes, and this provides an indication of changes in conditions on terrestrial landmasses. Lake sediments contain the remains of planktonic organisms, just like the ocean sediments, but they also contain a variety of other materials. Erosion of the catchment soils brings mineral materials into lake sediments and may also carry fossil fragments of plants and animals from surrounding terrestrial ecosystems. Stream water entering a lake also brings a suspension of pollen grains washed from the catchment, and this pollen load is supplemented by additional pollen

that falls from the air and lands upon the surface of the lake. Studies of pollen transport indicate that about 80 percent of the pollen entering a lake comes from stream transport and the other 20 percent directly from the air. The pollen component of lake sediments is of particular interest because it has proved so valuable as a source of information about the vegetation present in the area when each layer of sediment accumulated. The extraction, identification, counting, and interpretation of this pollen component is called *pollen analysis,* or *palynology* (see sidebar on page 188).

Pollen analysis has proved particularly valuable in the reconstruction of environmental history, but there are many problems in interpretation. It is not always possible to come to detailed conclusions about climatic conditions from a pollen assemblage. Quite apart from all the difficulties in determining how closely the pollen reflects the vegetation, many plants have quite broad climatic tolerance limits. Plants are also relatively slow to respond to sudden climatic change, especially from cold to warm, because it takes some time for them to immigrate into new areas that have become suitable for growth. Insects, on the other hand, are much more mobile and respond very rapidly to changes in climate, so the identification of insect fossils in lake sediment and peat is sometimes more informative about climate than the plant fossils. Beetles have proved particularly useful in this respect because their wing cases preserve well and can be identified to the species level. Fossils of larger animals, such as mammals and birds, can also be valuable in helping to interpret the environment of the time, but these are usually scarce in comparison to plant and insect fossils.

In order to trace the history of the tundra biome through the glacials and interglacials of the Pleistocene, therefore, geologists and paleoecologists (scientists concerned with the reconstruction of past ecological conditions) have turned to fossil plants and animals for evidence that they can compare with the climatic records of ice cores and marine sediments. The ice sheets and ocean sediments are of limited value for fossil study because the ice is very poor in fossils, even pollen grains, while the ocean sediments are dominated by the local marine organisms and contain few records of terrestrial origin. Paleoecologists must therefore rely on land-based materials.

Study of the tundra regions has been hampered by the fact that glaciations often scour the landscape clean of all former fossil-bearing materials. Lake beds and former peat bogs are swept away by advancing ice, so only sites that by chance or favorable location have escaped a particular ice advance contain evidence of earlier times. The numerous glacial episodes of the Pleistocene have ensured that very few such sites have survived, and there are none in the higher latitudes that have escaped all glacial activity and provide a full, uninterrupted record of Pleistocene terrestrial history. There are, however, numerous fragmented records,

Pollen Analysis

Pollen grains are produced by seed plants (gymnosperms and angiosperms) in considerable numbers as a means of genetic transfer during sexual reproduction (see sidebar "Pollination," page 126). Plants that use the wind as a means of pollen dispersal, which includes most gymnosperms (for example, conifers) together with many trees, grasses, sedges, plantains, and so on, produce larger quantities of pollen than insect-pollinated plants because the mechanism is a very chancy one compared to insect transfer. The tough outer coat of pollen grains survives well in wet conditions, where bacterial activity is partly inhibited, so these coats accumulate as fossils in lake sediments and peat. In these materials they become stratified over the course of time as new sediments accumulate above them, so they reflect the changing vegetation of the past. The relatively inert chemical structure of pollen grains enables scientists to extract them from sediments by dissolving away the surrounding matrix. They are so inert that it is possible to boil them for a limited period in hydrofluoric acid, which is strong enough to dissolve glass and which removes sand and silt from the sample. The structure and the delicate sculpturing of the pollen coats allow analysts to identify them, sometimes even to the taxonomic level of species, but more often to generic levels, such as oak or pine. The accompanying illustration shows the distinctive features of pollen grains when viewed with a scanning electron microscope. The abundance of the fossil pollen in most sediments is useful because the different types can be counted and their proportions in the total assemblage calculated. In this

Pollen grains viewed under a scanning electron microscope. The pollen grains are first coated with a thin layer of gold and then photographed under bombardment by electrons.   *(Peter D. Moore and Margaret Collinson)*

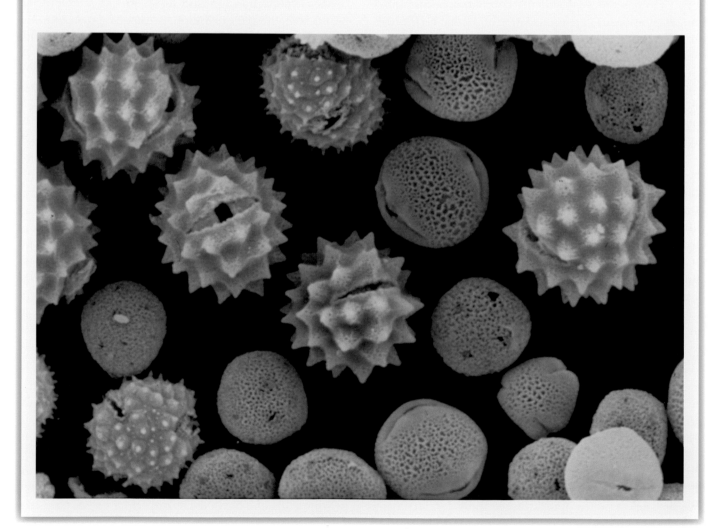

way it is possible to follow the changes in various plants over the course of time, reflecting changes in vegetation, which are themselves related to environmental changes such as climatic changes or in more recent times changes associated with human land management. The interpretation of pollen assemblages is fraught with difficulties, however. Different plants produce different amounts of pollen, so some species are underrepresented and others overrepresented. Some pollen grains disperse more effectively than others, and some are more delicate, so they are damaged or decomposed more easily. Then there is the problem of how far away the source plant was from the lake or peat bog where the pollen is finally preserved. Wetland plants are more likely to be growing nearby and so will be better represented than distant dry-land plants. All of these factors need to be taken into consideration when interpreting a fossil pollen assemblage in terms of the vegetation that gave rise to it.

and geologists are faced with the difficult task of fitting these fragments together, which requires accurate dating and correlation.

Alaska and the Northwest Territories of Canada have been intensively studied, and they can be taken to illustrate the problems of reconstructing the history of the tundra. Parts of this region were not covered by ice sheets during the Pleistocene, so some old sediments have survived. In the early Oligocene, about 35 million years ago, plant fossils from the area consist mainly of broad-leaved evergreen trees growing in subtropical conditions with an average annual temperature between 55 and 68°F (13 to 20°C). By the end of the Oligocene, 23 million years ago, conditions had become colder, and temperate deciduous forest was present, suggesting an annual average temperature of 46 to 54°F (8 to 12°C), with a likelihood of some frost in winter. Between 22 and 13 million years ago, in the Miocene, much of Alaska was dominated by coniferous trees such as spruce, pine, and hemlock. This forest was very rich, with more than 100 species of woody plants, but gradually declined through the Miocene. The forest extended far north into what is now the Arctic tundra, reaching the latitude of 80°N. By 6 million year ago, in the Pliocene, the diversity had fallen very considerably, indicating climatic cooling. At the close of the Pliocene, 2 million years ago, the Bering Sea region had lost its forest cover, and the trees were replaced by herbaceous and shrubby species. Grasses, sedges, and willows were the predominant plants, which suggest that conditions were now much colder and that the tundra had been born. There are no deposits from this time on the Arctic slope of Alaska, but it is reasonable to suppose that tundra extended through this region also by 2 million years ago. This was a time of active mountain building in the northwest of North America, so the development of high mountains must have also encouraged the emergence of a distinctly tundra and treeless landscape at that time.

The evolution and the coming together of the vegetation associated with alpine and polar tundra thus seems to have taken place within the last 3 million years. The origins of its various components are a subject for debate, but clearly the coniferous forests supplied a number of these. Many of the grasses, sedges, evergreen heathers (Ericaceae), and dwarf willows had undoubtedly evolved within the boreal forest biome, especially in clearings and disturbed areas. Their life forms gave them greater resilience and survival potential in the increasingly harsh conditions of the high latitudes, so they persisted after the trees were eliminated. The coastal regions were also undoubtedly a source of suitable plants for the tundra environment. Plants of cliffs and salt marshes may well have benefited from the reduced competition from more robust species as the climate cooled. Then there were the alpine species evolving on the emergent mountains. Alpine saxifrages and *Dryas* species may well have originated on the upper parts of mountains as the tree line was pushed down by icy winter winds.

The Pleistocene has been a time of rapid and severe climatic change, with cold conditions periodically interrupted by warm episodes. The oxygen isotope curves suggest that cold conditions have predominated during the last million years and that warmth has been the exception rather than the rule. But fossils do not always fully reflect this because during cold times there is less organic productivity and so less biomass available for fossilization. Trees, which produce large quantities of pollen, are fewer, so the overall pollen fallout is less. Lakes become filled with eroded soils brought down from the catchments by turbulent streams in the spring snowmelt, so any fossil material is diluted by a heavy load of sediment. The materials available for the paleoecologist, therefore, are often those of the warm intervals, the interglacials. In the early Pleistocene sediments of Alaska and northwest Canada fossils of spruce cone fragments are found, indicating that boreal forest was still present in the area, at least during the warmer times. But in the upper layers of such sediments there are also increasing amounts of pollen of herbaceous species, particularly *Dryas,* saxifrages, and members of the dock family, all of which indicate tundra conditions. The final, Wisconsinian glaciation covered parts of this region, but geologists have found evidence for a short warm period within the main glaciation, an interstadial. This is thought to date from about 40,000 years ago but is at the limit of radiocarbon dating. During this interval spruce forest was once more able to invade north of the Arctic Circle in the region of the Old Crow River to the west of the Mackenzie River. Open tundra conditions then returned as the Wisconsinian glaciation reached its maximum extent at about 22,000 to 20,000 years ago.

More detailed information is available for the final stages of the Pleistocene and for the more recent Holocene history of the region because there are several sites with sediments that date from these times. After 14,000 years ago the open tundra hillsides were invaded by shrub tundra, with birch and juniper, and the low-lying wetter areas bore willow scrub and sedge meadows. Spruce trees invaded the region 11,000 years ago, becoming the dominant vegetation between 9,000 and 6,000 years ago. Both white spruce (*Picea glauca*) and black spruce (*P. mariana*) were present in this community, together with paper birch (*Betula papyrifera*) and tamarack (*Larix laricina*). Alder (*Alnus crispa*) became increasingly frequent from 7,000 years ago onward, probably as an understorey species in the spruce forest, but between 5,000 and 4,000 years ago a major change took place in which all of the tree species went into a sharp decline. The forest retreated south, and tundra took over the coastal region, with shrub tundra on the low-lying regions and dwarf shrubs such as crowberry and bearberry, and such herbaceous species as Arctic lupine (*Lupinus arcticus*) and pasqueflower (*Anemone patens*).

The history of the Arctic tundra in this region of North America, therefore, is one of constant advance and retreat as climate fluctuated and the tree line moved north and south in response. The present distribution of the tundra has established itself in the course of the last 4,000 years, advancing as the spruce forest withdrew, and this picture is representative of all the northern tundra regions. The movements of the tundra are in accord with what is known of the climatic history of the last 12,000 years from oxygen isotope and other evidence. The climate became warmer from 14,000 years ago, leading to the deglaciation of North America and Northwest Eurasia, and the forest spread north. But the peak in warmth occurred early in the Holocene so that the climate was becoming cooler once more by 5,000 years ago, leading to the changes observed in the fossil record. The Arctic tundra of the continental mainlands is a very young biome and one that has proved very resilient in the face of rapid and frequent climatic changes.

■ THE END OF THE ICE AGE

In fact, there is no incontrovertible evidence that the Pleistocene Ice Age has finished. The last 20,000 years have seen great changes in climate, but they are no different in kind from those of previous interglacials. So it is quite possible that the present warm episode is the latest part of an ongoing cycle of cold and warm stages (see "The Threat of Climate Change," pages 222–224). For the last 14,000 years, however, annual average global temperatures have been between about 11 to 18°F (6 to 10°C) higher than in the preceding 10,000 years. The geological evidence used by paleoecologists and paleoclimatologists, including lake sediments, peat bogs, and their fossil contents, have not suffered from glacial erosion following their accumulation, so there is much greater opportunity for detailed reconstruction of environmental changes over this time. One of the important facts emerging from such studies has been the climatic instability that accompanied the final stages of the last ice age.

At the beginning of the 20th century Danish geologists discovered a sequence of sediments in which an organic layer containing fossils of birch and pine was both underlain and overlain by gray clays containing rock fragments and fossils of tundra plants, especially the mountain avens (*Dryas octopetala*). They rightly concluded that this sequence represented a climatic fluctuation in the final stage of the last glaciation in which the climatic warming that resulted in a change from tundra to boreal forest vegetation was interrupted by a cold phase in which there was a return to periglacial conditions in Denmark. The three layers of sediment, from the lowest to the highest, were named the Older Dryas, the Allerød, and the Younger Dryas. What was not clear at that time was whether this was a local climatic fluctuation in Denmark or whether it was more widespread. Over the subsequent years geologists have accumulated increasing evidence to show that the underlying climatic fluctuation was far more extensive than Denmark. Similar sedimentary sequences have been found throughout northwestern Europe, west into Britain and Ireland, north into Scandinavia, and south and east to the European Alps and parts of the Mediterranean. Researchers noticed, however, that the effects of the climatic changes of this time were less marked in southeastern Europe than in the western coastal fringe. On the western side of the Atlantic Ocean evidence for the fluctuation was found in Newfoundland and in the coastal region of New England but became weaker farther inland. Geologists have searched for evidence from other parts of the globe and have detected some changes in the Southern Hemisphere and the North Pacific region, but the main effect of the change evidently centered on the North Atlantic.

Radiocarbon dating of these sediments (see sidebar "Radiocarbon Dating," page 180) has revealed that evidence for warming began at 14,000 years ago and continued to accelerate until 12,700 years ago, when the Younger Dryas stadial (glacial advance of relatively short duration) interrupted the warming process. The fossils, especially those of beetles, which are excellent evidence for climate change because of their rapid response, indicate a very sudden downturn in temperature, perhaps falling by 18°F (10°C) or more in a few decades. The stadial lasted about 1,200 years but then ended as abruptly as it had begun, with temperatures in the North Atlantic region becoming similar

The changing drainage patterns of North America as its ice sheet melted at the end of the last glaciation. In the early stages of glacial retreat the ice cap extended south of what are now the Great Lakes, and the meltwater drained south along the Mississippi valley into the Gulf of Mexico. As the ice retreated northward, meltwater began to drain east along the St. Lawrence River and into the North Atlantic. The discharge of cold freshwater into the ocean interrupted the cycling of the oceanic conveyor belt (see the diagram on page 12) and caused the breakdown of global energy redistribution. Very cold conditions then returned for several centuries.

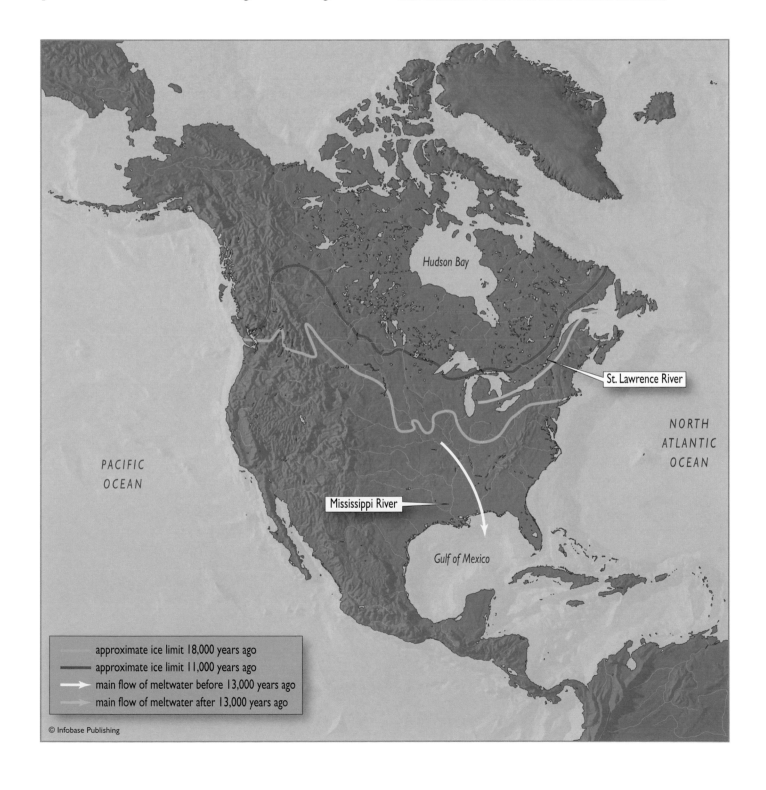

Hudson Bay

St. Lawrence River

PACIFIC OCEAN

NORTH ATLANTIC OCEAN

Mississippi River

Gulf of Mexico

— approximate ice limit 18,000 years ago
— approximate ice limit 11,000 years ago
→ main flow of meltwater before 13,000 years ago
→ main flow of meltwater after 13,000 years ago

© Infobase Publishing

to those of today, again within a matter of decades. Beetle remains from northern Europe at the end of the Younger Dryas stadial indicate that the climate changed from Arctic to Mediterranean extremely rapidly.

Geological and paleontological evidence thus indicates a very unstable climate at the end of the glaciation, and many ideas have been proposed to account for this fluctuation. Models based on the orbital variations of Milankovitch (see "Causes of Glaciation," pages 183–186) could not account for the climatic fluctuations underlying the geological record because the changes were relatively restricted to the North Atlantic, so other explanations were needed. One of the most convincing involves the patterns of circulation of the world's ocean currents (see the diagram on page 12). The North Atlantic receives warm water from the Caribbean as a result of the Gulf Stream. As this water disperses its heat to the atmosphere and land masses of northern Europe it cools, becomes denser, sinks, and returns south as a deepwater current. When the climate began to warm at the end of the final Wisconsinian glaciation, waters from the melting ice sheet over northern North America flowed south into the Gulf of Mexico. As the ice sheet retreated, however, a new outlet for the meltwaters became available in what is now the St. Lawrence River, as shown on the map. Cold freshwater from the melting ice sheet began to flow out into the North Atlantic in such abundance that it affected the oceanic conveyor belt, diluting the northern waters and cutting off the flow of the Gulf Stream from the Caribbean. Immediately the whole of the North Atlantic was deprived of the tropical warmth that the Gulf Stream supplied, and the entire region was plunged into cold conditions. Glaciers began to advance once again in Scandinavia and Scotland, and tundra vegetation replaced the boreal forests that had begun to colonize the north.

Once the flow of meltwater slowed, the oceanic conveyor belt rapidly reestablished itself, and the return of the Gulf Stream brought sudden warmth to the North Atlantic once again. This incident at the close of the glaciation demonstrates how fragile is the balance of the Earth's climate and how easily and unexpectedly it can be disrupted. The global extent and distribution of the tundra biome is closely dependent on such swings in climate.

Resurgence of the ice during stadial episodes such as the Younger Dryas created complex deposits around the edges of glaciers. Old stagnant ice was occasionally buried by new advances of fresh ice carrying its load of detritus that it deposited as till. With the retreat of the new ice, portions of the older ice lay buried beneath the till and were insulated from warming by the blanket of clay and rock debris. Buried ice can take a long time to melt even under warm conditions, and some of these ice burials may have lasted another 3,000 years before they finally collapsed and left behind kettle-holes in which wetlands would subsequently develop

(see "Glacial Debris and Its Deposition," pages 44–49). At last, however, by 11,000 years ago the ice was in permanent retreat, and the tundra biomes followed the ice withdrawing toward the poles and up into the mountains.

■ AFTER THE ICE

The retreating ice sheet was surrounded by a band of tundra about 60 miles (100 km) in width. As the ice withdrew, so the tundra belt followed, invading the land left bare and being invaded by trees from warmer locations that were extending their range as the climate rapidly changed. Pollen analysis has helped to reconstruct the movements of the biomes during the last 10,000 years, which geologists call the Holocene epoch. By examining the fossil pollen preserved in lake sediments, the arrival of trees and the loss of tundra vegetation can be detected. If researchers take a series of lakes aligned on a north-south transect and calculate the date at which the tundra-forest transition took place, it is possible for them to trace the speed of movement of the vegetation belts. Obviously, this varies considerably with local conditions and the availability of trees for the invasion, but the northward movement of trees in North America and in Europe following glacial retreat was about 1,000 feet (300 m) per year. This is quite a rapid spread for trees because it involves seed dispersal, germination, establishment, growth to flowering age, and the production of pollen before the arrival would be recorded in the pollen record of nearby lake sediments. Successive generations of trees thus moved north, and their spread led to the retreat of the tundra.

Birches and pines have light, wind-borne fruits and seeds, and they spread rapidly. The white spruce (*Picea glauca*), for example, completed the journey from its glacial survival center in Pennsylvania to Labrador in 7,000 years, averaging 600 feet (180 m) per year. In its final spread to the shores of the Arctic Ocean, however, it accelerated, effectively crushing the tundra biome from the mainland of North America. More warmth-loving trees followed, such as the oaks. These have larger fruits that are dispersed by mammals and birds, but their speed of movement was not hampered by this. Indeed, they moved somewhat faster than the initial pioneer trees, achieving rates of spread often exceeding 1,200 feet (350 m) per year. These rapid movements of the temperate forest resulted in severe contraction of the tundra biome, which had been particularly widespread during the glacial stage.

The open tundra of the glacial stage had provided grazing for many large mammals, including mastodons, mammoths, many species of deer, bison, bighorn sheep, musk oxen, and caribou. Some of these still survive in the polar and alpine tundra, but others have become extinct (see "Early

Tundra People and the Megafauna," pages 197–201). At least one species of horse and one camel species roamed what are now the Great Plains at the end of the Wisconsinan glacial. All of these grazers were preyed upon by a variety of carnivores, including wolves, cougars, bobcats, bears, and the infamous saber-toothed tiger. The mammal fauna of the late glacial tundra landscape in North America was thus more diverse than that of the more restricted tundra of the present day. It is possible that the plants of the tundra were also more diverse, but the fossil record is less complete because the delicate remains of plants have not survived as well in glacial detritus as have the bones of large mammals. But the pressures that sent many mammals to extinction, especially human hunting, did not affect plants, so extinction may well have been less widespread.

Although plant extinction may not have been common, the rapid changes of the early Holocene often resulted in plant populations becoming fragmented. The yellow-flowered mountain avens (*Dryas drummondii*), for example is now found in the mountains of western North America and in the Gaspé Peninsula and Newfoundland in the east. A fragmented distribution pattern of this sort is said to be *disjunct* and often results from changing environment causing a split in what was once a continuous distribution. In this case the invasion of forest to the south of the main area of ice in North America (the Laurentide ice sheet) split this species into two separate populations. The hairy willow (*Salix vestita*) is another example of a plant with a disjunct distribution pattern. This species occurs as an alpine in the Rocky Mountains of Montana and British Columbia and on the coastal tundra around the shores of Hudson Bay and in Labrador.

Mountaintops are often sites where tundra species with disjunct distributions survive because forest vegetation is unable to penetrate to high altitude as a result of cold conditions. Mountaintops are fragmented and scattered habitats because of their very nature, but they have often provided tundra species with a location in which they can survive through the present interglacial warm stage. Often several disjunct species are found together in such locations, which can be termed *refugia* because they have offered places of refuge to species under climatic and competitive pressures. Such species are sometimes called *relict species* because they have been left behind. The word *relict* is an old English word for a widow, someone who has literally been left alone as a result of changing circumstances.

It is possible that some mountaintop refugia have an even longer history. Areas occupied by ice sheets at the present day often have jagged mountain peaks called *nunataks* emerging from the ice cover as interruptions to the uniform ice landscape (see sidebar "Nunataks," page 38). It is possible that some plants and animals may have survived on these nunataks through the entire course of a glacial stage.

These refugial organisms might then continue to occupy the sites when the climate became warmer and would have disjunct distributions that resulted from being isolated by former glaciers rather than current forest. This possibility is perhaps most plausible for relatively low-latitude sites, where the microclimate on nunataks might have been more favorable than is the case for the modern nunataks of the ice sheets on Antarctica or Greenland. With no modern equivalents to study, it is difficult to reconstruct the conditions experienced on a low-latitude nunatak.

Coastal areas may also have carried nunataks with relatively mild climates. Off the coast of what is now British Columbia, for example, the Queen Charlotte Islands remained ice free through the last glacial maximum and undoubtedly provided many plants and animals with an ice age refugium. Perglacial refugia, those that had offered opportunities for survival through the ice age, would then form the focal points from which tundra could spread out when the ice retreated. Some biogeographers have located ice-free refugia by examining how many endemic species are found in an area, arguing that an area rich in endemics may have been a center of spread following glacial retreat. The southern part of Alaska, for example, has more than 30 species of endemic plants, which indicates that some of this region remained ice free through the glaciation.

Disjunct distribution patterns on a continental mainland, where there are no obvious physical barriers to dispersal, are most common among organisms that are limited in their mobility, such as plants, flightless insects, and small mammals. But the glaciations of the past 2 million years have also affected more mobile creatures, including birds. The northern flicker (*Colaptes auratus*), for example, is a common North American woodpecker that is found throughout the continent except the extreme north, but eastern and western populations are quite distinct in their plumage. The eastern, or yellow-shafted form, has yellow beneath its wings and a brown face and throat. The western, or red-shafted form, on the other hand, has orange underwings, and its face and throat are gray. When a species shows separation of this kind it usually implies that some barrier has prevented interbreeding and that the two isolated populations have developed along different lines. In the case of the northern flicker, the most likely explanation is that the species was split by the Wisconsinian glaciation into eastern and western populations, and now that they have come together again as a result of glacial retreat they still prefer to mate within their own groups, so the plumage differences persist. There are many examples of this type of differentiation within bird species, including the yellow-rumped warbler (*Dondroica coronata*), which was once regarded as two separate species, the Audubon's warbler in the west and the myrtle warbler in the east. The plumage, especially of the males, is very different, but they are now considered variants of the same species. In Europe the pied flycatcher (*Ficedula hypoleuca*) and the collared flycatcher (*F. albicollis*) are also very similar in plumage, the latter species, as its name suggests, differing by having a white collar around its neck. These two species differ in their migration routes, the pied flycatcher taking a western route through Spain into Africa for the winter and the collared flycatcher flying around the eastern end of the Mediterranean. It is probable that these were once a single species but were split by the last (Weichselian) glaciation and have since persisted in their different behavior patterns despite the fact that the old ice age barriers are now gone. The two groups interbreed only rarely, so ornithologists now regard them as separate species.

There is one other way in which the glaciations have affected distribution patterns and evolutionary processes: by altering global sea levels. When glaciations were at their height the global sea level was reduced by about 330 feet (100 m), leaving exposed much of the land that is now covered by shallow seas (see sidebar "Glaciation and Sea Level Change," page 183). Animals and plants were able to spread over these land areas and occupy new regions, but as climate warmed during an interglacial sea levels rose once again and land bridges were cut, leaving islands isolated by the sea. Beringia consists of a series of islands forming a chain between Alaska and Russia across the North Pacific, and these islands were linked together by land when the sea level was lower during the last glaciation. Rising sea level has now fragmented this former land bridge and isolated the different islands and their resident populations of mammals, especially shrews and voles, which have subsequently begun to diverge in their evolution.

Some islands have become isolated by rising sea level before invasion was possible for animals and plants migrating north with the warming climate. The island of Ireland to the west of the British Isles, for example, has no snake species because the water body that now forms the Irish Sea between Britain and Ireland had risen to an impassable level by the time these reptiles had reached the British mainland. The fact that they managed to get that far, however, shows that they crossed what is now the English Channel between France and England before the rising sea had cut off their migration route.

In some respects mountaintops resemble islands that have been cut off from one another by rising sea levels. In many mountain areas within the temperate and tropical latitudes, the summits were linked to one another by alpine tundra so that plants and animals could move between them. As climate warmed, forest invaded from lower altitudes, and, depending upon factors such as the height of the mountain range and the latitude, trees invaded much of the former tundra, eventually forming a closed forest canopy isolating the higher mountain peaks. For the less mobile organisms this meant fragmentation of populations, genetic isolation,

and independent evolutionary development. Mountaintops, like islands, are thus centers of evolution.

The glaciations of the last 2 million years have thus had a great influence upon the biogeography of the Earth. Not only have there been great shifts in the distribution patterns of individual species and entire ecosystems, but the disruption caused by glaciation has also stimulated evolutionary development. For some species, however, the challenge proved too great, and fossils are all that remain to indicate their former existence.

■ CONCLUSIONS

Ice ages, when polar ice caps develop and extend and the mountaintops of the world bear glaciers, are relatively rare events in the history of the Earth. The first ice age recorded in ancient rocks occurred about 2.3 billion years ago, before the advent of life, and others followed during the long Proterozoic era, ending 542 million years ago. None of these early ice ages were accompanied by the development of tundra ecosystems because there were no living things apart from bacteria at those times. Approximately 450 million years ago, in Ordovician times, the Earth entered another ice age, which was subsequently repeated 150 million years later. Ice caps are known to have been present on Earth at that time, and forest vegetation in high latitudes gave way to herbaceous life forms, the first true tundra. Warmth took over once more, but there were signs of cold conditions during part of the Cretaceous period, about 140 million years ago, but no signs of ice caps developing. In the last 65 million years the Earth's climate has been in decline once again, intensifying in the last 5 million and resulting in the widespread glaciation of the most recent ice age during the past 1.75 million years of the Pleistocene.

Geologists have only recently come to appreciate the severity of the changes in climate experienced during the Pleistocene. In the past 200 years scientists have been able to calculate the great age of the Earth, 4.6 billion years, and have come to realize that over this long history the landscape has been formed as a result of relatively slow but persistent processes, such as erosion and sedimentation, rather than catastrophes, such as the biblical Flood. The idea of interpreting the past in the light of the present gave rise to the principle of uniformitarianism, and this provided the basis for reconstructing the glacial events of the past 2 million years. Using this approach, the Swiss geologist Louis Agassiz first proposed that ice had been widespread in relatively recent times and had affected landscape development throughout Eurasia and North America.

Radiocarbon dating together with other techniques based on radioactive decomposition have greatly advanced the understanding of patterns in glacial advances and retreats, but the discovery of oxygen isotope methods for reconstructing global temperature changes has been the most valuable tool available to paleoclimatologists. It is now clear that there have been many glacials and interglacials even within the past million years. As the ice sheets developed water was taken out of the oceans, so global sea levels fell, perhaps by about 300 feet (100 m), as a result of these eustatic changes. Sea level studies, however, are complex because ice masses depress land surfaces, lowering them in relation to sea level, and the crust then rebounds upward when the weight of ice is lost. So some sea level changes are a consequence of these isostatic processes. The relative position of sea level in relation to land level is a consequence of both these effects and varies considerably with locality, depending especially on past ice distribution.

What causes the Earth to enter ice ages is still a mystery. There are approximately 150 million year intervals between global ice ages, but what underlies this cycle is still unknown. Within an ice age there is a shorter periodicity, glacial episodes recurring roughly every 100,000 years, and the cause of these short-term cycles appears to be determined by astronomical factors. Milutin Milankovitch proposed three different types of cyclic phenomena that, when superimposed, can account for the main pattern of glacial and interglacial episodes during the Pleistocene. Additional factors, such as major volcanic eruptions and the abundance of some greenhouse gases, such as carbon dioxide and methane, in the atmosphere may play a role in modifying these cycles.

The biological consequences of recent climate changes are best studied by extracting fossils from stratified sediments in the oceans or in lakes and peat bogs. Microscopic fossils, such as foraminifera and diatoms in ocean sediments and pollen grains in freshwater sediments, provide a record of oceanic phytoplankton and terrestrial vegetation, respectively. Pollen together with the fossils of beetles and other animals have proved a very illuminating source of information about terrestrial changes during the glacial cycles of the Pleistocene. On this basis paleoecologists have been able to reconstruct the history of the tundra biome and study its expansion during glacial episodes and its contraction in the interglacials.

The transition between the last glacial and the present interglacial was a time of extreme climatic fluctuation in the North Atlantic region. Signs of climatic warming and ice retreat began about 14,000 years ago, but the warming was interrupted by 1,200 years of extreme cold between 12,700 and 11,500 years ago, a period called the Younger Dryas,

during which both ice sheets and the tundra biome as a whole expanded once again. The effects of this cold episode diminished on moving into the continental areas of Europe and North America, so its cause evidently lay in the North Atlantic Ocean itself. One theory is that the great ice sheet of North America, the Laurentide ice sheet, gradually decayed, and at some point, perhaps around 11,000 years ago, its pattern of meltwater drainage changed and, instead of draining south into the Gulf of Mexico, drained east into the Atlantic. This would have created a large surge of cold, low-density freshwater into the North Atlantic, which interrupted the flow of the warm Gulf Stream and caused a sudden cooling of the entire region. The sensitivity of the ocean currents and their impact on climate change is revealed in this relatively recent event.

The pattern of change since the last glacial retreat indicates that warming was sudden, reaching a peak in the early Holocene (the last 10,000 years) and then declining. The northern continental areas of the Arctic, including northern Alaska and Canada, were covered by boreal forest in the early Holocene. Tundra expanded south from the High Arctic islands during the last 4,000 years as global climate began to cool. Much of the current polar tundra, therefore, is a relatively young ecosystem.

The mark of glaciers is found not only in features of the landscape but also in the current distribution patterns of plants and animals. Ice-free areas acted as refugia for organisms, and they subsequently spread from these pockets of survival as conditions became warmer. These refugia can be located by mapping the abundance of endemic organisms. The distribution of many tundra species has more recently become fragmented, isolated, and confined by the spread of forest, and many disjunct distributions are due to the tundra species becoming mere relicts of their more widespread pattern in former times. Some species of plants and animals became split into separate populations by the development of ice caps, and these disrupted populations followed independent evolutionary paths, leading to changes in appearance or behavior that have persisted even though the separating effect of the ice caps has now gone. Rising sea levels have also isolated oceanic islands that were formerly linked by land bridges, and changing climate has placed barriers between mountaintop tundra habitats, leading to their fragmentation and isolation. The aftereffects of the last glaciation are thus widespread and persistent in the biogeography and evolution of many plant and animal species, including those of the tundra and of other biomes.

8

People in the Tundra

Human beings are part of tundra biodiversity, particularly in the Arctic. *Homo sapiens* is a species that occupies most parts of the planet but with a very patchy distribution as far as population density is concerned. In parts of Europe, India, China, and Japan the population density of humans exceeds 800 people per square mile (300 km^{-2}), but the hot deserts and the tundra usually support less than 0.4 people per square mile (1 km^{-2}). As far as people are concerned, the tundra is a relatively inhospitable place. Just like the other animals and plants that endure the hardships of the tundra, the human inhabitants need to be hardy and resilient if they are to survive.

■ THE EMERGENCE OF MODERN HUMANS

Primates, the animal group to which biologists assign the human species, are largely tropical forest animals, although some members have proved very adaptable to other types of habitats. The primates are divided into Old World and New World groups, and these are thought to have separated some 40 million years ago as the Atlantic Ocean widened, separating Africa from South America. By 7 million years ago the Old World group had split into different lines of development, including one that led to the great apes and another that led to humans. These organisms were still essentially forest creatures, and it was not until about 4 million years that the ancestors of humanity showed an undoubted bipedal stance and adaptations to the hand that indicate life in a wooded grassland or savanna habitat.

The genus *Homo*, of which we are a member, is first found in the fossil record from deposits of about 2.5 million years ago, and our own species, *Homo sapiens*, first occurs in the geological record 260,000 years ago. Paleoanthropologists, scientists who study the early history of humans, still argue about the place of origin and the pattern of spread of the human species. One member of the *Homo* genus, *H. erectus*,

had spread out of Africa and into southern Asia by 1 million years ago, and this species is regarded as the direct ancestor of *H. sapiens*. Neanderthal man, *H. neanderthalensis*, was a species that diverged from the main line of ancestry about 200,000 years ago but became extinct about 28,000 years ago, and did not survive through the final cold episode of the last Ice Age. Both species adapted to life in the cold conditions of the Ice Age in Europe and Asia and assumed the ecological niches of top predator, scavenger, and gatherer of plant food in the tundra ecosystem. It was during that final Ice Age, around 50,000 years ago, that our species began to develop new techniques and habits, such as the production of more refined stone tools, the creation of artistic images depicting possible religious rites and deities, and the wearing of decorative jewelry.

Adapting to the cold conditions demanded a number of changes in culture and behavior. The general lack of body hair in humans is clearly a disadvantage for life in tundra environments, so humans took to using the better adapted skins of other animals. Evolving a hairy pelt would take a very long time, so it was clearly advantageous to steal someone else s. Fire proved extremely important in the tundra environment. A knowledge of fire among species of the genus *Homo* can be traced back into the times of *H. erectus*. Tropical fires, especially in the savanna environments, would have been common, and early peoples would use fire for cooking, habitat management, and social purposes. With the coming of the ice ages, this knowledge of fire was a key to survival, providing an artificial source of warmth unavailable to other animals. Humans in the tundra thus created their own microenvironments, first around their immediate bodies by creating insulating clothing and also within their cave habitations by using fire. These people were predators, but not very well-equipped predators. They were not particularly fast in pursuing their prey and had poor teeth and claws for killing and dismembering any animals they caught, but they compensated for these weaknesses with brain power. Early humans hunted in packs and developed techniques of stampeding game over cliffs or into confined spaces, where they could slaughter them with stone weapons. They even developed cooperative hunting with another intelligent pack animal, the wolf, and eventually this relationship led to full domestication.

The new surge in human cultural development was accompanied by geographic spread through Europe and Asia and, by 40,000 years ago, into Australia. The New World remained unoccupied throughout this time. The invasion of Australia is particularly remarkable because it must have involved sea travel, and there is no direct archaeological evidence for boats until relatively recent times, 13,000 years ago in the Mediterranean region. The invasion of Australia and the numerous islands of the Southeast Asian region may well have taken place by chance dispersal on rafts. From a biological and ecological point of view this dispersing species was particularly remarkable for its adaptability. In Africa it met with semiarid scrub and desert, in Europe it found temperate forests of deciduous and evergreen character, in India it found savanna, and in the southeastern regions of Asia it discovered both dry deciduous and tropical rain forests. No other species displayed such a capacity to cope with the challenges of so many different habitats and was able to prove so invasive.

The tundra of the far north proved less attractive, but the prospect of hunting large mammals, such as mammoth, giant elk, and caribou, drew the expanding populations of humans even into these testing regions. By 20,000 years ago people had spread into northern Europe and Asia, including the far eastern corners of Siberia. Global sea levels at that time were low because so much of the Earth's water was locked up in the ice sheets and glaciers, so there was a land bridge between Siberia and Alaska (see sidebar "Glaciation and Sea Level Change," page 183). The environment of this bridge was harsh, unforested tundra, but it was not covered by ice, so hardy mammals, including humans, were able to cross into the New World on foot. The vegetation, as revealed by fossil studies, consisted largely of grasses and sage (*Artemisia frigida*) forming an open grass tundra ideal for large herbivores. Exactly when the migration of people took place is still uncertain, but Alaska was certainly occupied by humans by 12,000 years ago. The subsequent spread of people through the Americas was remarkably rapid. By 11,000 years ago they had reached what is now Mexico, having spread right through North America, and within a few hundred years were present in Amazonia and all the way to Patagonia.

The study of ancient languages may help in sorting out the problem of when North America first received its human immigrants. There are three distinct groups of Native American languages, and their distribution through the continent corresponds closely to the pattern of maximum ice extent. One group lies to the far northwest, where there was an ice-free region; one group was along the north-west Pacific coast, which was also ice free even during the maximum of the glaciation; and the third group was in the south, beyond the ice limits. This pattern suggests that people were present throughout North America at the time of maximum ice 20,000 years ago and were isolated in the three regions, where languages developed independently of one another. The conclusion from this type of work is that North America was colonized relatively early, prior to the final glacial advance.

Some anthropologists feel that the Beringian view of the invasion of the Americas is too simple to explain the facts. They claim that the people arriving in the New World were not all from eastern Siberia but that some came from southern Asia and the southern Pacific Rim. Archaeologists have examined some of the most ancient American skeletons and feel that they more closely resemble the southern Asian or even the early Australian people than the folk of Siberia. Perhaps there were different waves of invasions involving different groups of people, and perhaps the Native American populations of the Baja and Mexican regions had a different origin from those of the north. There are also arguments about dates, because some South American settlements have been dated to 12,500 years ago, which does not fit in with the pattern of spread previously accepted unless the early arrival hypothesis is accepted. The arguments will undoubtedly continue until additional evidence is available, but one thing is certain: By soon after 11,000 years ago the peopling of the Americas was complete.

The spread of human cultures through North and South America illustrates the cultural adaptability of the human species to very different climates and environments. The people who had entered Alaska were of Siberian descent and had lived in a tundra environment. Some of these remained within the Arctic tundra of northern Canada, but others traveled south and encountered coniferous and temperate deciduous forest, prairie grasslands, dry scrub, and deserts. In Central America they first met up with the tropical forests and from there spread rapidly into the tropical rain forests of Amazonia. Within relatively few generations humans had made the change from living in tundra to living in the rain forest.

■ EARLY TUNDRA PEOPLE AND THE MEGAFAUNA

The people inhabiting the southern regions of North America 11,000 years ago have been named Clovis people after a town in New Mexico where their archaeological remains have been described in detail. They are characterized by their finely chipped stone spear points and other tools and are often associated with the bones of large mammals, such as mammoths, on which they preyed. Such a lifestyle would also have been appropriate in the alpine tundra habitats of tropical mountains, and one can imagine that these people were able to advance southward down the mountain chain of the Andes hunting game. Open savanna and prairie habitats would also have been suitable for this type of hunting.

The invading humans therefore encountered a tundra ecosystem in which they adopted an omnivorous role, with the capacity to act as a major predator. Plant products are very scarce in the tundra, so it was inevitable that humans adopt a mainly carnivorous diet. Large predators were already present, however, including the American cheetah (*Miracinocyx trumani*), the American lion (*Panthera leo atrox*), the saber-toothed tiger (*Smilodon fatalis*), and the dire wolf (*Canis dirus*). The last three of these may well have regarded humans, especially the very young and the very old, as a new source of prey, but to the invading people these large carnivores were in competition for the flesh of large herbivores. All of these species became extinct in North America about 11,000 years ago, but some of the other large carnivores from those times survived, including the polar bear, the grizzly bear, the mountain lion, the timber wolf, and the wolverine. It is possible that conflict with human invaders led to their extinction.

Some other species of large mammals are also believed to have became extinct at about the time of the human population expansion in North America, including the Shasta ground sloth (*Nothrotheriops shastensis*), the giant beaver (*Casteroides ohioensis*), and species of horses, camels, mastodons, and mammoths (see sidebar below). Overall, there were about 55 species of large mammals belonging to 35 genera that became extinct during the closing stages of the Wisonsinian glaciation in North America. These large mammals, all potentially over 110 pounds (50 kg) in weight, have become known as the *megafauna*. Undoubtedly, all these animals were on the list of human prey, but it is very difficult to prove that people were directly responsible for their extinction. One of the most compelling arguments for human involvement is the fact that the dates of human arrival and large mammal extinction very often coincide. The extinction of the megafauna was not restricted to North America but took place in many other parts of the world, including Europe, Asia, and South America. The circumstantial evidence is clearly strong and perhaps conclusive.

Mammoths

The mammoths are a group of large extinct elephants that were closely associated with tundra environments. Perhaps the best known is the woolly mammoth (*Mammuthus primigenius*), some of the largest of which were 17 feet (5.2 m) high at the shoulder and weighed almost 10 tons (10,000 kg). By the end of the last Ice Age mammoths were only about half this size, so a process of shrinking had gradually taken place during the Pleistocene epoch. They resembled elephants in their general form but were covered in long reddish brown hair and carried long curved tusks, far larger than those of modern elephants. Their backs sloped gently upward from their rear haunches to the head because their shoulder region was padded with massive accumulations of fat, presumably a food storage device. Thick blubber beneath the skin also provided insulation, allowing them to survive in temperatures as low as −50°F (−46°C). Like all herbivores, the mammoths needed to eat large quantities of vegetation to keep up their energy levels, and this meant hard wear on their teeth. To compensate for this, the mammoth jaw had only four grinding teeth on either side, and as these were ground down they were pushed forward and replaced by new teeth at the back. The tusks were the only frontal teeth present.

The word *mammoth,* or *mammut* as it was originally spelled, is derived from the languages of northern Asia, where it means an animal that lives underground (compare the word *marmot*). The mammoth was not, of course, a subterranean animal, but its bodies and remains have often been found buried in ice and permafrost in the Arctic, where the flesh is extremely well preserved. There are even records of mammoth steaks being served and consumed in Russian restaurants. The ivory of mammoth tusks has long been sought by hunters and traders in the tundra. Some zoologists have speculated that intact DNA might be recovered from frozen mammoth cells and be used to recreate the mammoth using elephants as surrogate mothers, but DNA is a relatively unstable molecule and has not yet been found in sufficiently good condition for this to be attempted. An alternative that may prove possible is to recover the frozen sperm from mammoth corpses and use these to fertilize the eggs of modern elephants. As yet these plans remain speculative.

The circumstantial evidence for the time of extinction, on the other hand, is not as strong as it may seem. It is very difficult to establish a date for the disappearance of a species in the absence of historical documentation. In the case of the Shasta ground sloth, for example, the main evidence for its extinction is that fossil fecal material suddenly becomes very scarce and then vanishes about 11,000 years ago. But is this firm evidence of extinction? Certainly, the megafauna became much rarer around this time, but did they disappear completely? Romantic hopes that somewhere the great mammals may have survived blossomed in the days of exploration, when pioneer adventurers, including Lewis and Clark in the American West, sought lost worlds in unexplored regions where they hoped to find monsters from the past. All of these failed to establish any survival of the lost megafauna into the present day, but geologists in Siberia have found remains of woolly mammoths dating from just 4,000 years ago, contemporaneous with the Bronze Age Xia Dynasty in China. This does not disprove the hypothesis that human hunters caused their decline, but it is a warning that circumstantial fossil evidence has its gaps and its weaknesses.

There are other considerations other than that of human predation to explain the death of the megafauna. The closing stage of the last glaciation was a time of rapid climate change (see "The End of the Ice Age," page 190–192) and hence of vegetation change as the glacial episode gave way to the current interglacial. Scrub and forest was invading the open tundra landscape and spreading northward through North America as the ice sheet retreated, and this environmental change could have been detrimental to these large, open-habitat grazers. But there had been several previous interglacials during the preceding 2 million years, and the megafauna had largely survived through these. The one major difference between the transition to interglacial conditions on this occasion was the presence of humans with newly devised hunting cultures. There is also, however, evidence that other animals, particularly birds, went through a phase of extinction at this time, and it is very unlikely that humans played any major role in their demise. It is possible that the loss of the megafauna resulted in consequential changes in the entire ecosystem. Grazing intensity would have been lowered, and there would have been less dung and fewer decomposing corpses for scavengers, so the loss of a large grazing animal could have induced a cascade of other extinctions.

One other piece of evidence has recently been unearthed that makes one hesitate before laying the entire blame for the loss of the megafauna on humankind. Paleontologists examining the fossil bones of megafaunal species just prior to their extinction have noted that the animals seem to have been getting smaller. There is an overall reduction in size of about 14 percent in most species between 20,000 and 12,000 years ago. This could be an effect of climatic warming,

and Bergmann's rule (see "Body Size in the Tundra," pages 113–115) may here be seen in operation, or perhaps it is an effect of human predation, whereby large animals are being selectively culled. These questions concerning the ecological effects of the entry of people into the tundra ecosystem may never be fully answered.

Although so many large mammals became extinct at the close of the last glaciation, there are some that survived, such as the bison and the caribou. Fossil and other evidence suggest that these animals were hunted by the early peoples of the tundra, so why did they not suffer the fate of the other megafauna? The fact that bison were the prime prey animal of the North American region and the caribou, or reindeer, were the preferred prey of the Asian people would suggest that they should have been even more likely candidates for extinction than most. But this was not the case, and the reason may lie in the way human predation was carried out. Most large predators concentrate their attentions on the very young, the old, the crippled, and the sick. Taking the young has a particularly strong impact on the population structure of a prey animal because it means that fewer animals survive until they are old enough to breed. Human predation, particularly when using methods of stampeding herds over cliffs or into gullies, does not select in the same way and can lead to enhanced survival of young and therefore an increase in the proportion of breeding animals. Humans are also rational creatures, and tribes dependent on particular species of herbivores may well have developed traditions of protection and taboo for breeding individuals, perhaps even instituting a no-hunting policy for the breeding season. Being a favored prey animal, therefore, could lead to protection from other predators and expanding populations.

The lives of the early tundra people were hard and evidently closely linked to their prey animals. This is clearly indicated in their art, which has survived in a number of locations, mainly in southern Europe (see sidebar on page 200).

Cave shelters were ideal for relatively long-term settlements, probably being occupied during the winter, but the northern migration of the herds in summer meant that at least part of the tribe would need to follow to ensure a continued supply of meat. The camps erected by summer hunting parties were less substantial and have been difficult for archaeologists to detect and study, but in southern Russia and Moravia traces of huts made of animal skins have been found. These were oval and dug into the ground around their edges. No timber was available in the tundra landscape, so the tusks of mammoths provided the simplest tool for stabilizing the skins and weighing down the edges. The mammoth hunters of Mezhirich in Russia went to even greater extremes and built the entire hut out of mammoth bones laid over the top of skin. Some of these huts were very large, generally oval in shape, and up to 98 feet (30 m) long.

Early Tundra Art

As the glaciers of northern Europe retreated around 15,000 years ago, human hunters began to invade the extensive tundra landscape they left behind, preying upon the herds of large herbivores and probably following them on their seasonal migrations. These people took shelter in caves and spent some of their time documenting the animals they hunted, covering the walls and ceilings of certain caves with powerful pieces of art. It is very unlikely that such art was purely recreational. The representations of animals, most of which are fully identifiable, were deeply symbolic and were in all probability linked to hunting ritual. Making a picture of the prey seemed to give the artist power over the subject, because the artist can control the outcome of events in a picture. It is probable that different tribes saw themselves as linked to specific totemic animals, as is still the case among many primitive peoples. The animals shown, such as the bison in the cave of Altamira in Spain, first discovered in 1879, are not cringing, defeated beasts but are magnificent in their strength and defiance. Undoubtedly these creatures had religious and magical significance because the future of the tribe depended upon the success and continued procreation of the prey populations as well as the outcome of the hunt. Some animals are shown with spears and arrows penetrating their bodies, so the artists were attempting to use their skills to ensure that the human predators were finally successful. People are rarely represented, and then only by sticklike drawings. The cave art of Spain, southern France, and the southern Ural Mountains in Russia depicts bison, mammoths, wild cattle, birds, deer, and horses. The techniques employed involved engraving outlines on the rock with stone tools, then blowing pigments derived from minerals and charcoal into the crevices using hollow reeds. Finally, the artists rubbed the pigments into the rock using their hands. All of this was conducted deep within selected caves using the light of small stone lamps.

In addition to cave art, the early tundra people also created carvings, some of which are delicate and sophisticated, such as that of a deer carved on the handle of a spear-throwing implement made of reindeer antler found at Mas d'Azil, France. Unlike the cave art, people figured strongly in the carvings of these people, with an emphasis on plump, evidently pregnant females. The artists were displaying their magical powers in granting fertility to their subjects.

There were several hearths in the larger huts, suggesting that several families shared the shelter and occupied their own space within them. The oldest recorded house in the world, dated to about 50,000 years ago, is a small version of one of these mammoth bone dwellings in Ukraine.

It is fairly obvious that the early prehistoric people of the tundra must have worn clothes in order to endure the harsh conditions, and some information about clothing is contained in their art. The tundra people from near Lake Baikal in Russia and parts of Siberia carved slender human figures, mainly female, from mammoth ivory, and these are represented as fully clothed with jackets, long leggings, and hoods over their heads, all apparently made from a single skin. The fur of the skin is turned toward the outside, but it is not possible to determine what kind of animal was used in the construction of this clothing. Bone needles are present in their settlement sites, so the manufacture of clothing and tents evidently involved some skilled stitching.

As the climate became rapidly warmer, the tundra people were pushed farther north into the narrow belt of land beyond the reach of the expanding coniferous forests. The southern continent of Antarctica, however, remained free from any human occupants until the great voyages of discovery much later in history.

The alpine tundra regions of the world did not prove very attractive for prehistoric settlement, especially after the advent of agriculture. Valley situations usually proved much more attractive as locations for settlement and the growth of crops, but grazing animals migrate in summer to the high tundra pastures, where herbaceous productivity is relatively high and blood-sucking insects are often less abundant. It was natural, therefore, that the prehistoric people of high mountain areas developed a pastoral system known as *transhumance*. Shepherds and cowherds took their grazing flocks to the high pastures during the summer months, establishing temporary camps in which to live, eventually returning to the valleys for the winter, where the animals were kept confined and fed on hay and tree-leaf fodder that had been cut and dried in summer. This system is still practiced in some mountainous areas such as the Swiss Alps in Europe, as shown in the photograph on page 201.

A chance discovery in 1991 has provided a great deal of information about the way of life among the alpine people of Europe. Two German mountaineers were climbing in the high Italian Tyrol, in the European Alps. They descended northward, crossing the Italian-Austrian border as they made their way across an ice field when they saw a brown object protruding from the melting ice. It was the head and shoulders of a human body, and the climbers immediately assumed that it was the remains of some other mountaineer who had met with an accident. They reported their find to the authorities in Austria, and police returned to the site to investigate. What they found was not a victim of a recent mountaineering accident but a very ancient body that had long been preserved within the ice. Pathologists and archaeologists were immediately called in to recover and examine the body, and it proved to be a man who had met his death about 5,300 years ago. The state of preservation was extremely good, so it was possible to examine in detail his clothing and equipment. The "ice man" wore clothing consisting of layers of chamois leather and grass padding to provide additional insulation. His boots were packed with grass, and on his head he wore a bearskin hat. He carried a longbow and a quiver full of arrows, a copper ice ax, a sheathed dagger, and a wooden-framed backpack. This prehistoric traveler carried fire-lighting equipment with flints, charcoal, and dried fungus for tinder, all contained within birch bark pouches. It is difficult to discern what he was doing on a high mountain pass, but the cause of death is likely to have been hypothermia, possibly as a result of his being caught in a sudden snowstorm or avalanche. The ice man's equipment, particularly the weapons, suggest that either he was a huntsman or a traveler who was well prepared for trouble on his journey. His tissues are so well preserved that it has proved possible to analyze the DNA of his cells, and this has shown that he originated north of the high alpine ridge, from Austria rather than Italy. Whether this early mountaineer was heading north or south at the time of his death will never be known, but his sacrifice has brought a wealth of new information to those who study the people of the alpine tundra and their history.

Cows graze freely over the high alpine pastures of Switzerland in summer, are kept in stalls at lower altitudes through the winter, and are fed on hay that is gathered from protected meadows. *(Peter D. Moore)*

MODERN TUNDRA PEOPLE

The study of tundra archaeology is extremely difficult because the instability of the soils, with constant frost heaving (see "Soil Formation and Maturation," pages 54–59), disturbs the evidence that can be used by archaeologists. Campsites with post holes and arrangements of rocks are soon made virtually undetectable by the regular freezing and thawing of soils. Only occasionally do the remains of stone chippings and implements provide a clue to the former occupation and use of a site. This is the case at Cape Denigh, close to the Bering Strait in Alaska, where archaeologists discovered quantities of small flint blades dating from 5,000 years ago. It is likely that these small blades would have been mounted on bone shafts to make harpoons, spears, or arrows that were used by the northern hunters of that time. Archaeologists call this the Arctic small tool tradition, and similar materials began to turn up throughout Alaska and northern Canada.

The people responsible for this stone tool culture had evidently spread through the far north of North America, eventually reaching Greenland by about 4,000 years ago. They were the founders of the modern peoples of the North American Arctic. These people were originally named Eskimos, a name derived from the Cree Indian language meaning "flesh eaters," but these people themselves prefer their own name of Inuit, meaning simply "the people." They were originally far less coastal in their zone of occupation than is the case with modern Inuit and ranged far inland in Alaska and Canada. They began to assemble along the coasts from about 3,000 years ago. It is possible that this shift in distribution was related to the increasing rarity of the musk ox, which was once a major source of food. From then on the Inuit people became essentially maritime in their lifestyle, building boats and hunting marine mammals, as shown in the photograph below. They developed intricate and highly imaginative styles of art, carving bone and soapstone often in the form of animals and resembling in many ways the ancient traditions of stone age art. These people managed to survive as far north as Ellesmere Island, within 470 miles (756 km) of the North Pole, which illustrates how hardy and resilient they were.

In Eurasia the mammoth-hunting people gave way to cultures that concentrated on the reindeer (caribou) as their prey. Their main hunting period was in the fall, when large herds of migrating reindeer headed south from their tundra breeding grounds. They were most vulnerable when crossing large rivers, and the development of a boat technology was important for the hunters to ensure a good harvest. Hunting was relaxed during the breeding season, allowing the herds to recover from the autumn cull. Some archaeologists have estimated the ancient reindeer population of Eurasia at about 3 million, with an annual increase of 7 percent. If this is correct, then these animals could have supported a human population of about 11,000 people. These were nomadic people, following the herds and becoming so dependent upon them that the human population was limited by the success of this one animal. Meat storage through the Arctic winter was not a problem, but fuel for fires was very scarce in the far north.

At some stage in prehistory the Eurasian Arctic people domesticated the reindeer, and this must have solved many problems. By maintaining tame herds, the pastoralists could ensure the safety of the animals from wolf predation and control their movements. This eventually led to some conflict between the reindeer herders and the wild reindeer, which persists to the present day. The pastoralists claim that wild animals graze the best pastures and thus reduce the productive potential for their own herds, and they also lead their animals astray, tempting them back into the wild. Currently there are about 2.5 million domestic reindeer in the Russian Arctic and less than 1 million wild animals, concentrated into three main herds. The biggest wild herd is on the Taymyr Peninsula of Siberia, which is beyond the range of the main range of husbandry, so there is no reason why wild and domestic reindeer should not continue to survive alongside one another.

As in the case of the Inuit people of North America, the northern Eurasian cultures were quite late in adopting a maritime life. Much of the central region of northern Eurasia has seas that are frozen for long periods, so the supply of marine mammals is poor. Only in the regions in contact with the Atlantic and Pacific Oceans is a maritime culture sustainable.

The diet of the northern peoples is thus extremely rich in meat, either the lean steaks of caribou or the extremely fatty flesh of seals, walrus, and whales. The flesh of the

A group of Inuit fishermen using kayaks; taken in 1905. (*Them Days Photo Archive, Happy Valley/Goose Bay, Labrador*)

polar bear is edible, but the liver is toxic because of its high concentration of vitamin A. Some vegetable material is available for the reindeer cultures because they also eat the stomach contents of the animals, which is rich in herbs and lichens, but the seal-based peoples are faced with what seems a very unbalanced diet. Some add herbs to their food as a means of improving its taste. The Chukchi people of Siberia, for example, make use of about 20 different plant species in their cuisine, and the Inuit people have long used bone hoes for cultivating and collecting herbs. Some saxifrages, willows, and docks are used in this way as well as certain marine algae, such as the large brown seaweeds of the genus *Laminaria*. Some of these plants, such as the docks (e.g., *Rumex arcticus* and *Oxyria digyna*) and nettles (*Urtica* species), are relatively nutrient-demanding and can easily be cultivated on dung heaps and middens, which is probably how these species became the subjects of a crude form of domestication among the northern peoples.

Construction of dwellings has remained a problem for the people of the far north right up to recent times. When explorers first visited and lived with the Inuit people of northern Canada, they were amazed to find whole villages built from ice taking the form of a series of rounded domes, the igloos. They carved blocks of compacted snow into the required curving bricks and fused them together with water that froze to seal them tightly. Snow has a high capacity for insulation and protected the inhabitants from the chill factor of the blisteringly cold winds with their suspended ice particles. The explorers described how the Inuit people illuminated the interior of the igloo with lamps of burning seal oil, and the effect of such light within an ice dome was extremely impressive. The flame was reflected from the vast array of ice crystals, filling the entire space with a soft and diffuse glow. The inhabitants constructed sleeping platforms within the igloo over which they spread caribou, bear, and musk ox skins.

■ ARCTIC EXPLORATION

The first nonnative people to travel to the tundra lands in historic times were the Norse, or Viking, people of Scandinavia. The Vikings had discovered Iceland by 850. They sailed from the European mainland and settled in Iceland between 870 and 930, reaching southern Greenland by 985 and establishing colonies there. The legends and myths of these people describe a land beyond these settlements, called Vinland, which historians have debated but could refer to the North American mainland or Newfoundland. Archaeological evidence has provided support for the idea that the Vikings did indeed reach Newfoundland and establish at least temporary settle-

ments there, so the "discovery" of North America can no longer be credited to Christopher Columbus (1451–1506), who reached the New World in 1492. Life in the northern lands proved hard and was based on fishing and pastoral farming, largely of sheep and cattle. The grazing season, especially in Greenland, was very short. Little hay could be gathered for winter fodder, so long-term survival of permanent settlements, especially as the climate steadily cooled through medieval times, proved impossible. Graveyard records suggest that some Norse people survived up to about 1500, but with the gradual collapse of these settlements, any knowledge of lands beyond the North Atlantic was relegated to the level of myth and saga.

The 16th century saw many great voyages of discovery, but most were concerned with establishing trade routes to the lucrative spice sources in the Far East and took the adventurers south around the Cape of Good Hope and Cape Horn. John Cabot (1450–98), an Italian-born Englishman, was one of the very few to look to the north for a route to the Pacific, sailing along the coast of Newfoundland in 1497, but he found no indication of a passage to India. In 1551 the attempt to find a northern route was renewed by an exploration company founded in London that financed an expedition by Richard Chancellor (unknown birth date–1556). This voyage of discovery set off with two ships in 1553 to seek a northeast route via Norway, but one ship was wrecked off Russian Lapland. The stranded sailors died there of starvation. Chancellor survived and set off from Norway once again, arriving in what is now the port of Archangel in northern Russia. Unable to continue the eastern journey by ship, he traveled by sledge to Moscow, a journey of 700 miles (1,100 km) and succeeded in establishing strong trade links with the Russian czar Ivan the Terrible (1530–84). So his expedition did not achieve what was hoped but did bring some rewards.

Queen Elizabeth I of England (1533–1603) then turned her attentions to the northwest in her search for a shortcut to India and began the long and fruitless quest for an anticipated northwest passage. In 1576 Sir Martin Frobisher (1535–94) sailed west beyond Greenland and discovered Frobisher Bay on the southern end of Baffin Island. He was amazed to find people there, and mistaking the Inuit for inhabitants of India, he captured an Inuit man together with his kayak boat and returned to London in triumph. But the Inuit man soon died of a cold. Subsequent attempts to settle a colony on Baffin Island failed miserably, so Frobisher's triumph was short lived.

Meanwhile, the Dutch, who were great trade rivals of the British at that time, concentrated their efforts on the possibility of a northeast passage. Willem Barents (unknown birth date–1597) undertook dangerous expeditions into the far north, discovering the archipelago of Svalbard and the island of Novaya Zemlya and giving his name to the Barents

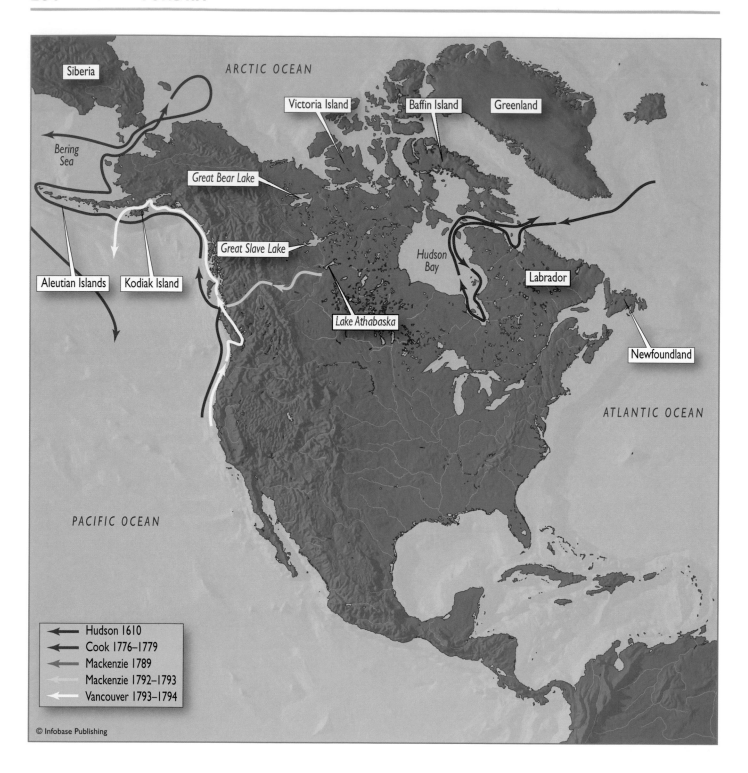

⬅	Hudson 1610
⬅	Cook 1776–1779
⬅	Mackenzie 1789
⬅	Mackenzie 1792–1793
⬅	Vancouver 1793–1794

© Infobase Publishing

Sea that lies between them. On one occasion his ship was trapped by the developing winter ice and was so badly damaged that it was forced to land on Novaya Zemlya. The crew salvaged timber to build a shelter for the oncoming winter. Somehow they survived the winter, the farthest north any Europeans had ever managed to endure this harsh season in the Arctic. Yet with the coming of spring there was still no hope of rescue, so the crew retrieved the lifeboats from their wrecked vessel and began to row the 1,600 miles (2,600 km)

Map depicting early European exploration of the tundra lands of North America. Oceanic surveys were conducted by Hudson, Cook, and Vancouver while Mackenzie first visited the northwestern tundra overland.

to the Kola Peninsula of Russia. Remarkably, most of the Dutch party managed to reach safety in Russia, but Barents died on the journey.

The English continued to look northwest, sending out expeditions under the command of Henry Hudson (unknown birth date–1611) in 1610 and William Baffin (1584–1622) in 1615. These men traveled widely around the Hudson Bay and Baffin Island region but were always turned back by ice. Baffin actually sailed into Lancaster Sound, which, although he never knew it, would have led him eventually into the Beaufort Sea and then on to the Bering Strait. He came very close to discovering a northwest passage, but even had he done so, the economic value of this hazardous route would not have been great. The voyages of Hudson and other early explorers of the North American tundra are shown on the map on page 204.

The Russians had the advantage of land connections with the Far East, but the lands of Siberia were as unexplored and unknown as the oceans. Czar Peter the Great (1672–1725) set up a program of exploration in 1724 employing the services of a Danish man, Vitus Bering (1680–1741), to lead the work of surveying, mapping, and claiming the lands for the Russian nation. He was extremely successful despite the hardships involved in crossing four major rivers and traversing a difficult landscape with no roads or trails. He eventually reached the Kamchatka Peninsula in eastern Russia. There he constructed ships and sailed through the Bering Strait to Alaska, claiming the land for Russia, and down the west coast of America to San Francisco. By 1784 the Russians had established settlements and trading posts in Alaska, which stirred the British, always nervous of the development of Russian imperialism, to renew their explorations in Arctic Canada.

The Pacific coast of North America was proving attractive to both the British and the Spanish in the latter part of the 18th century, and the arrival of the Russians in Alaska undoubtedly encouraged additional exploration in this region. In 1794 the British naval officer George Vancouver (1757–98) sailed his ship HMS *Discovery* into Glacier Bay in Alaska. He was impressed by the 10-mile- (16-km-) long ice wall where glaciers from the valleys entered the ocean. The American naturalist and conservationist John Muir (1838–1914) visited the region 85 years later and found that the glaciers had retreated 48 miles (77 km), possibly as a result of global warming over the intervening years. The main glacier of Glacier Bay subsequently became known as the Muir Glacier.

Exploration of the northern regions of North America also proceeded on land, particularly as a result of the companies with an interest in fur trading. In 1789 the Scot Alexander Mackenzie (1755–1850), who was employed by the North West Fur Company, set out with Native American and Canadian colleagues in birch bark canoes to seek a river route from the continental areas of North America into the North Pacific. He found a great river that led westward from the Great Slave Lake, but it gradually turned northward and eventually led into the Arctic Ocean instead of the Pacific. He named it the River of Disappointment, but it later became known as the Mackenzie River (see map on page 204). This waterway may not have provided a link to the West Coast, as was originally hoped, but it did represent a major advance in the exploration of the tundra regions. The discovery also opened up the region to fur trappers, whale hunters, and gold prospectors. More recently the region has become important for oil extraction.

The British had not lost all hope of an oceanic northwest passage, and in 1818 the British Parliament offered a financial reward to anyone either discovering such a passage or reaching the North Pole, which held a certain geographical prestige. There was an interest in exploration in general and the Arctic in particular, as demonstrated in 1818 by the publication of Mary Shelley's novel *Frankenstein, or the Modern Prometheus,* which depicted its now famous monster in an icy tundra setting. The challenge was taken up by John Ross (1777–1856), who first led a series of expeditions that managed to penetrate farther west than any previous attempts but still became blocked by ice short of the open waters of the Barents Sea. With his colleague William Edward Parry (1790–1855) he was forced to spend a winter in the High Arctic in 1820, which the explorers described as one of "deathlike stillness" and "dreary desolation" as they endured 84 days of total darkness. They maintained their morale by putting on a series of theatrical productions among the crew.

Parry then turned his attention to the other great goal of Arctic exploration, the North Pole. He decided to start from Spitzbergen, an island of the Svalbard group, and set out with 28 men carrying boats across the ice in the late spring of 1827. They walked over floating ice and used the boats to cross stretches of water, eventually reaching latitude 82°N, within 500 miles (800 km) of the North Pole, before they had to turn back. This was the farthest north anyone would reach for another 48 years. The pole was eventually reached in April 1909, when a U.S. naval officer, Robert Peary (1856–1920), led an expedition from Fort Conger on Ellesmere Island in Canada. Using relays of dogsleds and aided by the Inuit people, the team laid out a route with supply stations. Peary himself together with a companion covered the final 155 miles (250 km). The achievement was marred to some extent by a claim (later proved false) by another explorer that he had reached the North Pole the previous year, but Peary is now regarded as the first man ever to reach this remote and ice-bound location. The seasonal timing of such expeditions is critical, since they must take place after the worst of the winter weather and yet before the Arctic Ocean pack ice breaks up and becomes dangerous. Besides the dangers of collapsing or rafting ice, there is the ever-present risk of wandering and aggressive polar bears.

The successful discovery of a northwest passage did not occur until the voyage of the Norwegian Roald Amundsen (1872–1928), who was later to become the first man to reach the South Pole. His voyage took place between 1903 and 1906 and took him north of Baffin Island and Hudson Bay, through Baffin Bay, and then through the channels separating Victoria Island and Banks Island from the mainland. This passage is now known as the Amundsen Gulf. Eventually Amundsen arrived at the Mackenzie River delta. To travel from there to the Bering Strait, he had to wait until the ice had melted, and then he finally sailed down the west coast of North America to San Francisco. By the time he arrived, the news of his success had already become a little stale. Moreover, the excitement that might once have surrounded this great achievement had diminished because although the existence of the route had now been demonstrated, it clearly could never become commercially important. Somewhat ironically, jet aircraft from Europe to the West Coast of North America essentially follow the course of the long-hoped-for Northwest Passage through the Arctic tundra, so the route has become commercially valuable, but not as a trade route for eastern spices.

■ ANTARCTIC EXPLORATION

The existence of the Arctic tundra was never in any doubt. The Vikings were familiar with it, and tales of the bleak north were present in Greek and Roman mythology, probably based on tales brought back by travelers and traders from these distant regions. The Antarctic, however, was shrouded in mystery until very recent times. The Greek philosopher and naturalist Aristotle speculated in the fourth century B.C.E. that the great northern landmasses must be "balanced" by a landmass in the deep south. This supposed continent later became known as Terra Australis Incognita, the "Unknown Southern Land." Aristotle called it Antarktikos. The northern lands lay under the constellation Arktos, "the Bear," so the southern lands must be the opposite of this, he reasoned.

When the Portuguese explorer Vasco da Gama (1469–1525) sailed around the southern tip of Africa and eventually reached India in 1497, he proved that Africa was not joined to the Unknown Southern Land. Ferdinand Magellan (1480–1521), another native of Portugal, sought the southern tip of South America in 1519. He sailed through the strait that now bears his name, which separates the mainland of South America and the island of Tierra del Fuego to its south. But he was not sure whether this was truly an island or another great landmass, the Unknown Southern Land. The English sailor Francis Drake (1540–96) settled this issue when his ship was blown south in the Pacific by a great storm and ended up rounding Cape Horn, thus demonstrating that there was no connection between South America and the Unknown Southern Land. He also provided the first description of penguins, noting that they were easy to catch (being flightless) and good to eat.

By the mid-17th century Dutch traders were regularly sailing in the South Pacific, traveling along the west coast of Australia, and they considered it possible that this was the Unknown Southern Land of Aristotle. Among these Dutch explorers, Abel Janszoon Tasman (1603–59) was the first person to circumnavigate Australia and the first European to encounter the islands of Tasmania and New Zealand, but these lands were temperate in latitude and climate, not the equivalent of the Arctic wastes.

The notion that the Antarctic continent might exist continued to appeal to explorers, but it did not have the economic appeal of the Arctic in terms of the new possible sea routes it might offer. The fishing and whaling potential of the Southern Ocean, however, eventually led to its systematic exploration. The British navigator James Cook (1728–79) spent much of his life exploring the Pacific Ocean, including the southern regions, during the latter half of the 18th century. His voyages are shown on the map on page 207. He was the first to sail south of the Antarctic Circle, in 1773, but had to turn back on his southbound voyage because of pack ice. He little realized that he was only 80 miles (130 km) from the legendary continent. Cook succeeded in circumnavigating the fringing ice packs of Antarctica but was never convinced that the southern polar regions contained more than floating ice.

A Russian, Thaddeus von Bellingshausen (1778–1852), first sighted land in 1820. In unusual conditions of excellent visibility (something denied to Cook), he spotted high cliffs of rock, thus proving that a southern continent existed. The first people to set foot on the continent were the crew of an American boat, the *Cecilia,* who landed on the Antarctic Peninsula just a year later, in 1821. James Ross (1800–62), of Arctic tundra fame, was also eager to add the Antarctic to his polar explorations, and in 1841 he led an expedition that aimed to reach the South Pole. As he approached the pole, however, he found his passage blocked by an enormous mass of pack ice, now known as the Ross Ice Shelf, and he was unable even to approach the region of Antarctica in which the pole lies. He did see a great volcano that was active at that time, and he named it Mount Erebus. Erebus is the name the Greeks gave to the region of darkness through which the dead must pass, which gives some indication of the awe in which Ross and his fellow sailors held these inhospitable regions. The volcano remains periodically active, but its main claim to infamy is a tragic accident in 1979, when a plane carrying tourists became enveloped in a blizzard and crashed into Mount Erebus, killing 257 passengers and crew.

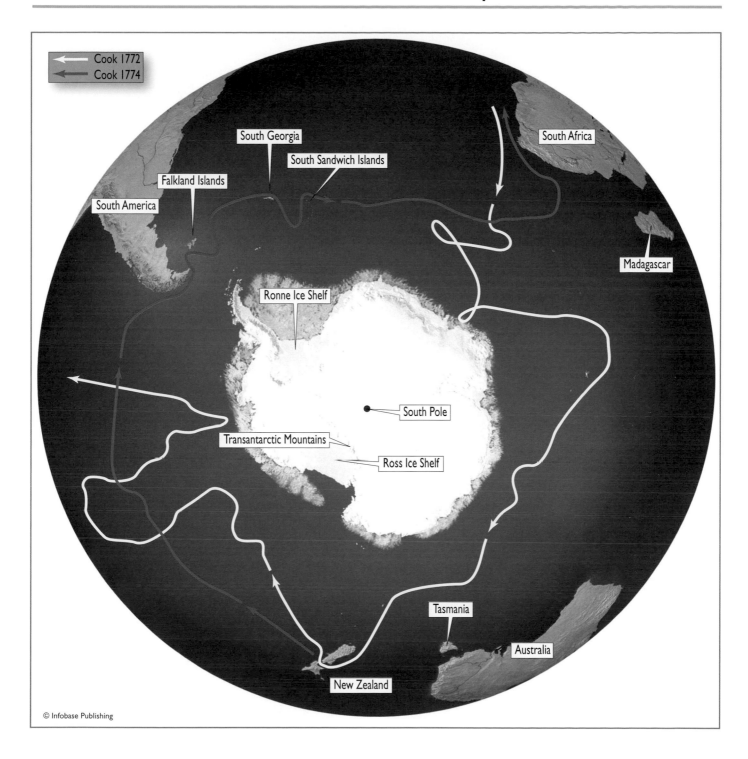

Map showing the voyages of Captain Cook around Antarctica in the late 18th century

The South Pole itself remained unconquered until 1911, when a dramatic and tragic race took place between two teams of explorers. From Norway came a team led by Roald Amundsen, and from Britain one led by Robert Falcon Scott (1868–1912). The Norwegians set out on October 19, 1911, using four sledges pulled by dogs. Their most difficult section consisted of the climb up to the polar plateau over the dangerous ice mass of the Axel Heiberg Glacier. From there the going became easier as Amundsen's team crossed the relatively flat plateau, even-tually reaching the South Pole on December 14 and rais-ing the Norwegian flag. The British team reached the pole on January 17, having hauled their sledges over the entire course without the use of dogs. Scott's comment on reach-ing the South Pole was "This is a terrible place." Ill fortune plagued the return journey of Scott's team. Despite acts of heroism and self-sacrifice, blizzards and frostbite took the lives of all members before they reached their base. Once again, the remarkable success of Roald Amundsen was somewhat overshadowed in the news media by the romantic yet tragic failure of Scott's team. In both his Northwest Passage expedition and in his attainment of the South Pole, Amundsen seems to have missed out on the public acclaim that was his due.

With the Amundsen and Scott expeditions the great romantic age of Antarctic exploration ended, and the period of commercial and scientific investigation took over. This was also the beginning of a period of human history in which human populations have expanded dramatically, and the need for food and raw materials led people to exploit even the most unlikely of the world's wildernesses.

■ ALPINE TUNDRA EXPLORATION

Mountains have always been a source of admiration and often religious awe for humanity. The people of the Old Testament looked to Mount Sinai as the place where God made known to the Israelites the Ten Commandments and the Greeks regarded Mount Olympus as the place of residence of their community of gods. High mountains provided few opportunities for human settlement or for the cultivation of crops, and even grazing domestic animals was possible only in the summer months in the grassy tundra zone between the tree line and the snow line. The snowy peaks and icy glaciers were an obstacle to travel, and few would have been tempted to climb them. The Inca civilization of Peru lived among the high Andes Mountains of South America, using the steep slopes to protect their cities, hunting over the alpine tundra, and cultivating the lower slopes of the mountains, but the very high peaks held few attractions.

Early travelers' tales tell of the dangers of mountain travel, where weather conditions change quickly and freak snowstorms can prove fatal. The Islamic explorer Leo Africanus (1485–1554) writes of the dangers of travel in the Atlas Mountains of North Africa. He tells of a three-day ordeal trapped by blizzards in a mountain cave from which he was fortunate to escape alive, but in the process he lost an entire wagon train of goods he was taking through the mountain chain.

The 19th century was a time of imperial expansion, and with the conquest and exploitation of distant nations came the need for detailed geographical survey. The British East India Company, for example, controlled trade in southern Asia and became interested in making accurate maps of India, which included surveying the Himalaya Mountains. One of the most skilled surveyors of the time was George Everest (1790–1866), who developed trigonometric methods of surveying that included the calculation of the heights of mountains. His successors applied his techniques to a survey of the high Himalayas, which could not be surveyed directly because the team was forbidden entry to Nepal. The surveyors worked from northern India, about 160 miles (260 km) from the main mountain chain, and calculated the heights of all the major peaks. The highest reached an altitude of 29,002 feet (8,839 m), and they considered it to be the highest mountain in the world. George Everest was honored in the naming of this mountain, Mount Everest, which is indeed the highest of all mountains and which actually rises to 29,028 feet (8,848 m). The trigonometric methods of George Everest had provided a remarkably close estimate.

Although geographical science provides one great incentive for exploration, there is also within the human spirit a desire to take up physical challenges and overcome the obstacles of nature. The journeys to the North and South Poles illustrate this driving force, which has subsequently taken humanity to the Moon. The high mountains of the world have always presented people with this type of challenge, and many have died in their attempts to succeed. One of the most visually impressive mountains in the world is the Matterhorn in Switzerland (see the photograph on page 41), and its steep sides and overhangs produce a constant rain of falling rock, which present serious hazards to all who attempt the climb. Despite various attempts to conquer the mountain, it remained unclimbed in the middle of the 19th century. In 1865 two teams, one English and one Italian, attempted the climb from opposite sides. The English team, led by Edward Whymper (1840–1911), was first to the top, but disaster struck during the descent. One climber lost his footing and fell, knocking his companion over and pulling an additional two, who were attached by rope, off the rock face and down 4,000 feet (1,200 m) onto the surface of a glacier. The remaining two braced themselves for the strain, but the rope snapped. Although four climbers lost their lives, Whymper and two companions survived. For two hours these climbers were unable to move because of their shock and grief, which drained their confidence and their nerves. At last they were able to move and begin their perilous descent into the Alpine village of Zermatt, where they had to announce their success in the conquering of the Matterhorn and their grief at the loss of their comrades.

The greatest mountain challenge of all, Mount Everest, remained unconquered for almost another century. Many gave their lives in the attempt, one of the most persistent climbers being another English mountaineer, George Leigh Mallory (1886–1924), who made a total of three climbs on Everest. In 1921 he succeeded in mapping the north face of

Mont Blanc, at 15,781 feet (4,810 m), is the highest mountain in the European Alps. A local doctor, Michel-Gabriel Paccard, from Chamonix, France, first climbed it in 1786. He, thus, initiated a wave of enthusiasm for mountaineering in Europe.   *(Peter D. Moore)*

Everest, and in the following year he climbed to 27,000 feet (8,230 m), higher than anyone had previously ascended. In 1924 he made a final attempt on the summit with one colleague and was last seen heading for the top, but he never returned. His body was not found for another 75 years. It is still unknown whether Mallory may have been the first man to conquer Everest, meeting with an accident on his return, or whether he failed in the attempt. The first to succeed in reaching the summit and returning alive were the New Zealander Edmund Hillary and the Nepalese Tenzing Norgay, who finally reached this last terrestrial frontier in 1953.

Of all Earth's biomes, the tundra has proved the most resistant to exploration and scientific study. It occupies a climatic zone that is least tolerable to human physiology and has held few attractions for agriculture and settlement, but the tundra has exerted a magnetic impulse on humanity and has inspired acts of remarkable bravery and sacrifice.

■ CONCLUSIONS

Human population levels are low in the tundra, but the Arctic tundra has long supported people. They must be regarded as part of the tundra ecosystem. Modern humans, *Homo sapiens*, have been present on the planet for more than a quarter of a million years, during which time there has been a major glaciation, when tundra was far more extensive than in the present day. Humans have a long history of tundra dwelling. The first people in North America were likely to have been of tundra origin, crossing the land bridge of the Bering Strait and spreading through the continent, adapting to different climates and ways of life.

The arrival of people in North America during in the closing phases of the last ice age, approximately 12,000 years ago, coincided with the extinction of many large mammals, the megafauna. The relative roles of climate change and human hunting in the decline of the megafauna is still debated, but it is likely that the presence of a new and efficient predator in the form of humans put additional pressure on these large mammals that were already suffering from loss of habitat as the climate warmed and forests spread. Some of the earliest examples of graphic art depict these great beasts, including mammoths and bison.

The ancient people of the tundra kept warm by dressing in the skins of their prey animals and also made shelters and tents out of these skins. Where caves were available, they readily used them, but on the open tundra they used whatever building materials were available, including the bones and tusks of mammoths or compacted ice from which they built igloos. In the alpine tundra regions the people practiced transhumance, taking flocks of domestic animals from valleys to the high tundra pastures in the summer, but in the Arctic hunters followed the migrating herds of caribou, eventually domesticating them. The diet of tundra people has a high protein and fat content because they concentrate on animal sources, but the gut contents of herbivores, such as caribou, are an important source of vegetable matter.

The Arctic tundra has been inhabited throughout human history, and Viking settlement of Greenland and probably Newfoundland dates back more than a thousand years. Through much of history, however, these northern lands have been of no economic consequence and so were largely ignored until the sea traders and explorers of the 16th century began to seek trade routes around the north of North America, the much sought after Northwest Passage to the Pacific Ocean and the spice islands of the Far East. The British spent much effort and many lives seeking this route, while the Dutch headed east and sought a passage around the north of Siberia. Both failed. It was a Norwegian, Roald Amundsen, who eventually sailed from the Atlantic to the Pacific via a northwest passage in the early 20th century, but by this time it was widely appreciated that it could never become a viable commercial sea route. The North Pole itself, which is situated on permanent floating ice, was first reached in 1909 by the American Robert Peary.

The existence of the Antarctic continent was no more than a vague possibility fueled by legend until the Russian explorer Thaddeus von Bellingshausen first sighted land in 1820. Its commercial importance increased with the growth of the whaling industry, but most of the people who have spent time there have been explorers and scientists. Roald Amundsen was the first man to succeed in reaching the South Pole in 1911, his triumph being partly overshadowed by the disaster that hit the rival British team under Robert Falcon Scott, all of whom perished on their return journey.

Alpine tundra has been occupied by people throughout history, including the Incas of South America and the Swiss pastoralists of Europe. The challenge of conquering mountains reached its height in the 19th century, when some of the highest peaks in the European Alps were finally conquered, including the Matterhorn, which cost four lives. George Mallory may have climbed Mount Everest, the highest mountain in the world, located in the Himalayas, in 1924, but he died on the mountain. It will never be known whether the accident occurred before or after reaching the summit. Edmund Hillary and Tenzing Norgay finally conquered the peak and returned intact in 1953.

The tundra has always been a difficult and unattractive environment for permanent human settlement, but some hardy cultures have achieved it and manage to survive under its testing conditions. For all humans, the tundra has offered seemingly irresistible challenges for exploration, discovery, and conquest. The call of the wild is nowhere stronger than in the vastness of the open tundra and among the snowy mountain peaks. But the highly romantic and attractive nature of the tundra wilderness may lead to its downfall. As more people learn of the tundra and are drawn by its wildness tourism develops and pressures on the wild environment increase. The influence of continued exploration of the tundra will play an important part in the future of the biome, as will be examined in chapter 10.

9

The Value of the Tundra

The word *value* immediately generates ideas of finance. How much money would this particular object or commodity fetch in the marketplace? In many areas of life this approach is perfectly appropriate, but there are other areas where value cannot be assessed in financial terms. There are some things that money simply cannot buy. The problem faced by conservationists and ecologists is that they often argue for the protection of parts of the environment that could be exploited in alternative ways, and the economic arguments for exploitation may be very convincing. The question of evaluation in environmental matters is therefore one that requires careful thought and consideration.

■ THE PROBLEM OF EVALUATION

Conservation is more than simply preservation. It involves the active management of an ecosystem to ensure that it continues to function, and sometimes this means that human activities need to be sustained. Landscapes have often been modified by a long history of human intervention, such as in the use of fire, or farming, or the harvesting of certain products, including hunting, providing a financial return from the land. The conservationist, therefore, is not always averse to the sustainable use of an ecosystem for human support or recreation, all of which add value from a human point of view, but conservation is opposed to activities that will damage the ecosystem and reduce its biodiversity. It is possible that conservation can accommodate economic requirements in the form of a sustainable harvest from a managed ecosystem. In the case of tundra, for example, local human populations may graze domestic reindeer, or harvest a proportion of wildfowl, or take fish and marine mammals for food. These activities can be sustained as long as they are kept within the limits of the ecosystem's productivity. For the local human population this harvest is of considerable value, being essential for the maintenance of life in a harsh environment.

Biodiversity is itself extremely valuable. Every species of organism on the planet has its own unique genetic makeup that has been honed over millions of years to fit the environment in which it lives. Even with a species there are variations in genetic constitution that confer adaptability on the organism, and all of these variants are of value because they could not be reconstructed if lost. Humans place great economic value on certain works of art because they are unique and could never be replaced if lost, and species of plants, animals, and even microbes are similarly irreplaceable. Extinction is indeed forever.

But are all species of equal value? Is the giant panda worth the same as the smallpox virus? This is a difficult question to answer, but until the possible value of each species is known, it is wise to avoid steps that cannot be retraced. Many obscure and unlikely species have proved unexpectedly valuable as the sources of drugs and chemicals or as laboratory research tools, so discarding species before they are fully appreciated is extremely unwise. There is also an ethical argument. Many people believe that species apart from humans have certain rights, and the right to existence is basic among these. Even if the dodo or the passenger pigeon had little economic value, their loss at the hands of human mismanagement is something that most people would regret. The world is a poorer place without them.

The same argument can be applied to whole ecosystems and landscapes. Human psychology is extremely complex, but many people feel uplifted by natural environments and wilderness. For those who may never have the opportunity to visit truly wild places, it is still reassuring to know that they exist. Tourism to the world's wildernesses (see "Ecotourism and Recreation," pages 216–219) is greatly on the increase, and television wildlife documentaries are a source of pleasure to many. So the existence of global biodiversity can have a positive impact on human health and well-being, even though this is difficult to evaluate in economic terms.

Many of the Earth's great ecosystems, the tundra among them, have significant effects upon the global environment, including climate (see "Tundra and the Carbon Cycle," pages 219–221). Scientists have great difficulty in trying to quantify such effects because of the scale of the exercise and the problems experienced in obtaining precise measurements, but most are agreed that conservation is necessary to ensure the continued functioning of global processes, including the interaction between the atmosphere and the biosphere. The economic implications of this argument are so vast as to be almost incalculable.

The global aspects of conservation are very considerable and may be difficult to conceive, but at a local level it is often possible to appreciate more easily the importance and the value of natural ecosystems to the people who are a part of them. In the case of the tundra human populations are often quite low, but these people are dependent upon the continued productivity of the ecosystem. Any assessment of the value of tundra must therefore begin with the needs of the local people.

■ HUNTING AND TRAPPING

Hunting animals has been the support system for human life in the tundra since our species first invaded these realms. Tundra plant life offers few opportunities for food gathering apart from some berries in the late summer. Arable agriculture, as the Norse people who invaded Iceland and Greenland were to discover, was of little use in these northern latitudes, but herds of large grazing mammals, from mammoths to caribou, provided a reliable source of food for human cultures prepared for a largely carnivorous diet. The Inuit people of Alaska and northern Canada have traditionally been coastal in their distribution and marine-based in their culture. Fishing and hunting seals and other marine mammals became an important and sustainable source of food for them. The most northerly settlement of people in North America is on Ellesmere Island, where Inuit settlers have been present for at least 4,000 years, and even here the population is able to survive on a largely animal diet.

Later, when explorers and pioneers of European extraction penetrated north through the boreal forests of Canada, they discovered lands rich in mammals whose skins were extremely valuable in the fashionable cities of the south. Similarly, the wastes of Siberia in northern Asia provided a supply of furs for the citizens of Moscow and St. Petersburg in Russia, so trappers lived hard lives but made good profits. The formation of trading companies such as the Hudson's Bay Company developed more formal trading routes for the movement of furs from the outback to the consumers in the cities. For many decades the wealth of the tundra lands north of the forests lay mainly in their mammals.

Hunting marine mammals is still permitted for the indigenous people of the Arctic, but the Canadian authorities have imposed strict quotas, especially for the rarer animals, to ensure that they are not brought to extinction. In 2005, for example, the Inuit hunters were given permission to kill 518 polar bears for their own consumption. This represented a 29 percent increase on the previous year based on statements by the hunters that the bears were becoming more common. Conservationists are concerned that such evidence is purely anecdotal and could be biased, so they are demanding that in the future quotas should be based on scientific evidence. There are additional factors, such as climatic warming causing the early breakup of pack ice, that could affect the survival of the polar bear, so any population control measures need to take many variables into account.

Fishing in the seas that surround the tundra lands has already resulted in the collapse of some species, such as the cod, and the decline in Arctic fish stocks will have consequences for the local people and the marine mammals of the region. In the Bering Sea to the north of the Aleutian Islands of the North Pacific Ocean, the Steller's sea lion (*Eumetopias jubatus*) has declined catastrophically since 1960. This animal was abundant when it was first described in 1741, but its rapid recent decline has reduced its population to about 35,000. It is regarded by conservationists as "threatened." There are many possible causes for the decline, including pollution, disease, predation by killer whales, and climate change, but the most likely cause is commercial fishing, particularly focused on the walleye pollock (*Theragra chalcogramma*), a fish resembling cod, which has become very popular with the Japanese. This is an important food source for the Steller's sea lion, and it may be necessary to conserve stocks by limiting fishing in order to save this mammal.

In Europe one animal, the caribou, or reindeer, formed the main source of sustenance for the Sami (sometimes called Lapp) people of northern Scandinavia and western Russia. They followed the migratory herds over the tundra landscape, using the animals as a source of meat and hides. They protected the herds from other predators, such as wolves, thus ensuring that young calves survived and resulting in an increase in the size of the herds. Thus began a kind of symbiotic relationship between the Sami people and the reindeer, resulting in domestication of this docile mammal. Herds are now carefully protected and managed in northern Europe and Asia and remain the foundation on which Sami culture is based.

In Antarctica uncontrolled hunting of marine mammals for their skins became a major industry within a few decades of Captain Cook's first penetration of the Antarctic Circle in 1773 (see "Antarctic Exploration," pages 206–208). Seals, including fur seals, were abundant in the subantarctic islands at the beginning of the 19th century, but many of these island colonies had been completely exterminated by 1830. When the seal populations had been depleted, hunters turned to penguins, which provided a source of oil from their stored fat. The protection and conservation of seal and penguin populations in recent times has fortunately led to the recovery of many species, including fur seals and king penguins.

Whales and other cetaceans, such as dolphins and porpoises, abounded in the southern oceans, and whaling stations were set up on the Antarctic islands early in the 19th century to assist in the harvesting and butchering of these marine mammals. The major impact on whale populations began in the early 20th century with the development of sophisticated hunting methods, powerful harpoons, and robust ships. The major problem with whale harvesting is the very slow rate of reproduction, which means that populations recover from losses only very slowly. Since 1949 the International Whaling Commission has attempted to regulate whale harvesting, and since the 1960s some species have been fully protected. A recovery in whale populations is now apparent among some species, such as the minke whale (*Balaenoptera bonaerensis*), but the slow-breeding species will take a long time to recover from the impact of hunting. Perhaps even more important for the survival of the whales is the conservation of the crustaceans of the southern oceans, especially the tiny krill on which many whales depend for food.

The Arctic regions have a long history of hunting seals and cetaceans by their native peoples. For many species hunting by traditional methods involves a limited harvest that is sustainable. In the case of the beluga (*Delphinapterus leucas*), for example, a regulated harvest is sustainable, but for some scarcer species, such as the bowhead whale (*Balaenoptera mysticetus*), the largest of the Arctic sea mammals, even a small harvest could damage the limited population. This means that even the traditional hunting by native tundra peoples needs to be monitored and controlled. Seal hunting for skins that can be exported, which has traditionally played an important part in the economy of the Inuit people, has suffered a major downturn in recent years. The reason is that the demand for seal skins around the world has declined as people have moved from animal skin to synthetic materials for their clothing. Ecological concerns among the consumers of the world are having a substantial impact on the economy of the people of the far north.

Large bighorn rams are prized by trophy hunters, but regular removal of the largest males from a population could weaken the genetic constitution of the species. *(Ron Shade, Yellowstone Digital Slide File, Yellowstone National Park)*

The yellow-bellied marmot (*Marmota flaviventris*) is frequently found on the rocks of tallus slopes of the Rocky Mountains. Some individuals sit upon a high rock and produce a penetrating whistle when a predator, such as an eagle, appears. Its European equivalent has suffered persecution by hunting for its fat.   *(Christine Duchesne, Yellowstone Digital Slide File, Yellowstone National Park)*

Hunting in alpine tundra has also resulted in drastic falls in the populations of some mammals. Bighorn sheep have suffered in North America, and chamois and ibex have undergone considerable declines in the Alps of Europe. All of these species have increased as a result of protection in recent years. Hunting is still permitted, but it is controlled and is often directed at large males, which are prized as trophies. Even this activity could have an impact on the genetic constitution of the species because it involves the consistent elimination of the biggest and perhaps the fittest males. Conservationists have become worried about the repeated removal of the largest bighorn rams (see photograph on page 213), which could deplete the genetic strength of the population.

In Europe marmots have been subjected to a great deal of hunting pressure because their considerable reserves of fat are regarded highly. The fat is considered a valuable insulator when applied to human skin. Fortunately, the yellow-bellied marmot of the North American mountains (see photograph above) has escaped such persecution.

As biotechnology becomes an increasingly important part of modern industrial research, the exploitation of tundra animals may become focused on their biochemistry rather than their value for food or fur. In this industry microbes often provide the most useful sources of new materials, and there is currently much interest in the microbes of Antarctica, where some organisms have extremely efficient antifreeze chemicals that allow them to grow even at very low temperatures (see "Microbes in the Tundra," pages 141–143). But conservationists are concerned that unregulated "bioprospecting" in the Antarctic could cause untold damage to ecosystems, and they are pressing for international agreements on the control of such activities, similar to those in place relating to mineral prospecting (see "Mineral Reserves," pages 214–216).

■ MINERAL RESERVES

Mineral prospecting in the tundra regions developed alongside hunting and trapping in the early history of tundra exploitation. In the early days the remoteness and the transport difficulties meant that only the most valuable of geological resources were worth pursuing, and chief among these was gold.

On July 17, 1897, a steamship arrived in Seattle harbor from the extreme northwest of Canada. It carried news of a considerable strike of gold on the Klondike River, a tributary of the Yukon River of Alaska. The news became exaggerated when journalists claimed that the boat contained a ton of solid gold, and the outcome was a crazed rush into the northern lands of men lured by the prospect of untold riches. The strike was indeed a rich one, but the Klondike River lay in an extremely remote region of mountains. Those who set off faced extreme hardships of travel and survival in the inhospitable tundra. Despite these problems prospectors arrived not only from North America but from as far away as Europe and Australia, and the settlement of Dawson City soon arose as a boomtown to service the gold rush of the Klondike.

The extraction of mineral wealth inevitably involves geological disturbance and results in the contamination and silting of streams and rivers. This was not a primary concern in the late 19th century, when environmental contamination was given little attention, but has become an increasingly important consideration in the environmentally conscious days of the 21st century. A modern understanding of the fragility of the tundra ecosystem has added to these concerns. The retrieval of gold from rocks is now more sophisticated and efficient than in the days of the Klondike gold rush. This efficiency means that larger masses of rock can be treated, and as a consequence more waste is produced. Extraction of gold for a pair of wedding bands can generate a truckload of waste rock. The environmental damage associated with gold mining, however, has now largely moved from the tundra regions to the Tropics, in Brazil and Africa.

The tundra habitats of high mountains are also subject to the impact of mining activities. The mountains of Utah contain the largest hole that has ever been created by human mining activities, Bingham Canyon. More than 2.5 miles (4 km) in diameter and half a mile (0.8 km) deep, it was created to permit copper extraction from the rocks. Open-pit and strip-mining activities of this type create the most visually destructive impacts on the landscape, but even subsurface mining creates the problem of waste disposal and results in the accumulation of heaps of discarded rock that is often slow to be colonized by plants because of its high metal content. Colonization in the cold tundra habitats is even slower than that associated with low-latitude spoil heaps. Apart from metals, some mountain areas are mined for the rock itself, which is used for building or decorative purposes. Granite, limestone, and slate are particularly in demand, and many mountain regions of the world are scarred by quarries for rock harvesting.

The possibility of commercial mineral exploitation in Antarctica has raised many difficult questions about the management of this wilderness continent. In 1959 all interested governments, including the United States and the then Soviet Union, signed the Antarctic Treaty. Its main concern was that the continent should be used only for peaceful purposes and scientific research, denying any country the right to establish military bases or to conduct weapons testing on the continent. But environmental protection required more than this, and the question arose of whether and how to exploit the likely mineral resources of Antarctica, a subject not covered by the treaty. During the cold war the Antarctic provided an important point of contact for the great world powers, and an Antarctic Treaty System evolved, which sought to develop international cooperation on management strategies for Antarctica. During the 1980s the interested nations constructed a Convention on the Regulation of Antarctic Mineral Resource Activities, which assumed that mining would occur but tried to ensure that environmental damage should be kept to a minimum. Then in 1989 France and Australia took a tougher stand and demanded a ban on all mining activities. The other states involved eventually agreed to this policy, but only on the condition that the ban should have a time limit permitting future reconsideration of the exploitation of mineral resources in Antarctica. The date agreed upon is 2048, providing 50 years of protection after the agreement came into force. So for the present the Antarctic is safe from mineral extraction, but the battle for the protection of the Antarctic tundra is not yet won. Within the lifetime of the present generation of young people there is likely to be a renewed scramble for mining rights in the Antarctic wilderness.

In the northern tundra regions an additional resource has been discovered that is creating even greater environmental problems than the gold of the past, namely oil and gas reserves. Major consumers of oil, in particular the United States, have been anxious to discover resources within their own national boundaries and thereby avoid the expense and vulnerability associated with depending on imported fossil fuels. The tundra regions of Alaska and Canada have proved to contain important reserves of oil that are now being extracted. It is important to bear in mind, however, that these reserves are limited and will be exhausted in a relatively short period of time. The North Slope of Alaska, for example, is estimated to contain only enough oil to supply all the demands of the United States for three years. Other reserves will undoubtedly be discovered. There are major fields of oil beneath the Beaufort Sea, along the coastal region of northeastern Alaska and northern Canada. But this region of Alaska comprises the Arctic National Wildlife Refuge, so two very laudable aims, the provision of the nation's energy needs and the conservation of its wildlife heritage, have come into conflict. It remains to be seen whether the two goals can be reconciled. Meanwhile, additional fields are being discovered among the islands of the Canadian Arctic, so the debate and the problem will continue for some time to come.

The exploitation of oil reserves in the tundra can expose the landscape and wildlife to many harmful environmental impacts. The most obvious of these is spillage and pollution, both at the point of extraction and during transportation. Oil is particularly harmful in marine situations. Having a low density, it floats upon water, and, depending on its viscosity and stickiness, it can produce either masses of thick, coagulated rafts or a thin film distributed over a large area. Floating oil is especially harmful to seabirds because it can coat their feathers and leave them unable to fly or to dive. These oil-contaminated birds are then in danger of starvation. In addition, as they preen their feathers the seabirds ingest toxic chemicals from the oil and become poisoned. The oil may also wash onto shores, where it damages marine life along cliffs and beaches. Cleaning up after oil spills usually involves the use of detergents that emulsify the oil, dispersing it into very small globules that eventually decompose in the water. But the detergents are often more harmful to wildlife than the oil itself, and many conservationists believe that the cleaning operations do more damage than good. Detergents, for example, remove the natural oil from a bird's feathers, leaving them easily wettable and destroying their thermal insulating properties.

Oil extraction in the tundra also involves the establishment of human settlements and the development of roads

A gas pipeline carries fuel over the tundra. Conservationists fear that long pipelines in Alaska may disrupt the migration of caribou herds, but the permafrost prevents such pipelines from being placed below ground. *(BASF Group)*

or other transport systems. Settlements generate waste, and, as has been described, waste matter decomposition is slow in the tundra, so waste mountains can develop. Apart from being unsightly, waste heaps attract pests, from rats and gulls to polar bears. Road development on tundra soils is also difficult because of the freezing and thawing that take place each fall and spring. Hard surfaces in the winter turn into wetlands in the summer, and roads quickly break up under the strain. Oil pipelines provide an alternative to the use of truck transport, especially where long distances are involved, as shown in the photograph below. The Trans-Alaska Pipeline runs from Prudhoe Bay on the North Slope south to the Gulf of Alaska and avoids the need for either fleets of tankers passing through the Bering Strait or lines of trucks traversing the Alaskan wilderness, but it nonetheless creates certain problems. Pipelines fragment the landscape and can create barriers to the movement of large mammals, such as migratory caribou. These barriers could affect the survival of some herds, isolating them genetically and perhaps exposing them to new and higher levels of predation.

ECOTOURISM AND RECREATION

Television, books, and films have provided a wealth of information about the Earth, making the wonders of the natural world increasingly accessible, even those wilderness regions remote from human habitation. It is perhaps inevitable that people should increasingly desire to visit such places and experience the wilderness at firsthand. As wealth, leisure time, and the availability of global transport increase, more and more people are indulging in an activity that has become known as *ecotourism*. Essentially, this is a type of tourism in which the focus of attention and interest is the natural world and its associated human cultures. Ecotourism sets out to be environmentally friendly, seeking to avoid any ecological damage to the area visited and aiming to look but not to touch. In some respects the approach of ecotourism resembles the game-hunting spirit of former times except that the gun has been replaced by a camera. The natural wilderness found in the tundra regions of the world, both polar and alpine, provides an ideal target for the development of such ecotourism.

In the late 19th century the prospect of visiting the Arctic tundra as a tourist was almost as ambitious as space tourism is today, but in 1892 a German shipping company began running tourist trips to the Arctic island of Spitsbergen. These and similar tourist voyages continued until 1975, interrupted only by two world wars, when an airport was opened on the island. Cruise ships remain the most popular means of visiting the polar regions of both Russia and North America, but the services now provided usually include education and information supplied by expert lecturers and guides. The North

A cruise ship carries tourists into the polar tundra regions of Alaska. By creating a minimum of disturbance to the ecosystems being visited, this type of undertaking can be classed as ecotourism. *(Celebrity Cruises, Inc.)*

Atlantic provides the best opportunities for approaching the polar regions, concentrating on Greenland, Spitsbergen, and Baffin Island in Canada. Voyages in the North Pacific usually focus on Alaska, as shown in the photograph above, particularly Glacier Bay, but some trips pass north through the Bering Strait. In recent years Russian, Canadian, and American icebreakers have even succeeded in taking paying passengers to the North Pole.

One great advantage of using cruise ships as a means of ecotourism is that they can carry all the supplies needed for the tourists and remove all the waste generated. Waste disposal is a particularly important consideration in such cold climates, where natural decomposition is slow. Ecotourism can, however carefully organized, create some environmental and ecological problems. Tourists often wish to visit sites rich in wildlife, and this can create problems of disturbance. In the Antarctic, for example, visits to seal and penguin colonies are very popular, and the animals often show little fear of humans. But this can encourage very close approaches, which cause stress to seals and penguins with young. There is also the constant danger that visitors will inadvertently bring diseases that endanger wildlife populations. Heavy trampling of fragile tundra soils and the possibility of contamination from spills of fuel and toxins add to the dangers inherent in wilderness tourism. Strict controls need to be imposed on both the numbers and the activities of visiting parties in the Arctic and the Antarctic to prevent environmental overload and consequent damage.

Alpine habitats have an even longer history of ecotourism. Mountaineering, climbing, and hiking became popular activities in the 19th century among those rich enough to afford the travel, the equipment, and the local guides. People who visited the mountains were often naturalists who were eager to collect both the plants and animals they discovered, thus destroying some of the wildlife that had attracted them to these locations. Others were "sportsmen" whose main concern was to kill and collect larger specimens of mammals and birds as trophies. As in the case of polar ecotourism, the modern emphasis of alpine ecotourism is upon observation rather than destruction. Despite this, the greater accessibility of the mountains to a larger proportion of the world's population places new kinds of stress upon alpine tundra habitats. Excessive trampling can erode trails. Waste disposal becomes a particular problem because it is not always possible for visitors to take all of their waste products away with them. In areas of the Himalaya Mountains in Asia, for example, regions of picturesque mountain landscape have become scarred by deposits of litter and other waste materials along the well-used trails. Ecotourism in the mountains therefore requires careful planning and investment. Tourist sites must provide properly equipped camping areas, including latrines and waste-disposal systems, or they rapidly become littered and spoiled. As in the case of polar ecotourism there is a limit to the number of people that fragile alpine tundra habitats can accommodate, and it is necessary to impose a system of permits to ensure that such limits are adhered to.

The economics of ecotourism are an important consideration. In areas such as the Himalaya, the arrival of visitors can result in economic gain for the local people, and this is essential if the local populations are to appreciate the financial value of wildlife conservation, encouraging them to maintain the natural environment. But if this leads to the development of sophisticated hotels to house the visitors, then the money may flow into the pockets of developers from outside the region rather than the indigenous population. Even food may be imported because local produce may not be to the tastes of foreign visitors. There is also the danger that local communities will be disrupted and cultures changed by the presence of tourists. In many instances local cultures effectively become converted into museum displays for the sake of visitor entertainment, a development that is seen both in the Himalaya and in Canada. The dangers of ecotourism, therefore, are considerable. It is very difficult to observe either wildlife or local cultures and customs without having a disturbing effect upon them, and the regulation of the numbers of visitors is essential for the maintenance of meaningful ecotourism.

Apart from their attractions for ecotourists, tundra habitats can also offer opportunities for recreation. Mountain tundra has become an important location for winter sports, especially skiing, and this has led to new threats to this

Tourists gather for whitewater rafting, a popular sport in mountain areas but one that can create disturbances for local wildlife and so cannot be classified as ecotourism.   *(Keystone Canyon Whitewater Rafting)*

wilderness habitat. The skiing industry now attracts very large numbers of people to locations that were once deserted or occupied only by small farms and villages. The influx of people has led to extensive development of roads and residential complexes. Even more serious for the tundra habitat has been the construction of ski runs and ski lifts that greatly modify the landscape. Most of the activity, of course, takes place in winter, when the tundra is covered with snow, but the major damage occurs early and late in the season, when the snow cover is thin and bare areas of grass are visible. These brown patches on ski runs are the sites of major damage to the underlying vegetation and soil as pressure, friction, and physical damage degrade them. In summer the ski run is usually easy to recognize because of its poorer vegetation cover and much lower plant diversity, caused by this damage from the skis. Ecologists have examined in detail the effects of establishing ski runs over tundra vegetation and concluded that many alpine plants suffer under such conditions. Woody species of plants and those that normally flower early in spring are adversely affected. The use of artificial snow to prolong the skiing season causes the vegetation to resemble that of late-melting snow patches. Ski runs have a much lower diversity of arthropods than natural tundra habitats, and both the abundance and the diversity of birds is also lower. Many attempts have been made to mitigate the impact of skiing on vegetation, but the dam-

age done by establishing ski runs cannot be repaired even after many years of abandonment. Recreation is important for humans, and it is inevitable that mountain landscapes should attract winter sports. As in the case of ecotourism, however, excessive use can damage the very resource that provides the pleasure, and both tourists and those responsible for the management of recreation areas must bear this in mind. Additionally, reserves for wildlife conservation need to be established to ensure that disturbance and damage do not extend into all available habitats.

One summer activity that attracts tourists to mountain tundra habitats is whitewater rafting on the turbulent rivers of the alpine areas. This activity can disturb wetland animals, but its main problem lies in the need for infrastructure in the form of roads to provide access to remote areas (see photograph above). Like so many tourist activities, numbers need to be controlled to avoid destruction of the habitat that is the focus of tourist enjoyment.

The value of all tundra habitats, like that of other wilderness areas, is very apparent to all who enjoy mountain scenery but is difficult to translate into economic terms. Wildness has a great appeal to the human spirit, which is why artists, writers, and poets have long extolled the beauty and the appeal of remote locations free from the impact of human society. Most people are inspired by pictures or descriptions of wild places, and many will go to great lengths in order to visit such locations and experience silence and solitude at firsthand. But even people who may never have the opportunity to visit the wilderness may gain pleasure from the simple knowledge that it exists and that there are still relatively uncontaminated places on the Earth. Polar and alpine tundra, therefore, are of great value to all humanity as a source of inspiration and inner peace.

■ TUNDRA AND THE CARBON CYCLE

The tundra, like all other biomes, has a close relationship with the atmosphere. Plants fix atmospheric carbon dioxide during photosynthesis and release oxygen as a by-product. Plants, animals, and microbes use the products of photosynthesis as an energy source for their growth and activities, and their respiration releases carbon dioxide back into the atmosphere. Carbon is thus constantly moving between the atmosphere and the living members of the ecosystem. Dead tissues of plants and animals are a source of energy for detritivores and decomposers, but in cold and wet conditions the process of decay can take a long time because bacteria and fungi are slower in their activities. The dead organic matter thus forms a temporary reservoir of carbon in the soil or even as a layer on top of the soil, where it accumulates as peat (see "Soil Formation

The carbon budget of the atmosphere expressed in gigatons (billion tonnes) per year. The metric tonne is roughly equivalent to the imperial ton. There are sources and sinks for carbon that either add or remove carbon to or from the atmosphere. Sources of carbon currently exceed the sinks to the extent of 3.2 billion tonnes per year, so the carbon content of the atmosphere is steadily rising. This is important because carbon dioxide is a major greenhouse gas. The "missing sink" consists of the carbon that is evidently leaving the atmosphere but is currently unaccounted for. Some of this lost carbon may be accumulating in the regrowth of forests, and perhaps soils are taking up carbon, especially the peaty wetland soils of the Arctic.

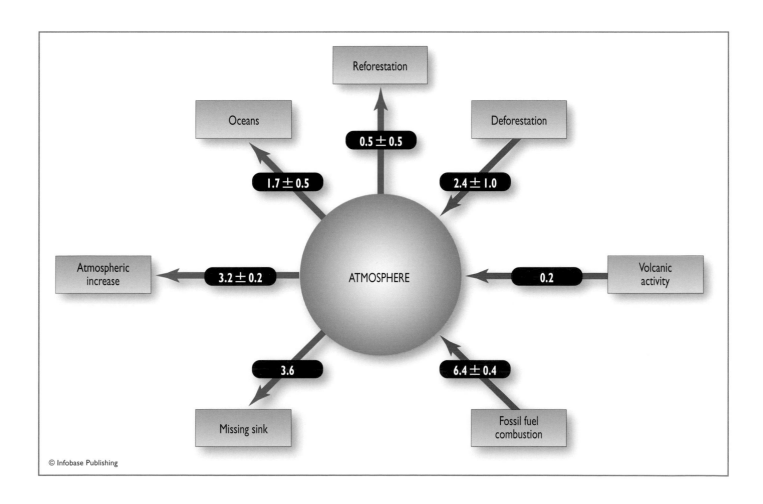

Carbon Cycle and the Greenhouse Effect

The atmosphere is a layer of mixed gases surrounding the Earth and held there by gravitational forces. The gases are translucent, so sunlight passes freely through the atmosphere and brings light energy to the surface of the planet, apart from that which is reflected back into space by clouds or by other reflective surfaces, such as the white cover of the ice sheets. When light energy strikes a dark surface, however, such as soil or vegetation, it is absorbed and causes that surface to become warmer. Heat energy is radiated at a longer wavelength than visible light and passes back into the atmosphere. But some atmospheric gases that do not hinder the passage of short-wave light energy are absorbent when exposed to long-wave, infrared heat energy. These gases include water vapor, carbon dioxide, methane, ozone, some oxides of nitrogen, and a group of compounds called chlorofluorocarbons. They are collectively known as *greenhouse gases.*

The term *greenhouse gas* refers to the way in which these gases behave, rather like the glass of a greenhouse, which also allows light energy to enter but prevents heat energy from leaving, thus elevating the temperature within the greenhouse. The greenhouse gases in the atmosphere have the same effect upon the Earth, retaining heat and causing its temperature gradually to rise. This is the so-called *greenhouse effect.* Increasing quantities of any of these gases in the atmosphere enhances the greenhouse effect and will steadily change the climate of the planet.

and Maturation," pages 54–59). Carbon dioxide also dissolves in water, and the oceans that surround the tundra landmasses provide a sink for atmospheric carbon. There are thus four main reservoirs for carbon in the tundra regions, the atmosphere, the oceans, the plant and animal biomass, and the soil.

It is possible to view the entire Earth as an ecosystem and to construct a carbon budget for all its different components. A global view of carbon cycling shows how important the atmosphere is as a carbon reservoir, as shown in the diagram on page 219. It also shows how the forest clearance and industrial activities of people are playing an important part in the global carbon cycle, taking carbon from ancient reserves in the ground and returning it to the atmosphere.

The tundra ecosystem plays a significant role in the global carbon cycle.

Atmospheric carbon dioxide has become the subject of considerable debate in recent years for two reasons. First, it acts as a greenhouse gas (see sidebar at left), and second, its concentration has been steadily increasing over the past 200 years. Evidence for the increase comes from monitoring stations around the world, especially from a long-term study based in Hawaii, where scientists take regular measurements of the levels of atmospheric carbon dioxide. More ancient values have been calculated from the analysis of gases trapped in the ice caps. Bubbles of atmospheric gas enclosed within the compacting ice provide fossil samples of past conditions so that very long term trends in atmospheric conditions can be analyzed.

The period over which atmospheric carbon dioxide levels have been rising corresponds to that of human industrial development powered by fossil fuels containing carbon. The industrialized nations have been burning geologically ancient reserves of carbon in the form of coal, oil, and gas, once buried in the rocks, to generate the energy used in daily living. The increase in atmospheric carbon dioxide over the last 200 years may not seem very great, from around 0.028 percent to 0.038 percent by volume of the atmosphere. This is a very small concentration of gas, hardly more than a trace, and the increase in concentration is also relatively small, but most atmospheric scientists believe that its impact on the climate has been considerable. Global temperature has risen by about 3°F (1.5°C) during those 200 years, and it seems likely that the rise in atmospheric carbon dioxide has been an important contributory factor to this increase.

If atmospheric carbon dioxide concentrations continue to rise at the current rate, we can expect additional increases in global temperature and considerable changes in the Earth's climatic patterns, most of which would be harmful for human populations and for wildlife conservation. Any ecosystem that takes up more carbon dioxide from the atmosphere by photosynthesis or by any other chemical process than it releases to the atmosphere by respiration is called a *carbon sink* and must be regarded as valuable and worthy of protection. Growing forests are carbon sinks, as are wetlands. The forest is storing up carbon as new tree biomass, while the wetland creates organic mud and peat in which carbon is stored. The tundra biome, especially the tundra of the Arctic region, is also currently a sink for carbon. Decomposition in tundra is slow (see "Decomposition in the Tundra," pages 98–99), and organic materials can accumulate in soils. In wetter sites, which are frequent in the tundra regions of the Arctic, peat develops and forms a store of carbon. The tundra, therefore, is acting as a sink for atmospheric carbon and will continue to do so while conditions remain cold and damp. One of the major concerns about the future of the tundra, however, is whether this will continue to be the case.

Of the various carbon reservoirs in the tundra, therefore, it is the soil carbon, especially the peat, that has the potential to make this biome act as a sink. Any change in climate that stimulates decomposition, such as warmer or drier conditions, could lead to a reversal of the current situation. The tundra could change from being a sink for carbon into a source as the store of soil carbon is converted into atmospheric carbon dioxide by soil microbes.

■ CONCLUSIONS

As in the case of all natural biomes, it is difficult to place a financial valuation on the tundra. The biodiversity of the tundra ecosystem is not as high as that of many other biomes, but the range of animals, plants, and microbes that inhabit these harsh regions have adaptations that could provide a rich source of scientific information. The genetic variability and flexibility of tundra organisms indicates that there is a great resource of material in this biome that could one day be of inestimable value to agriculturalists and medical scientists.

The climate of the tundra currently limits the extent to which human food can be farmed there, but the caribou, or reindeer, has proved a useful domestic grazing animal that is adapted to the cold conditions and is able to support human populations. Other animals, including marine mammals, are hunted as food by local people, but the value of the tundra as a source of skins has diminished as human tastes in clothing and fashion have changed.

The mineral wealth of the tundra has been an attraction to people through the course of history, especially during the times of gold prospecting in Alaska and Canada. The interest in gold has now largely been replaced by the concern to exploit buried deposits of fossil fuels, which are particularly rich in Alaska and the Northwest Territories of Canada. The extraction of this resource, however, has brought industrialists and conservationists into a degree of conflict. Antarctica is protected by international agreement from any attempt to exploit its mineral resources until 2048, but it is impossible to predict what the future may hold for the hidden wealth of that continent.

One rapidly developing resource for all tundra habitats is ecotourism. Visitors who wish to experience the wilderness without causing damage in the process are a possible source of revenue for local populations as well as being a financial justification for wildlife conservation. Both the polar and the alpine tundra have developed ecotourist industries, but the habitat is fragile. Those concerned with the expansion of ecotourism need to regulate the numbers visiting sites because any tundra location is easily damaged by human presence. Recreational use of tundra, as in the case of skiing and other snow-based sports, is rarely compatible with wildlife conservation, so the two must be kept separate, with some areas protected from disturbance.

At a time when rising levels of atmospheric carbon dioxide are causing global concerns, one of the greatest attributes of the tundra is its capacity to act as a carbon sink. This results from the accumulation of dead organic matter in the soil as a consequence of slowed decomposition in the cold and sometimes wet conditions. The role of the tundra in the global carbon cycle is currently the focus of much research. Any change in climate could easily convert the tundra from a sink to a source of atmospheric carbon.

10

The Future of the Tundra

The tundra is a valuable ecosystem, but it is also a fragile one. It is low in biodiversity, so it would be at risk of collapse if any of its component species were to become extinct. Its nutrient capital and its productivity are relatively small, so the ecosystem is sensitive to disruption, whether by changing climate or human disturbance. Its soils are unstable, easily damaged, and easily eroded. The tundra, therefore, is a biome that is placed in danger by environmental changes, and the problems that face this biome need to be studied carefully if they are to be avoided.

■ THE THREAT OF CLIMATE CHANGE

The climate of the Earth is not static but is constantly changing. Some of these alterations in climate are due to regular astronomical cycles and so can be predicted to a certain extent (see "Glacial History of the Earth," pages 176–178), but other changes are less predictable, especially those associated with human activity. The entire world is currently becoming warmer, and most climatologists agree that human industrial activity, particularly the burning of fossil fuels, is contributing to this temperature rise. Over the past 150 years global temperature has risen by at least 3°F (1.5°C), and the results of this rise are apparent from many sources. The retreat of the Glacier Bay ice in Alaska, as shown in the diagram on page 223, provides a direct historical record of the effects of climate change in the Arctic. It is impossible to predict with accuracy the climate of the future, but if the present warming trend continues, the global mean temperature is likely to become between 2° and 4°F (1° and 2°C) warmer in 2050 than it is at present. This may not seem very much, but it could have considerable geographical and biological consequences. These consequences will not be evenly spread over the face of the Earth because the warming process will not be the same in all locations. Observations of the increased

temperature over the past 50 years indicate that the high latitudes, which include the regions of polar tundra, have become warmer much faster than the equatorial latitudes. If this trend continues into the future, the polar latitudes are expected to undergo a greater change in climate than anywhere else on Earth, as shown in the diagram on page 224. It is possible that by the end of the current century the Arctic tundra will be 14°F (8°C) warmer than it was prior to 1990, while the increase in equatorial regions may be only 4°F (2°C).

The tundra biome contains the world's reserves of ice, and a warmer climate will mean that some, perhaps most, of this reserve will melt and join the oceans. This is already causing considerable problems in some alpine tundra regions, such as the mountains of Bhutan, where glaciers are retreating by up to 330 feet (100 m) per year, causing dams to burst and rivers to flood downstream. The melting of the Greenland ice sheet, which is almost 10,000 feet (3,000 m) deep at its center, would raise the level of the world's oceans by 23 feet (7 m). Low-lying countries, such as the Netherlands and Bangladesh, would be flooded. Many of the world's great cities, including New York and London, would be swamped. The behavior of the Antarctic ice sheet is difficult to predict. The current evidence shows that the surrounding sea ice is breaking up extensively, but in the long term much will depend on future levels of precipitation. Warmer sea temperature means faster evaporation of water, and this could bring more precipitation over Antarctica. More snow would fall over the vast ice sheet, so the ice volume might not decline as fast as might be expected on the basis of future temperature calculations. So there are still some important unknown factors in the equation. When water is heated it expands, so increasing the temperature of the water in the world's oceans will itself cause expansion. This will also add to sea levels globally. Taking into account the combined effects of melting ice and expanding water, observers expect the level of the Earth's oceans to rise by four to 10 inches (10 to 25 cm) by 2050. A sea level rise of this magnitude would flood many coastal

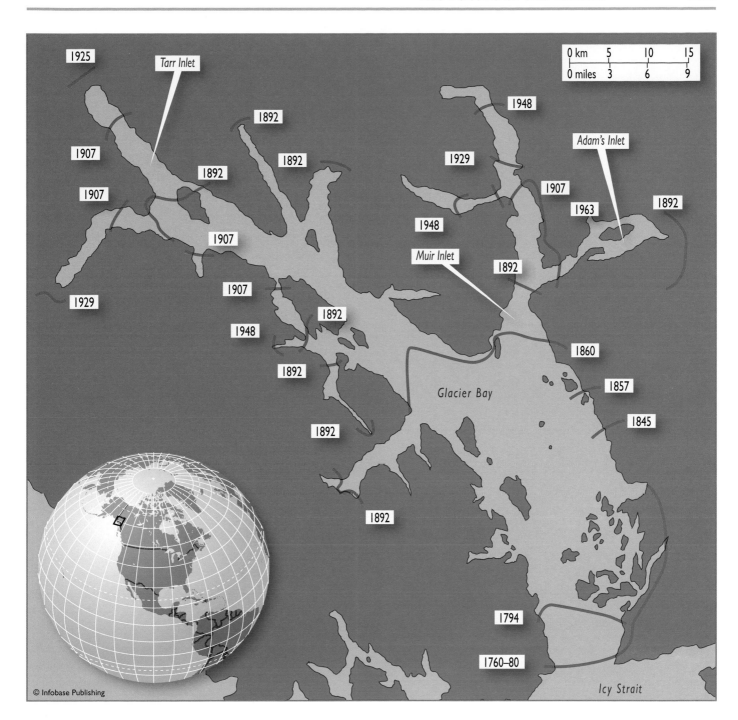

Glacier Bay in Alaska showing the progressive retreat of ice since the 18th century, when Captain Cook first observed the glacier

areas of Arctic tundra, but on the other hand, the loss of ice and snow over the tundra landscape will expose more elevated land areas where tundra vegetation can invade. The replacement of snow and ice with dark vegetation changes the reflectivity, or albedo (see sidebar "Albedo," page 8), of the ground, leading to greater absorption of solar energy and more rapid warming. This is a positive feedback mechanism, in which climatic warming results in environmental changes that induce further warming.

There are additional factors that complicate the picture. If the Greenland ice sheet melts, the entire pattern of oceanic circulation could be changed because very large volumes of freshwater would be released into the North Atlantic and might disrupt the oceanic conveyor (see sidebar "The Oceanic Conveyor Belt," page 13) that distributes low-latitude warmth to the high-latitude regions. This could cause cataclysmic and relatively rapid climatic change, lead-region expected to experience >11°F (6°C) rise in annual mean temperature by the end of the century

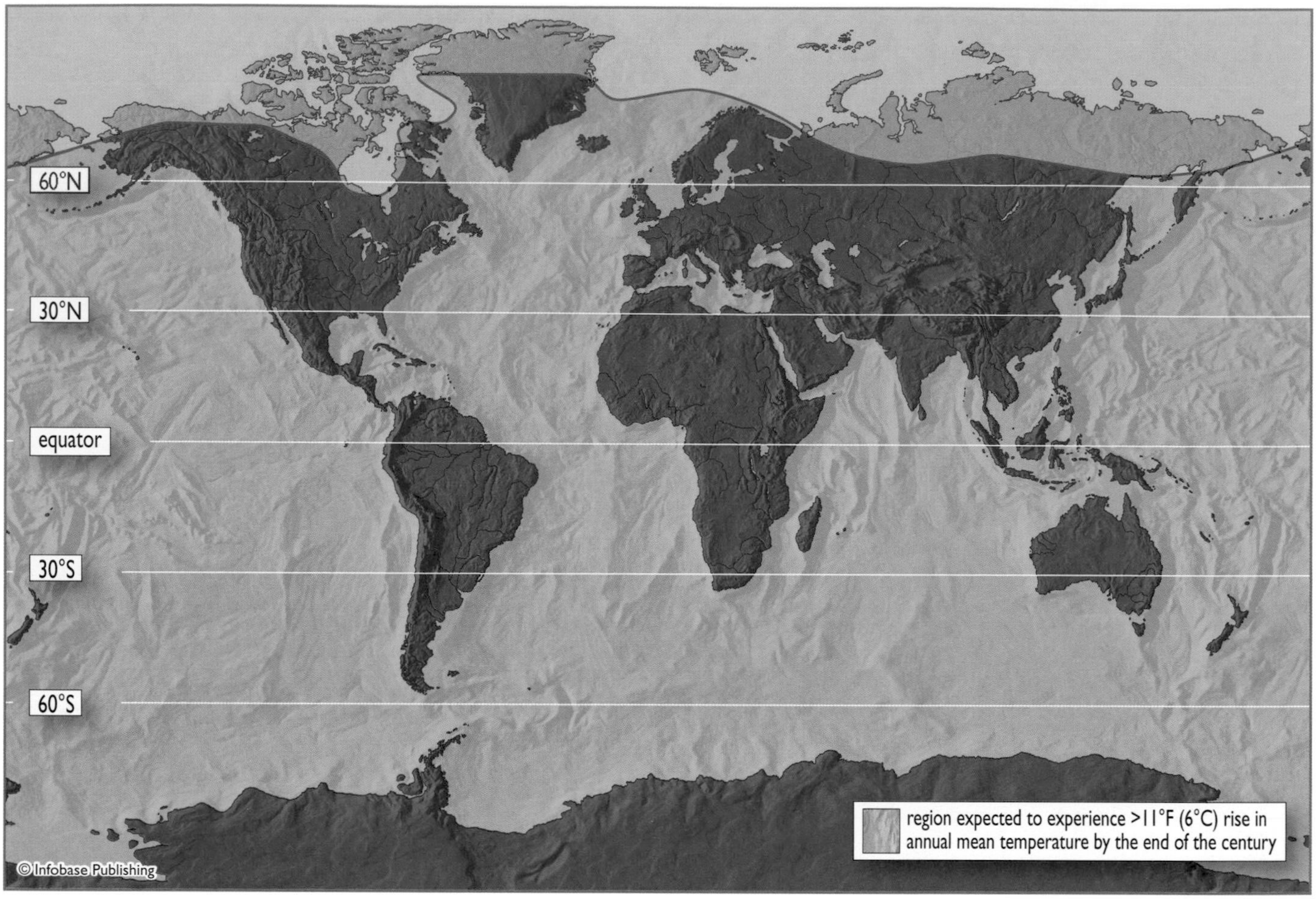

© Infobase Publishing

Scientists all agree that the Earth is currently becoming warmer, but predicting the future with accuracy is very difficult. A number of models have been constructed, and most agree that the Arctic regions will heat up faster than any other part of the world. The projection shown here indicates that the tundra lands in the Northern Hemisphere may become warmer by more than 11°F (6°C) by the end of the current century. This would have many implications for tundra wildlife, forest invasion, and wetland changes, which in turn would have a feedback on the global carbon cycle.

ing to a rapid fall in temperature in the Arctic as the supply of warm water is cut off.

Another great unknown is the impact of climatic warming in the Arctic tundra on decomposition processes. At present the Arctic tundra, especially the wetlands, act as a sink for atmospheric carbon (see "Tundra and the Carbon Cycle," pages 219–221) because some of the plant material created by photosynthesis fails to decay and accumulates as soil organic matter, or peat. Warmer, drier conditions could increase the rate of microbial activity in the soils, leading to the release of carbon from soils and peat, and the tundra

could be converted from a sink to a source of carbon. This is another example of positive feedback. Drying peat lands are likely to release more methane, which is a strong greenhouse gas (see sidebar "Carbon Cycle and the Greenhouse Effect," page 220), adding to the positive feedback.

Patterns of cloud formation and precipitation, however, could also alter in a warmer Arctic and might bring negative feedback to climate change. Current climate models predict that precipitation will increase considerably in the high latitudes under a warmer climatic regime, and this will lead to more saturated soils and could favor the development of peat-forming wetlands. Higher levels of cloud cover would also increase albedo and reduce the amount of energy reaching the surface of the Earth. So both of these effects would act as negative feedbacks to climatic warming.

■ BIOLOGICAL RESPONSES TO CLIMATE CHANGE

Animals and plants have climatic limits within which they are able to survive. They also have more confined climatic boundaries within which they grow most rapidly, reproduce most effectively, and compete most strongly with their neighbors for the limited resources of the environment. These inner limits define the optimal conditions for the organism. When climate changes this alters the geographical locations of both the optimal zone and the survival limits. The warming of climate for a tundra organism, therefore, should lead to a spread poleward and a retreat from the lower latitudes as conditions become less optimal and competition from other warmth-adapted species strengthens.

In the polar desert and semidesert zones (see "Polar Desert," pages 65–66) the vegetation is open, and there is ample opportunity for plants to invade and spread if the climate warms. Many plants are able to expand vegetatively, so they can rapidly colonize bare soils and form a complete canopy cover. Animals take advantage of the new conditions by their mobility. Invertebrates, including flies, butterflies, springtails, beetles, and mosquitoes, all have very effective dispersal mechanisms, so they respond rapidly to changing conditions. Birds and larger mammals similarly have no problem in extending their geographical ranges into new regions where the expanding plant populations raise the primary productivity and offer new feeding opportunities. In the Arctic the presence of the Arctic Ocean limits the spread northward, and rising sea levels could reduce further the area of land available for colonization by tundra organisms. Changes in the pattern of pack ice formation, breakup, and movement are likely to have severe effects upon polar bears, which use the ice as a base for hunting seals. Some species could become geographically squeezed by climate changes and might even become extinct.

While the rising sea level will erode the polar tundra from the Arctic Ocean in the north, vegetation changes will also take place from the south. Tundra scientists in the Alaskan Arctic have studied vegetation changes using aerial photography, and they have shown that the abundance of shrubs in a large study area has doubled in the past 50 years as shrubs, particularly alder and willow, have invaded the herb-dominated tussock tundra. At the boundary between the shrub tundra and the boreal forest, or taiga, in the Low Arctic, warmer conditions will enable trees to fruit and spread their seeds into new regions. The forest will advance northward into regions that are now covered by low shrubs and cushion herbs. There is already evidence of spruce invading beyond its former limits in Canada and extending the boreal forest into the tundra zone, and in Scandinavia trees have moved north by up to a quarter of a mile (400 m) over the past 50 years. The tundra biome, therefore, will become crushed between the forest and the sea, and it is likely to occupy an increasingly narrow zone as climate warms. Some forest pests are also increasing in warmer conditions, however, such as the spruce bark beetle, which

is thought to have killed some 40 million spruce trees in Alaska during the past 15 years. Such pests may help to hold the spread of forest at bay.

Botanists in Canada have also examined the higher hills and mountains where tundra vegetation survives above the tree line, and they find that the extension of trees into higher altitudes is not taking place as fast as the spread into higher latitudes. Perhaps the hilltops are exposed to more severe frosts, higher winds, or more prolonged snow cover, all of which restrict invasion by trees. It is possible that there is some hope in the hills for tundra and that fragments of the biome will survive on the tops of high-latitude mountains.

In the low-latitude mountains of the world the tree line will probably extend upward as the climate warms, and the line of permanent snow at high altitude will retreat. This means that the belt of tundra will gradually occupy higher altitudes. But this process can only continue until the permanent snow zone is completely lost, and this snow loss will depend upon the latitudinal position of the mountain (equatorial mountains will lose their snow first) and the overall altitude of the mountain (higher mountains will retain their snow longer). The Quelccaya ice cap in the Peruvian Andes is the world's largest tropical ice mass, covering 17 square miles (44 km²). Its main glacier, the Qori Kalis, is currently retreating at a rate of 195 feet per year (60 m y⁻¹). Some estimates suggest that this ice cap, along with other tropical glaciers, such as that of Kilimanjaro in East Africa, will have disappeared by 2012. In the case of lower mountains in warm, low-latitude climates, increased warmth will eventually eliminate alpine tundra vegetation. As the trees spread upward the tundra plants will have nowhere to retreat and will eventually become shaded out by the expanding forest. As in the case of the high-latitude mountains, however, the tree line is often determined by local microclimatic factors. High winds and precipitation may hold back some tree lines and permit the survival of alpine tundra. In the Tropics the development of cloud banks and permanent mist shrouds can greatly modify the development of forest.

The Antarctic region is likely to see a greater availability of bare rock and soil that is left behind by melting ice and is available for plant and animal colonization, so tundra vegetation may expand on this continent. Some animals that are dependent on permanent ice, however, may be adversely affected. Populations of Adélie penguins on the western fringes of the Antarctic Peninsula have almost halved in the past 30 years, during which the midwinter period has become up to 7°F (4°C) warmer. Ice has retreated, but there is more snowfall; the birds are less successful in rearing their young in snowy, slushy conditions. The populations of emperor penguins have also declined by 50 percent in the last 50 years, and this is associated with warmer seas and a reduced extent of sea ice. Climatic warming does not bode well for the penguins.

The prospects for the tundra biome in the event of continued climatic warming may therefore appear rather bleak, but the tundra has survived as a biome for several million years despite episodes in the Earth's recent past that were much warmer than our present world. This suggests that the tundra has a higher degree of inertia, or resistance to change, than one might expect. Two possible reasons have been put forward to explain this. First, the very low levels of nutrients in tundra soils may actually help in the survival of this biome. Low nutrient reserves might make an ecosystem fragile, but poor soils can also make the tundra difficult to invade. Trees generally need more chemical nutrients than smaller plants, and most find it difficult to establish themselves in poor soils. Even the tough birches, pines, and spruces that occupy the border regions of the Arctic tundra find it difficult to germinate and survive in very poor soils, especially when they are being grazed upon by voracious herbivores such as Arctic hares and caribou. In this way nutrient poverty may actually protect the tundra from tree invasion.

The second feature that the tundra biome has in its favor is a high level of genetic diversity in its flora and fauna. The number of species of plants and animals in the tundra is low, but among the species present there are many genetically distinct subspecies and races. The environment itself is quite diverse, with wet locations and dry ones, salty and fresh areas, high altitudes and low, exposed and protected sites, snow-covered and open locations. The species of plants and animals present, though low in number, have developed a whole range of genetically adapted forms, called *ecotypes,* that are able to cope in each of these different microhabitats. This genetic diversity will greatly assist the tundra as it faces the challenge of climate change. Whatever new sets of conditions are generated by the changes, it is likely that many existing species will possess the right set of genetic adaptations to take advantage of the new challenges and opportunities. Perhaps this high genetic diversity is the clue to how the tundra has managed to survive those many occasions when climate change has threatened this biome in the past.

◼ OZONE HOLES

Other changes are taking place in the atmosphere of the polar regions apart from those due to climatic change. Scientists have become deeply concerned about the annual appearance of a gap in the ozone layer of the upper atmosphere above the North and South Poles. This could have considerable implications for the living organisms that occupy these regions.

It is easy to become confused about the gas ozone. It is a very reactive material, having the chemical formula O_3

and possessing great powers of oxidation. It attacks many types of materials and can even decompose rubber. When people breathe it in, therefore, it damages the lungs and can prove fatal to those with weak respiratory systems. Ozone is one component of automobile exhaust fumes, and, when coupled with strong sunlight and other products of combustion, it can contribute to the formation of photochemical smog. So ozone in our immediate environment, especially in sunny cities, is a pollutant gas that is a health risk and needs to be avoided.

But ozone also occurs naturally high in the stratosphere, which is a layer of the Earth's atmosphere lying about nine to 30 miles (14 to 50 km) above the ground. Even in the stratosphere ozone is only a trace gas, and if all the ozone present in the stratosphere were concentrated at ground level under normal atmospheric pressure, it would make a layer only 0.1 inch (3 mm) thick, but this small amount of ozone in the stratosphere performs a vital function for all the living things on the land surface of the Earth: It absorbs the harmful ultraviolet radiation (UV) from the Sun. In the absence of such protection the UV radiation would cause genetic damage among all organisms, apart from those living deep in the oceans, where they are protected by the ozone-screening effect of water. When scientists discovered that the ozone layer was becoming very thin over the polar regions (see sidebar on page 227), it naturally gave rise to considerable concern.

Scientists have conducted many experiments to determine how additional ultraviolet radiation might affect the plants and microbes of the polar tundra. Using fluorescent tubes that emit enhanced ultraviolet rays, they have tested the effect of exposure on bog mosses and found that growth was depressed by 20 percent. If this proves general among plants, it would greatly reduce the primary productivity of the tundra and would affect its role as a sink for atmospheric carbon. Soil microbes are even more sensitive to ultraviolet radiation, so healing of the polar ozone holes is evidently a high priority if the tundra is to survive.

◼ POLLUTION OF THE TUNDRA

One might suppose that the tundra, which has low human populations, would be free from the blight of pollution, but this is sadly not the case. Pollutants are carried into the polar tundra regions by the ocean currents, leading to the contamination of food chains based on marine organisms and then leading to terrestrial ones. Pollutants are carried by the atmosphere, and these find their way into both polar and alpine tundra habitats, being deposited along with rain and snow on the land surfaces. Finally, some pollution results from mining activities in the tundra. Fossil fuels are the main target of mining in the Arctic, but other minerals and

History of the Ozone Holes

Research into the changes in atmospheric ozone began in 1957, when atmospheric scientists set up an international network of stations, including one in Antarctica, to monitor the level of ozone in the stratosphere. By 1985 they had accumulated sufficient data to demonstrate that the overall quantity of ozone in the stratosphere, particularly over the Antarctic, had severely declined during the period of monitoring. Its concentration had halved since 1957. Detailed studies showed that the loss of ozone was seasonal, with the strongest decline occurring in the Antarctic spring. Two possible causes were investigated, first the increasing abundance of stratospheric jet traffic, and second the release by humans of increasing quantities of a group of chemicals called chlorofluorocarbons (CFCs), used in aerosol sprays, refrigerators, and the production of plastic foam. Little evidence could be found to support the jet plane hypothesis, but CFC release became an increasingly likely explanation for the development of an "ozone hole" over the Antarctic. CFCs are released at the Earth's surface and then diffuse upward through the atmosphere, gradually breaking down to release chlorine, which then interacts with ozone, causing it to decompose and form oxygen. Due to circulation patterns in the atmosphere coupled with low temperatures (destruction begins below −112°F [−80°C]), ozone destruction is greatest in the South Pole region, but this process has also been taking place above the Arctic in recent years. Destruction is worst in the spring because increased light intensity also assists in the process.

In 1987 an international meeting in Canada led to the Montreal Protocol, an agreement to restrict the production and use of compounds such as CFCs that are known to affect the ozone layer. The signatories put its provisions into effect, especially the developed nations, which are responsible for the bulk of CFC production, and the outcome has been promising. Concentrations of CFCs have reached a plateau and are beginning to decline. Scientists predict that the worst of the ozone destruction should be over by 2010, after which there will be a gradual recovery of the ozone layer, reaching 1980 levels by the middle of the century, but the size of the ozone holes varies greatly from year to year. The Antarctic hole was exceptionally large in 2000 and 2003, was small in 2002. But in 2006 the Antarctic ozone hole was bigger than ever previously recorded, extending over an area of 11 million square miles (29 million km²). The Arctic hole reached its greatest extent in 2005 following a particularly cold winter, so it is difficult to follow trends in this potentially very harmful atmospheric feature.

---

geological materials prove attractive in some mountain areas and may one day tempt people to exploit the reserves of the Antarctic (see "Mineral Reserves," pages 214–216). Mining leads to spillage and to waste dumping that pollute the environment in which it takes place, so future developments will undoubtedly create new hazards.

The ice of the glaciers and ice sheets in the tundra has accumulated in sequential layers over many thousands of years, and it is formed from the precipitation, mainly in the form of snow, that falls upon its surface. This snow contains contaminants it has gathered from the atmosphere, and these chemical compounds and particulate fragments become incorporated into the developing ice mass. Snowflakes often begin their formation around dust particles, called nuclei, and carry these to the ground as they fall. So the ice preserves a record of past conditions, and when scientists analyze its chemistry they can reconstruct the gradual process of atmospheric contamination that has accelerated over the past few centuries. The tundra, therefore, not only suffers from pollution but conserves an archive of pollution history.

As the snow is compacted into ice, bubbles of gas from the contemporaneous atmosphere become trapped, and they, too, provide a record of past conditions from which a picture of changing atmospheric conditions can be constructed.

Some of the particles that accumulate in tundra ice are formed from heavy metals generated by industrial activities in lower latitudes. Heavy metals include all those metals with an atomic weight greater than iron, including lead, zinc, mercury, nickel, copper, and cadmium. Lead is one of the most common pollutant heavy metals, and it is an accumulative poison in plants and animals. The ice in the Greenland ice sheet that accumulated prior to 1800 contained only very small traces of lead, but concentrations then rose steadily through the 19th and early 20th century, finally accelerating abruptly after 1950. Current lead concentrations in Greenland snow are five times those of 1930. This increase in heavy metal fallout over the tundra has an influence on food webs, where such pollutants will accumulate, especially in top predators. The slow growth of many plants, especially lichens, means that tissues live long enough to collect heavy metals, which are then passed on to grazers, such as caribou, and will then be accumulated by wolves. The long residence time of elements within the tundra biota means that pollution of this kind is particularly serious.

Radioactive fallout is an even greater cause for concern in the tundra. Scientists have analyzed the ice layers at the South Pole to trace the level of radioactive fallout over the course of time. The profile of radioactivity shows a steady growth through the 1950s as above-ground testing of nuclear bombs proliferated, peaking in 1966. Subsequently, such tests have been abandoned, and the level of radioactivity in snow has steadily decreased and has returned to the background levels similar to those prior to 1950. The disastrous accident at the Soviet nuclear power station in Chernobyl in 1986 led to clouds of radioactive cesium 137 dust being forced into the upper atmosphere and drifting over northern Europe. Northern Scandinavia was badly affected by fallout, and local people were advised not to eat reindeer meat while the contamination was present in the vegetation. But the radioactive isotope ^{137}Cs is very long lived, and the background radioactivity of soils remains elevated as a result of this pollution event.

One of the most widespread and serious of the aerial pollutants is not a toxin but a fertilizer. Oxides of nitrogen are emitted from burning fuels, and these enter the atmosphere, where they dissolve in rainwater to form nitric and nitrous acids. Acidic rain has direct harmful effects on plants, stripping the wax cuticles from their leaves and leaving them open to desiccation and infection. The acids also affect wetlands and water bodies, lowering the pH (see sidebar "Acidity and pH," page 54) and causing the death of sensitive organisms, including fish. But when the acids are neutralized in the soil, they form nitrates, which are materials essential for plant growth. Scientists studying the aerial fallout of nitrogen compounds in the Arctic have found that the quantities arriving are minute, usually between 0.0003 and 0.003 ounces per square foot (0.1 and 1.0 g m^{-2}) in a year. This may seem very small, but when it is compared with other sources of nitrogen in the ecosystem, such as the breakdown of soil materials by decomposition and weathering and the fixation of atmospheric nitrogen by bacteria, it becomes apparent that aerial deposition is as great as the combination of nitrogen from all these sources. Pollution, in other words, has roughly doubled the nitrogen supply to the tundra ecosystem.

The soils of the tundra are poor in nutrients, and the fertilization resulting from nitrate pollution causes considerable changes to the vegetation. In general, the more robust and fast-growing species, such as woody shrubs and grasses, are stimulated, while the slow-growing species, such as lichens and mosses, are overwhelmed by the competition. The above-ground biomass of the ecosystem thus increases until it becomes limited by other factors, such as a short-age of phosphorus or simply by the length of the growing season. But nitrate addition to a tundra soil also stimulates decomposition processes, and it is possible that the continued deposition of pollutant nitrates in the tundra could lead to an overall loss of soil organic matter. As in the case of climate change, the tundra could change from being a sink of atmospheric carbon into a source. This type of pollution could turn out to be one of the greatest threats to the future of the tundra biome.

Pesticides are extensively used throughout the world as a means of increasing the yield of crops or the health and productivity of domesticated animals. They are toxins that are selected to kill some organisms (such as insect pests) without causing harm to others, including people. But scientists and agriculturalists have often miscalculated and used pesticides that had a harmful impact beyond the target organism. The use of DDT to kill insects in the 1940s and 1950s, for example, was very successful in the control of many insect-borne diseases, including malaria, so it undoubtedly saved many hundreds of human lives, including those of soldiers during World War II. But the compound proved very durable in the environment. Its concentration built up in the fat reserves of birds and mammals until it eventually impaired their breeding and sometimes even caused their death. DDT also spread around the world, finding its way through marine food webs to the tundra regions and accumulating in the fatty tissues of seals and penguins. Fortunately, the harmful effects of this compound were recognized in time, and most developed countries have now banned the use of DDT, which has prevented the destruction of tundra mammals and birds as well as those of other biomes.

Organic pollutants, including DDT, accumulate in the polar tundra partly because of the oceanic currents, which carry them into the high latitudes, and partly because of the presence of large marine mammals, including whales and seals, that accumulate organic pollutants in their blubber. The polar tundra regions are thus a sink for organic pollutants, and as a consequence the native peoples of the Arctic are exposed to very high levels of pollution in their diet, which contains large proportions of animal fats. The Inuit people of Canada contain some of the highest levels of persistent organic pollutants and mercury recorded among any of Earth's inhabitants. The most serious health effect of this contamination is among pregnant women because elements such as mercury can cause irreversible neurological damage to the fetus, leading to learning difficulties later in life. Children also suffer from an impaired immune system, leading to abnormally high levels of infectious diseases. Nutrition scientists often recommend a shift from the traditional animal fat diet, especially in the case of pregnant women.

Pollution by human waste products and accidental spills creates a threat to all the Earth's ecosystems, but particularly to the sensitive tundra biome. The exploitation of mineral resources is often accompanied by environmental damage as a consequence of pollution. Oil spills at the site of extraction or during transport by pipeline or ship are statistically inevitable, given the large quantities taken out of the rocks beneath the tundra and the long distances the materials have to be carried for their treatment and consumption. Contamination of rivers and soils from mining waste is likewise an unavoidable outcome during the extraction of metals from the rocks beneath the tundra. Waste products from human settlements are difficult to dispose of and create pollution and health problems. All these sources of pollution are local to the tundra, and the best way to avoid them is to control the extent of exploitation of the habitat, possibly even denying humans the right to commence such exploitation, as in the case of mineral extraction in Antarctica. The conflict between economic exploitation and habitat conservation in the tundra will undoubtedly become a much-debated issue in the coming century.

■ TUNDRA CONSERVATION

If the tundra is to be retained as one of the Earth's major biomes, then conservation by humans is essential. Conservation is not quite the same thing as preservation. Preservation involves the isolation of an object from injurious influences and its protection to ensure that it remains unaltered by events. In the natural world this would be an impossible task because the natural environment itself is constantly changing. Climate changes partly because of natural cycles, and unpredictable events such as wind, earthquake, fire, and flood can disrupt natural ecosystems. Humans, as has been seen, have wide impacts far beyond the regions they inhabit, so even the most remote of wildernesses cannot be fully protected from the effects of our species. The pollution of the tundra by pesticides and airborne particles is a case in point. In conservation, therefore, one must protect but also manage the ecosystem in order to modify or control the pressures that come from outside. There was a time when ecologists believed that the best form of conservation was to leave an ecosystem alone and to avoid any form of human intervention. It is now recognized that this attitude is based on an impossible ideal and that people need to actively manage ecosystems in order to maintain levels of biodiversity on Earth.

In the polar tundra protection from the harmful human impacts is clearly the first priority. The hunting of mammals, for example, needs to be controlled if their populations are to remain stable. In the case of some animals, such as the great whales, this may involve a total cessation of all hunting, but for other animals, such as seals and caribou, there may be a case for limited harvesting of the herds. Indeed, there are

occasions when culling populations is a humane reaction to overpopulation. For instance, the red deer (elk) populations on the alpine tundra habitats of Scotland have recently expanded, with numbers increasing fourfold between 1960 and 2000. Additional culling is clearly required here. The alpine tundra habitat is being damaged, and each winter a proportion of the elk herd inevitably starves. The imbalance is a result of past human activity, namely eliminating the wolf from Scotland (the last wolf was shot in 1743). Now this large deer has no natural predators. There is, however, a new herbivore in the alpine tundra that people have introduced, namely the sheep, and the sheep herds contribute to overgrazing and vegetation destruction. In response, human management of the Scottish Highland tundra must reduce either sheep grazing, or deer numbers, or both. Currently hunters harvest about 50,000 red deer every year in Scotland, representing approximately 13 percent of the population. But this may not be enough to stabilize the population, and further culling will probably prove necessary.

The control of Scottish red deer is an example of the need to manage ecosystems, especially if they have already been modified by human activity and left unbalanced as a consequence. Sometimes it is necessary to try to correct errors of the past, such as the local extinction of a species. Reintroduction of a lost animal is always controversial for many reasons. For instance, it may bring problems for human populations. If the wolf were to be reintroduced to Scotland, sheep farmers would undoubtedly object, just as farmers have done near Yellowstone National Park, where wolves have been successfully reintroduced. The musk ox is a species that has attracted less controversy than the wolf but has suffered even more at the hands of human hunters. Once widespread through the Arctic, it became very restricted as a result of a long history of hunting from prehistoric to modern times. It was completely lost from Europe, Asia, and Alaska, but the surviving herds on Greenland have formed the basis for reintroduction to Russia and Alaska. In some sites the introduced musk oxen have found it difficult to compete with native caribou herds, but generally the species seems to be holding its own. In the case of the musk ox, for which humans have been responsible for local extinctions, it seems only reasonable that humans should try to set things straight by reintroduction. One of the dangers of this strategy, however, is that the stock may be taken from herds with slightly different genetic makeups. In the process of reintroduction the conservationist may be taking animals and plants to regions where they are not fully suited for survival.

People have long used the tundra for pastoral farming. The cold conditions of the polar and alpine regions make them entirely unsuitable for growing crops, but some grazing animals can survive and provide humans with

The yak (*Bos mutus*) is a shaggy Tibetan ox that has long been domesticated. Its hair is woven into cloaks, its milk is the source of a rancid butter, or *ghee,* and its dung is used for fuel. Wild herds still exist in the Himalaya Mountains.   *(Nepal Tourist Board)*

Albatrosses

There are 21 species of albatrosses, of which nine species are regarded as endangered. They are massive sea birds, with wing spans of up to eight feet (2.4 m), which allow them to soar over the waves, taking advantage of the air turbulence above the surface of the ocean. They have long life spans, up to 60 years, but the juveniles do not breed until 12 years old. Each year they lay only one egg that needs to be incubated for three months. Some species lay an egg only every two years. This slow rate of breeding leaves them susceptible to population crashes, from which it takes a long time to recover. Many populations of albatrosses are currently experiencing severe losses because of a system of fishing for tuna that was introduced during the 1980s in the Southern Ocean called "long-lining." In this method extensive fishing lines with thousands of baited hooks are trailed for up to 80 miles (130 km) behind fishing boats. Albatrosses are tempted by the bait, become caught on hooks, and drown. Conservationists estimate that 100,000 birds die in this way each year. When the lines are fully set, the baited hooks lie deep in the water, out of the reach of albatrosses, but during the early stage of line setting the hooks are close to the surface. That is when the birds are at greatest risk. There are ways in which the situation could be improved so that danger to albatrosses could be avoided. If additional weights were added to the line, it would sink faster, and as the birds feed by day, it is possible that the laying of the lines at night would also save some fatalities. International agreement to implement various strategies for albatross conservation has now been ratified by many countries. The United Nations Agreement on the Conservation of Albatrosses and Petrels (ACAP) came into force in February 2004, but some countries have not yet signed up to the agreement. There also remain many fishing vessels sailing under flags of convenience that do not consider themselves bound by it.

meat, milk, and skins. In the Arctic regions of northern Europe and Asia it is the reindeer that occupies this role, while in the Himalaya Mountain chain it is the yak (see photograph above). But cattle and sheep have also been used for alpine tundra grazing, sometimes with disastrous consequences. Alpine plants grow slowly and only for a short period in summer, so the amount of grazing they can sustain is very limited. Pastoralists have often tried to maximize the grazing pressure, and this has resulted in overgrazing and vegetation destruction. Sheep, especially merino sheep (see photograph on page 231), have often been the cause of upland erosion by excessive grazing. The conservationist John Muir, who spent much time in the Sierra Nevada of California, was once a shepherd, and he came to regard sheep as "woolly locusts" because of their great potential for vegetation destruction. The control of grazing, particularly in alpine tundra habitats, is therefore a conservation priority.

Just as many plants grow very slowly in the tundra, many animals have low rates of reproduction, and these are often among the most threatened when new environmental pressures arise. A fast-breeding animal can recover quickly from catastrophe, but one with poor powers of reproduction will recover only slowly and sometimes not at all. Many of the birds of the Antarctic fall into this category, having long periods as juveniles prior to breeding, laying only one egg, and sometimes not nesting every season. The penguins and the albatrosses fall into this category (see sidebar on page 230).

In alpine tundra habitats local extinction is a very common problem because of the isolation of mountain peaks. If an animal or plant is lost at one site, it may be difficult for the species to reinvade from surrounding but isolated mountains. Reintroduction of animals lost as a result of human persecution has therefore been extensively used in mountain regions. In North America the mountain goat has been reestablished in Oregon, Nevada, Utah, Colorado, Wyoming, and South Dakota from its residual populations in the northern Rocky Mountains and Cascades. The closely related chamois of the European Alps was similarly severely reduced by hunting but has recovered much of its former range as a result of reintroduction. Conservationists

have also helped some birds to spread back into mountain areas where they had previously died out. The bearded vulture (*Gypaetus barbatus*) of Europe and Asia is a massive bird that feeds on broken bones it drops onto rocks from the air, and it has benefited from human transport and care in bringing it back to several mountain ranges, including the Alps and the Pyrenees, from which it has been lost as a result of former human persecution.

Large birds of prey, including the golden eagle (see photograph on page 232), need a great deal of space for their survival, and this is becoming an increasingly scarce commodity. Human beings are now so numerous and so influential in their pollution of the atmosphere and the oceans that there is no part of the planet free from our impact. Even the remote areas of tundra, both polar and alpine, have experienced changes as a result of the arrival of humans on the scene. As discussed in

Many alpine tundra areas are grazed by domestic animals in summer. In this case merino sheep graze the high pastures of the Austrian Alps, near Obergurgl. Sheep are often responsible for overgrazing alpine tundra. *(Peter D. Moore)*

The golden eagle, a top predator of alpine tundra regions, is a symbol of wilderness. Like many top predators, it demands large areas of undisturbed land in which to hunt. *(Watson, Digital Slide File, Yellowstone National Park)*

chapter 9, the tundra is a useful biome for people because of its mineral wealth, its recreational opportunities, and its role in maintaining the overall balance of the planet. People, therefore, will need to play an increasing part in the conservation of tundra, both protecting it from further harm and putting right some of the problems this biome has already suffered as a consequence of human errors. The future of the tundra, as is the case for all biomes, now lies in human hands.

■ CONCLUSIONS

The tundra biome, both in the polar regions and in the mountains of the world, faces a number of problems, some of which threaten its very survival. Continued climatic warming is almost certain, and this will have a particular impact on the Earth's coldest biome. Tundra lies at the world's extremities, either close to the poles or near the highest parts of mountains. There is nowhere to which tundra can retreat if warming continues; it will be gradually eliminated from the tops of mountains and will be forced northward into the Arctic Ocean. Only in the Antarctic is there a strong likelihood of long-term survival for tundra if global warming continues. The probable rise in global sea levels in the warmer world will only make things worse, as

this will swamp low-lying tundra lands, particularly in the Arctic. Added to these problems, it seems probable that the polar regions will experience more rapid rises in temperature over the next century than most other places on Earth. Many of the highly adapted, cold-tolerant species of plants and animals will find themselves under climatic and spatial stress, eliminated by competition from warmth-loving and faster-growing species. The one hope for tundra biodiversity is its great genetic variability. The organisms of the tundra are not only highly adapted but also highly adaptable, and this may prove vital for their survival.

Ozone depletion in the polar stratosphere has posed a new threat to the tundra because it permits higher levels of ultraviolet radiation from the Sun to penetrate the atmosphere. This can cause genetic damage to exposed tissues and reduces the growth rate of some plants. The human causes of ozone thinning appear to have been identified in the form of chlorine-containing organic molecules, and efforts on the part of many nations to reduce this potentially serious threat to tundra survival seem to be meeting with some success.

Pollution, both in the form of direct deposition of harmful compounds, such as heavy metals and nitrates, and in the less obvious form of oceanic contamination by pesticides, is a further source of damage to the tundra. It is the responsibility of people to make pollution control a priority in order to protect this biome.

Conservation of the tundra will involve the enforcement of protective measures at an international level, and many of these are already agreed to and in place. But conservation demands more than just protection; it requires proactive management. Species whose ranges have been reduced by human activities may require human assistance in reestablishing themselves over their former ranges. Ski runs, ski lifts, and walking trails fragment the landscape into small patches, as shown in the photograph, left, and this leads to increasing disturbance of wildlife. Hotels and highways are built in the valleys, and the mountainsides of many alpine regions are becoming covered by funicular railways and viewing platforms. Reserves must be established to ensure the survival of some wilderness areas, and some habitats may need to be carefully managed to ensure the survival of sensitive species. The tundra is useful to us, as well as being valuable in its own right, and the future of the tundra lies very much in human hands.

The town of Chamonix in the French Alps viewed from the slopes of Mont Blanc. The valley is intensively developed, and a growing number of trails and ski lifts fragment the alpine tundra landscape into ever decreasing patches of wild habitat.   *(Peter D. Moore)*

11

General Conclusions

Few areas are left in the world that are true wildernesses and probably none that are totally unaffected by the presence of humans on the planet. The tundra has the appearance of wilderness. Its vast and bleak landscapes are almost devoid of human settlements, and one can travel for hundreds of miles over the polar tundra without encountering another person, a road, or a trace of habitation. But our species is responsible for an invisible yet insidious contamination of the tundra. Raised carbon dioxide levels in the atmosphere, the product of industrial consumption of fossil fuels, is changing the climate, increasing average temperatures and causing the early breakup of pack ice in the spring and the melting of glaciers and ice sheets. Chemicals made and emitted by people in the developed world have destroyed ozone in the stratosphere. Harmful ultraviolet radiation now bathes the sensitive organisms of tundra habitats. The oceans and the atmosphere around the tundra act as a global trashcan, collecting many human waste products as oceanic and atmospheric circulation bring refuse to the poles. People are increasingly looking to the tundra as a reserve of valued minerals and fossil fuels that can be extracted and transported to more populous lands, leaving behind the waste and detritus of a spoiled landscape. Even those people who value the tundra lands and wish to visit them as tourists or to enjoy the recreational and sporting activities they offer are in danger of destroying the very features they so ardently seek.

Many serious problems face the tundra biome. It may be the last great wilderness, but this fragile ecosystem is rapidly approaching a point where it will begin a slide into final collapse. It would be wrong to regard this process as inevitable. Although the tundra is one of the least inhabited, least visited, and least agriculturally productive parts of the world, its loss could have serious consequences for our species.

As in the case of tropical rain forests, wetlands, and all the other biomes of the Earth, it is much easier to destroy than to recreate, so any process that leads to the degradation of a habitat and the loss of species needs to be carefully considered before it is allowed to proceed. The organisms of the

Earth, classified by taxonomists into species, genera, families, classes, and so on, are each the product of thousands of years of gradual adaptation to specific environmental conditions. Each genetic makeup is unique, honed by natural selection until it best fits the role it has adopted in the great scheme of things. Put all the genetic variations, the range of species, and the variety of habitats and conditions together, and the outcome is biodiversity, one of the greatest assets the Earth possesses.

The tundra is not rich in species compared with most other biomes, but its species have been hardened to survive in one of the most testing sets of environmental conditions on Earth. The tundra ecosystem, from lowly microbes to carnivorous bears, consists of an assemblage of tough species that thrive under conditions that would spell rapid death to most other inhabitants of the planet. The secrets of this survival are held within their genes, and the organisms of the tundra are among the most genetically variable of all the Earth's species. They need to be because they live in a habitat that can bring catastrophic cold, flood, heat, and drought, all rapidly alternating with one another. To lose any of these remarkable collections of genes without having examined them for potential use to humanity would be irresponsible and foolish. The conservation of the tundra is an issue that must attract widespread concern.

Considered in the light of Earth history, the tundra can be regarded as one of the rarest of biomes. Only under certain unusual conditions of astronomy and continental arrangements do the poles of the Earth become sufficiently cold to sustain permanent ice caps, and there have been relatively few occasions in the last 4.6 million years when these conditions have been met. The arrival of human beings on the planet has taken place during one of these cold episodes, the ice ages of Earth history. People should count themselves as particularly privileged that they live on Earth at a time when there are glaciers and ice sheets adding to the great diversity of the world's landscapes and providing opportunities for the existence of such a remarkable range of organisms as the tundra sustains.

The tundra is already the focus of many different types of scientific research. Meteorologists and atmospheric scientists work in the tundra to examine the movements of the Earth's air masses and how they behave under polar conditions. Satellite observations are providing more detailed evidence of environmental changes that cannot be detected from the ground. Oceanographers work on the circulation of the oceans, how they redistribute energy around the world, and what factors modify their speeds and their courses. The future of the world's climate can be predicted more accurately only when more is known about the behavior of the atmosphere and the oceans in the polar regions.

Geologists have much to do in the tundra regions. Many of the oldest rocks on Earth lie beneath the tundra, and the quest for the first organisms to inhabit the planet concentrates on these ancient deposits. Ice and fire often meet in the tundra where plate boundaries and geological instability lead to volcanic ridges and subduction zones. From Alaska to Iceland some tundra regions display the thinness of the Earth's crust as molten rocks meet frozen, icy landscapes in spectacular geological displays. In many of the alpine tundra regions, such as the Himalaya, the thrusting of plates continues to heave mountains to new heights, revealing more information on the nature of tectonic movements and their effects on climatic and biogeographical patterns.

The ice itself is an archive of Earth history, preserving in accumulated layers a record of past events, such as volcanic eruptions that have cast their dust clouds around the world. The ice layers tell the story of developing atmospheric pollution as contaminant dust and chemicals are transported by air to the remote tundra regions. Isotopes in the ice crystals can provide information about changes in the Earth's temperature, and trapped bubbles of gas permit the reconstruction of the atmospheric history of recent times. The ice sheets of the tundra are thus a unique source of scientific research that will help in the elucidation of climatic history and thereby assist in the prediction of coming climate changes.

Biologists have much to learn from tundra organisms, studying their biochemistry, the ability of their cells to resist extreme cold, and their complex patterns of behavior that avoid the worst of the environmental conditions. How birds manage to navigate over these extensive landscapes, wintering in the Tropics and breeding in the tundra summer, is still something of a puzzle to zoologists. It is possible that their brains contain some kind of magnetic compass that enables them to establish their position at any time and to determine which way to go. The question of how an animal such as a marmot manages to lower its temperature during its winter dormancy and yet survive is an important one for physiologists to solve. If scientists were to discover how plant cells can continue to live at temperatures well below the freezing point of water, they could apply these findings to the development of new crops. The answers to these questions are of great theoretical interest to scientists, and they could also have important applications in industry, medicine, and society.

Microbiologists have a particular interest in the tundra because when conditions get tough it is usually the microbes that survive best. Lakes buried deep beneath the Antarctic ice for thousands of years still contain living microbes. These organisms have much to reveal about the very nature of life and survival, and their biochemistry will be of great interest to those concerned with the use of low temperature for preserving living tissues, whether for use in medicine or in the food industry.

Archaeologists, anthropologists, and paleozoologists need to combine their various skills to piece together the story of human invasion of the tundra, examining how people managed to survive in this harshest of all the world's landscapes. Since human populations were and remain small in the tundra, the question of the impact of people on the natural environment needs to be examined carefully. Archaeologists speculate about whether early human cultures were responsible for the extinction of some of the great tundra animals, such as the mammoth. If so, the process of megafaunal extinction illustrates just how fragile is the tundra ecosystem, where a small number of people were able to make such drastic changes to its species composition.

Climatologists concerned with predictions of the changes that can be expected in the next century are concentrating on the tundra. All the predictive models currently being developed agree on the fact that global warming will continue and that it will be most strongly felt in the high latitudes. The consequences for precipitation are more difficult to predict, but the general expectation is that the tundra may also receive more snowfall. Rising sea levels will restrict the area available for tundra in the Arctic, but the melting of the Antarctic ice sheet may expose more landmass for tundra organisms in the Southern Hemisphere. Alpine tundra is likely to become scarcer globally, and sites will become more widely scattered and isolated.

Under warmer and wetter conditions forest is likely to replace shrub tundra, and shrub tundra will replace herbaceous tundra and lichen-dominated polar desert. Precisely how these changes would affect the global carbon cycle is difficult to predict. Warmer conditions would increase microbial activity in soils, leading to faster decomposition and the conversion of much soil organic matter into gaseous carbon dioxide. This would be most strongly felt in the tundra wetlands, where extensive peat deposits could become oxidized, converting the entire ecosystem from a sink for atmospheric carbon into a source. This change would be made even more severe if methane production from tundra wetlands increased, which is another likely outcome of warming. But increasing

Two wolves in the snow, nose to nose. The wolf has figured strongly in myths and legends, always being cast in the role of villain, but wolf packs show a strong degree of social bonding and cooperation. *(Jim Peaco, Yellowstone Digital Slide File, Yellowstone National Park)*

precipitation could have a reverse effect, waterlogging soils and reducing microbial activity. Higher atmospheric carbon dioxide concentrations could also lead to greater biomass accumulation as photosynthetic production increased, and this effect would be enhanced by nitrate deposition from the polluted atmosphere. Tree invasion of the tundra would also increase biomass and therefore add to the carbon reservoir within the ecosystem. So the outcome of global change as far as the carbon balance of the tundra is concerned is still a matter for debate. More detailed experimental research in tundra habitats is currently underway and will hopefully help to answer some of these questions.

The tundra may seem remote from many people's lives, but what goes on in the tundra affects everyone in the world. The loss of the tundra would affect not just a small group of ecotourists and whale enthusiasts or even those addicted to winter sports but would have a global impact. Science, industry, medicine, and even political decision making would all suffer if the tundra were lost along with its great wealth of recorded information and highly adapted organisms. Indeed, the entire balance of the Earth's oceans and atmosphere, which control so much of global climate, could be severely disrupted if the tundra disappears. Tundra may not have the charismatic appeal of tropical forests. Some of its inhabitants, including polar

bears and wolves (see photograph above), have often been regarded as villains in myths, legends, and fairy stories, and this has added to the negative feelings many people have for the tundra. But this is the last great wilderness, and everyone needs to know that wild, uncontaminated places still exist on Earth.

The solutions to the problems of the tundra and all the other biomes of the world require two very considerable shifts in human thinking. The first is international cooperation, putting global survival before national economic interests, and the second is long-term planning rather than the seeking of short-term gains, which is a difficult goal even in democratic societies in which politicians need to make promises of immediate improvement if they are to be elected. Ultimately, global problems can be solved only by the dissemination of knowledge to the individual people who make up the nations, and it is to be hoped this account of the tundra biome will contribute to that end.

Glossary

ablation the loss of ice at the base of a glacier, where the warmth of the Earth causes melting

active layer the upper soil layers in PERMAFROST environments that melt in the summer and freeze in the winter

aerosol particles suspended in the atmosphere as a result of their very small size

aestivation a period of dormancy associated with the unfavorable conditions of summer drought (equivalent to HIBERNATION in winter)

albedo the reflectivity of a surface to light

allelopathy the capacity of a plant to produce and disperse chemical compounds that interfere with the germination and growth of potential competitors

allogenic forces outside a particular ecosystem that may cause internal changes; for instance, rising sea level can influence water tables in freshwater wetlands farther inland and is therefore considered an allogenic factor

alpine tundra vegetation dominated by herbs and dwarf shrubs found above the TIMBERLINE of high mountains where conditions are too cold for tree growth

anaerobic lacking oxygen

anion elements or groups of elements carrying a negative charge, e.g., NO_3^-, HPO_3^-

annual an organism (usually a plant) that completes its life cycle in a single year

anoxic lacking oxygen

aquifer a body of rock that is porous and permeable to water underlain by an impermeable layer, resulting in the storage of water beneath the ground

arctic-alpine describing an organism that is found in both Arctic and alpine tundra habitats

Arctic tundra a region with vegetation dominated by herbs and dwarf shrubs found in the polar regions where conditions are too cold for tree growth

aspect the compass direction in which a surface is oriented

ATP adenosine triphosphate; the molecule in cells that acts as a temporary storage system for energy

autogenic forces within an ecosystem that result in changes taking place, e.g., the growth of reeds in a marsh results in increased sediment deposition. *See also* FACILITATION

autotrophic organisms that are capable of constructing complex organic molecules from inorganic sources, such as green plants

biodiversity the full range of living things found in an area, together with the variety of genetic constitutions within the species present and the range of microhabitats available at the site

biological spectrum the proportions of different LIFEFORMS within an ecosystem

biomass the quantity of living material within an ecosystem, including those parts of living organisms that are part of them but are strictly speaking nonliving (e.g., wood, hair, teeth, claws), but excluding separate dead materials on the ground or in the soil.

biome a large-scale community that is defined on the form of its vegetation; biomes include tundra, boreal forest, desert, tropical rain forest, and others

biosphere those parts of the Earth and its atmosphere in which living things are able to exist

biota the sum of living organisms, plants, animals, and microbes

blue-green bacteria (Cyanobacteria) once wrongly called blue-green algae. Microscopic, colonial, photosynthetic microbes that are able to fix nitrogen. They play important ecological roles in some wetlands as a consequence of their nitrogen-fixing ability

bog   a peat-forming wetland in which the water supply arrives entirely by rainfall rather than groundwater

bog mosses   a distinctive group of mosses, all belonging to the genus *Sphagnum*. They have the capacity to hold up to 20 times their own weight in water and are also able to retain CATIONS. Most species are associated with acidic mires

boreal   northern, usually referring to the northern temperate regions of North America and Eurasia, which are typically vegetated by evergreen coniferous forests and wetlands. Named after Boreas, the Greek god of the north wind

calcareous   rich in calcium carbonate (lime)

capillaries   fine tubes, as in the structure of partially compacted peat and soils

carbon neutral   an ecosystem that generates as much atmospheric carbon dioxide through respiration as it captures by photosynthesis

carbon sink   an ecosystem that stores carbon, taking in more carbon than it releases to the environment

catchment   a term meaning a region drained by a stream or river system (equivalent to WATERSHED)

catena   of soils, refers to changes in the profile of the soil along an environmental gradient

cations   elements or groups of elements with a positive charge, e.g., Na^+, NH_4^+, Ca^{++}

cation exchange   the capacity of certain materials (such as peat and clay) to attract and retain CATIONS, and to exchange them for hydrogen in the process of LEACHING

chamaephyte   a plant that grows close to the surface of the ground, below a height of one foot (25 cm), and in this way escapes the effects of intense wind blasting in tundra habitats

charcoal   incompletely burned pieces of organic material (usually plant) that are virtually inert and hence become incorporated into lake sediments and peat deposits, where they provide useful indications of former fires. Fine charcoal particles may cause changes in the drainage properties of soils, blocking soil capillaries and leading to waterlogging

climate   the average set of weather conditions over a long period in a region

climax   the supposed final equilibrium stage of an ecological succession. Many question whether real stability is ever achieved

community   an assemblage of different plant and animal species all living and interacting together. Although they may give the appearance of stability, communities are constantly changing as species respond in different ways to such environmental alterations as climate change

competition   an interaction between two individuals of the same or different species arising from the need of both for a particular resource that is in short supply. Competition usually results in harm to one or both of the competitors

composition   the range of species associated in a community

consumer   an organism that relies upon other organisms for its food (heterotrophic). Consumers may be primary, secondary, tertiary, and so on, depending on their position in a FOOD WEB

continental climate   a climate characterized by hot summers and cold winters, often with low precipitation resulting from the weak influence of the world's oceans

Coriolis effect   the tendency of free-moving objects, such as air masses, to be deflected by the rotation of the Earth on its axis. Deflection is to the right in the Northern Hemisphere and to the left in the Southern Hemisphere

Crassulacean acid metabolism (CAM)   a photosynthetic mechanism in which carbon dioxide is temporarily fixed in the form of organic acids (often during the night) and is later released within the plant cells to be fixed again by conventional metabolic processes

cyanobacteria   *See* BLUE-GREEN BACTERIA

day length   the time period from surise to sunset. This may be nonexistent in a polar winter

deciduous   a plant that loses all its leaves during an unfavorable season, which may be particularly cold or dry

decomposer   a MICROBE involved in the process of DECOMPOSITION

decomposition   the process by which organic matter is reduced in complexity as microbes avail themselves of its energy content, usually by a process of oxidation. As the organic materials are respired to carbon dioxide, other elements such as phosphorus and nitrogen are returned to the environment, where they are available to living organisms once more. It is therefore an important aspect of the NUTRIENT CYCLE

deterministic   a process in which the outcome is predictable and does not allow for chance (STOCHASTIC) events

detritivore an animal (usually invertebrate) that feeds upon dead organic matter

diatoms a group of one-celled photosynthetic organisms (PROTISTS) that form an important part of the phytoplankton in wetland habitats

disjunct of populations of an organism that are widely separated geographically

dissociation the separation of two elements from one another in solution, forming charged IONS

diversity a term that includes both the variety of elements in an assemblage and the relative evenness of their representation

DNA deoxyribonucleic acid, the molecule that contains the genetic code

drumlin a deposit of TILL beneath a glacier that is often carved into a linear shape by the movement of the ice above

ecological niche *See* NICHE

ecosystem an ecological unit of study encompassing the living organisms together with the nonliving environment within a particular habitat

ecotone boundary region where one type of habitat gradually blends into another

ecotourism tourism to wilderness areas of the world that tries to avoid damaging the environment

emergent an aquatic plant that is rooted in the sediments below the water level but has shoots projecting above the water surface

endemic a species that is native to an area and confined to that area

energy sink a region of the world, such as the Arctic, where energy balance is sustained because of constant import of energy from another region, such as the Tropics

erosion the degradation and removal of materials from one location to another, often by means of water or wind

erratic a rock that is carried far from its original position by the movement of a glacier and is eventually deposited when the glacier melts

esker a ridge of glacial detritus running along the edge of a glacier or beneath the ice, aligned with the direction of glacial flow

eukaryote an organism with cells containing a distinct nucleus enclosed in a double membrane. *See also* PROKARYOTE

eustatic a change in sea level resulting from changing quantities of water being locked up in ice, or the volume of the oceans changing because of alterations in their temperature. *See also* ISOSTATIC

eutrophic rich in plant nutrients. *See also* OLIGOTROPHIC

eutrophication an increase of fertility within a habitat, often resulting from pollution by nitrates or phosphate or by the leaching of these materials into water bodies from surrounding land. This increase in fertility results in enhanced plant (often algal) growth, followed by death, decay, and oxygen depletion. Although usually applied to wetland habitats, it can also be used of terrestrial ones, such as the tundra

evaporation the conversion of a liquid to its gaseous phase; often applied to water being lost from terrestrial and aquatic surfaces

evapotranspiration a combination of evaporation from surfaces and the loss of water vapor from plant leaves

evergreen a leaf or a plant that remains green and potentially photosynthetically active through the year. Evergreen leaves do eventually fall but may last for several seasons before they do so

exfoliation the erosion of rocks by the flaking off of surface layers, often caused by frost

facilitation one of the forces that drives ecological succession. When a plant grows in a particular location, it may alter its local environment in such a way that enables other plants to invade. When a water lily grows in a lake, for example, its leaf stalks slow the movement of water and encourage the settlement of suspended sediments. The lake becomes shallower as a consequence, and other species of plant are able to invade, eventually supplanting the water lily

fiord (or fjord) a deep, steep-sided valley flooded by the sea

firn powdery or granular deposits of snow that accumulate on the surface of a glacier

floodplain the low-lying alluvial lands running alongside rivers over which the river water expands during times of excessive discharge

food web the complex interaction of animal feeding patterns in an ecosystem

forest tundra the ECOTONE (border region) of the forest and the tundra. Trees that survive here are usually dwarfs stunted by strong winds and are referred to as KRUMMHOLZ

fossil ancient remains, usually applied to the buried remnants of a once-living organism, but the term can be applied to ancient buried soils or even the organic remains called fossil fuels

fragility   the degree of ease with which an organism or a habitat may be damaged

fragmentation   the breaking of a habitat into smaller patches, often as a result of human land use

frost heaving   the mechanism of freezing and thawing of a soil that forces stones and unfrozen layers of soil to the surface

functional type   refers to organisms that perform certain ecosystem functions, such as nitrogen fixation or primary production

fundamental niche   the potential of an organism to perform certain functions or to live in certain areas. Such potential is not always achieved because of competitive interactions with other organisms. *See also* REALIZED NICHE

gene pool   the sum of genetic variation found within a population of an organism

geothermal energy   the heat energy released as the core of the Earth cools

glacier   a mass of permanent ice found in cold conditions, often occupying valleys through which the ice moves slowly under the influence of gravity

gley   a soil that forms under waterlogged conditions, which is therefore ANOXIC

greenhouse effect   the warming of the Earth's surface as a result of short-wave radiation passing through the atmosphere, then being converted to long-wave radiation as a result of interception and reflection by the Earth. Long-wave radiation is more likely to be absorbed by the atmosphere after reflection because of the presence of GREENHOUSE GASES

greenhouse gas   an atmospheric gas that absorbs long-wave radiation and that therefore contributes to the warming of the Earth's surface by the greenhouse effect. Greenhouse gases include carbon dioxide, water vapor, methane, chlorofluorocarbons (CFCs), ozone, and oxides of nitrogen

groundwater   water that soaks through the soils and rocks, as opposed to water derived from precipitation

habitat structure   the architecture of vegetation in a habitat

halophyte   a plant that is adapted to life in saline conditions as a result of its physical form, or its physiology, or both

hibernation   a period of dormancy that certain animals undergo as a means of coping with cold winter conditions

High Arctic   the northern regions of the Arctic, where the growing season for plants is less than two and a half months

homiothermic   warm-blooded

hotspot   a region with exceptionally high levels of biodiversity

hydraulic conductivity   a measure of the ease with which water moves through a material. A high hydraulic conductivity means that water moves easily through that material

hydrogen bonding   a bond between two molecules in which one of the components is a hydrogen atom, often linked to oxygen or a halogen. The hydrogen bonding between water molecules contributes to the great cohesive strength of this material, which is vital for the uplift of water in tall plants

hydrology   the study of the movement of water in its cycles around ecosystems and around the planet

ice sheet   an extensive cover of permanent deep ice. Only two ice sheets currently occupy the Earth, one covering Antarctica and the other based on Greenland

ice wedge   water that freezes in a tundra soil and expands, forming a wedge shape that forces its way down into the soil and may split the landscape into a series of polygons

igneous rocks   rocks created as a result of volcanic activity

inbreeding   a population in which genetic exchange from outside is severely restricted, resulting in lack of genetic variation. *See also* OUTBREEDING

inertia   the ability of an ecosystem to resist disturbing forces

insectivorous   an organism that feeds upon insects and other invertebrates. The term may be applied to certain plants that trap insects and digest them as a source of energy and nutrient elements

interception   the activity of plant canopies in preventing rainwater from reaching the ground directly. Intercepted water may continue on its way to the ground or may be evaporated back into the atmosphere

interglacial   a prolonged (tens of thousands of years) period of the Earth's history in which the climate is warm, preceded and followed by a cold glacial period

interspecific   taking place between different species

interstadial   a geologically short (less than five thousand years) period of warmth preceded and followed by STADIAL or glacial periods

Intertropical Convergence Zone (ICZ)   the band of the low-latitude zone of the Earth where air masses from the north and the south converge. Its position varies with season, migrating poleward during the summer season in each hemisphere. It is a region of low atmospheric pressure and consequently high precipitation

intraspecific taking place within a population of a given species

ion a charged element or group of elements. *See also* ANION *and* CATION

isostatic a change in the relative position of the sea level to land level as a result of crustal warping under the influence of such factors as ice volume during glaciation

invertebrate an animal lacking a backbone, including, for example, insects, mollusks, and crustaceans

jet stream a high altitude movement of air from west to east resulting from the rotation of the Earth

kettle hole a hollow in glacial detritus deposits resulting from the melting of a block of ice in that position. It may become filled with water to form a deep, steep-sided lake

krummholz vegetation dominated by trees that have been distorted and stunted by strong winds

K-selected species those species that expand their populations relatively slowly but are able to survive long term because of their competitive ability in tapping environmental resources

lapse rate the rate at which temperature falls with increasing altitude

latent heat of evaporation the energy needed to convert liquid into vapor at the same temperature

latitude conceptual lines running around the world that are named according to the angle subtended to the equatorial plane. Thus, the equator is regarded as 0° and the poles as 90° North or South. The equatorial regions thus lie in the low latitudes and the polar regions in the high latitudes

leaching the process of removal of IONS from soils and sediments as water (particularly acidic water) passes through them

lichen an organism consisting of a combination of an alga or a Cyanobacterium with a fungus. The combination may have a leafy form or may look like paint on a rock. Lichens are generally resistant to cold and drought

leaf area index the area of leaf cover divided by the area of ground overlain

lek a mating system in which males display as a group for the attention of females, which are responsible for mate selection

life-form a system of classifying plants according to where they carry their perennating organs

limestone sedimentary rocks containing a high proportion of calcium carbonate (lime)

litter the dead remains of plant and animal material that fall from the canopy and accumulate on the surface of the soil

loess windblown dust and sand carried by winds over bare, glacial terrain

longitude conceptual lines running from pole to pole and intersecting the equator. They are numbered from 0° at the Greenwich Meridian in southeastern England running east and west to 180° running through the Pacific Ocean

Low Arctic the southern regions of the Arctic, where the growing season for vegetation is generally between three and five months

macrofossils FOSSILS that are large enough to be examined without the use of a microscope; sometimes referred to as megafossils

macrophyte large aquatic plant that can be observed without the use of a microscope

management in the context of wetlands ecology, refers to the process of manipulation by humans in order to achieve a particular end (e.g., flooding, mowing, burning, harvesting, and so on)

megafauna extremely large animals, many of which became extinct at the end of the last glacial episode

megafossils *See* MACROFOSSILS

metamorphic rocks rocks that have been modified in their structure and composition as a result of high temperature and pressure in the vicinity of volcanic or tectonic activity

metapopulation a population of an organism that is naturally patchy or has been broken into separate patches as a result of habitat FRAGMENTATION

methane a gas produced by some bacteria as a result of the incomplete decomposition of organic matter. It is a GREENHOUSE GAS that increases the heat-retention properties of the atmosphere

methanogenic bacteria bacteria that produce methane gas as a result of their metabolism

microbes bacteria, fungi, and viruses

microclimate the small-scale climate within habitats, such as beneath forest canopies or in the shade of desert rocks

microfossils FOSSILS that can be observed only with the aid of a microscope, such as POLLEN GRAINS, DIATOMS, plankton remains, and so on

migration the seasonal movements of animal populations, e.g., geese, caribou, or plankton in a lake

mimicry the adoption of a form or coloration that causes an organism to resemble a different species, usually as a means of protection

minerals inorganic compounds that in combination make up the composition of rocks. The term is also used of the elements needed for plant and animal nutrition

mire general term for any peat-forming wetland ecosystem

mire complex a wetland area that consists of a series of different mire types

moraine an unsorted mass of glacial debris deposited at the lowest point of a melting glacier (terminal moraine) or beneath the ice mass (hummocky moraine)

net primary production the observed accumulation of organic matter mainly by green plants after they have used some of the products of photosynthesis in their own respiration and metabolism

niche the role a species plays in an ECOSYSTEM. It consists of both spatial elements (where the species lives) and the way in which it makes its living (feeding requirements, growth patterns, reproductive behavior, etc.)

nidifugous of young birds that leave the nest shortly after hatching

nitrogen fixation the porcess by which certain organisms are able to convert nitrogen gas into organic molecules that can be built into proteins

nunatak the peak of a mountain projecting from a mass of ice

nutrient cycle the cyclic pattern of element movements between different parts of the ECOSYSTEM, together with the balance of input and output to and from the ecosystem

occult precipitation precipitation that is not registered by a standard rain gauge because it arrives as mist, condensing on surfaces, including vegetation canopies

oceanic climate a climate in which summer temperatures are cool and winter temperatures mild, often accompanied by high precipitation. Such conditions are most often encountered in regions close to the oceans

oceanic conveyor belt the movement of the oceanic waters of the world in a pattern that distributes energy from the equatorial regions to the polar regions. Warm, low-density water moves northward into polar regions, where it cools and becomes denser, returning to southern regions as deep water currents moving in the opposite direction to those at the surface. *See also* THERMOHALINE CIRCULATION

oligotrophic poor in plant nutrients. *See also* EUTROPHIC

ombrotrophic fed by rainfall. BOGS are ombrotrophic mires, receiving their water and nutrient input solely from atmospheric precipitation

osmosis the movement of water molecules from low SOLUTE concentration to high solute concentration through a semipermeable membrane that prevents the diffusion of the solute

outbreeding a population in reproductive contact with other populations of the same species and able to exchange genetic material. Such populations generally contain more variety in their GENE POOL than INBREEDING populations

ozone hole extreme thinning of the ozone layer over the polar regions in their respective summers, which allows excessive ultraviolet radiation to reach the Earth's surface

paleoecology the study of the ecology of past communities using a variety of chemical and biological techniques

paleomagnetism the retention of magnetic alignment in a rock as a result of being formed within a magnetic field

palsa a wetland type found only within the Arctic Circle. Elevated peat masses expand as a frozen core develops within them. Palsas pass through a cycle of growth and then collapse, forming open pools. *See also* PINGO

paludification a process in which an ecosystem becomes inundated with water

palynology the study of pollen grains and spores

peat organic accumulations in wetlands resulting from the incomplete decomposition of vegetation litter

perglacial survival of an organism that has persisted in a region through a period of glaciation

periglacial describing the climatic conditions found around the edges of a glacier or glaciated region

permafrost permanently frozen subsoil. The upper layer (ACTIVE LAYER) thaws during the summer and freezes in winter

pH an index of acidity and alkalinity. Low pH means high concentrations of hydrogen ions (hence acidity). A pH of 7 indicates neutrality. The pH scale is logarithmic, which means that a pH of 4 is 10 times as acidic as pH 5

phanerophyte the LIFE-FORM of a tree in which buds are held well above the surface of the ground

photosynthetic bacteria bacteria possessing pigments that are able to trap light energy and conduct photosynthesis. Some types are green and others purple

physiognomy the general structure of vegetation

physiological drought a condition in which water is present in a habitat but is unavailable to a plant because of low temperature

pingo a structure formed in the tundra soil by water freezing under pressure, often fed by a spring. As the water turns to ice and expands, it forces the surface of the ground to rise into an extensive mound

pioneer an initial colonist in a developing habitat

plate tectonics the thesis that the crust of the Earth is divided into plates that move over the surface, occasionally colliding and buckling to form mountain chains

podzol a soil type, common in the boreal zone and in the ECOTONE with the tundra, in which LEACHING has resulted in the movement of iron, aluminum, clay minerals, and organic matter down the profile to be deposited in lower horizons

poikilohydric of plants that become desiccated under drought conditions but rapidly recover when water is available once more

poikilothermic cold-blooded

polar front the boundary where tropical air masses encounter cooler polar air masses, resulting in unstable weather conditions and the formation of depressions

pollen analysis the identification and counting of fossil pollen grains stratified in peat deposits and lake sediments

pollen grains cells containing the male genetic information of flowering plants and conifers. The outer coat is robust and survives well in wetland sediments. The distinctive structure and sculpturing of the coats permit their identification in a fossil form

polygon mire patterned wetlands of the Arctic region in which raised polygonal sections of land are separated by water-filled channels

polyphyletic a trait that has arisen a number of times in the course of evolution

population a collection of individuals of a particular species

precipitation aerial deposition of water as rain, dew, snow, or in an occult form

primary productivity the rate at which new organic matter is added to an ecosystem, usually as a result of green plant photosynthesis

prokaryote a simple organism having cells that lack a true nucleus

protist a simple organism belonging to the kingdom Protoctista, consisting of unicellular or colonial organisms, including amoebas, flagellates, diatoms, foraminifera, and many others

pyramid of biomass a principle that applies to terrestrial ecosystems stating that the succeeding trophic levels of an ECOSYSTEM have a lower total BIOMASS than preceding levels

quality of light, refers to the spectrum of different wavelengths present

radiocarbon dating a technique for establishing the age of a sample of organic matter based upon the known decay rate of the isotope ^{14}C (carbon-14)

raised bog a mire in which the accumulation of peat results in the formation of a central dome that raises the peat-forming vegetation above the influence of groundwater flow. The surface of the central dome thus receives all its water input from precipitation (OMBROTROPHIC)

realized niche the actual spatial and functional role of a species in an ecosystem when subjected to competition from other species. *See also* FUNDAMENTAL NICHE

reclamation the conversion of a habitat to a condition appropriate for such human activities as agriculture or forestry

redox potential a scale indicating the potential for oxidation and reduction in an environment

rehabilitation the conversion of a damaged ecosystem back to its original condition

relative humidity the quantity of water present in a body of air expressed as a percentage of the amount of water needed to saturate the air under the same conditions of temperature and pressure

relict a species or a population left behind following the fragmentation and loss of a previously extensive range

replaceability the ease or difficulty with which a particular habitat could be replaced if it were to be lost

representativeness the degree to which a site illustrates the major features characteristic of its habitat type

resilience the ability of an ecosystem to recover rapidly from disturbance

resource allocation the division of the products of photosynthesis among different parts of a plant, such as leaves, stem, and roots

resource partitioning the manner in which different species assume different roles (NICHES) in an ecosystem and thus divide the resources between them

respiration the oxidation of organic materials resulting in the release of energy. Waste products include carbon dioxide, methane, and ethyl alcohol depending on the availability of oxygen (*see* REDOX POTENTIAL) and the type of organism involved

rheotrophic a wetland that receives its nutrient elements from groundwater flow as well as from precipitation. In rheotrophic mires the groundwater flow is usually responsible for the bulk of the nutrient input

rhizopods one-celled microscopic animals resembling amoebas but with a protective shell around their bodies. These shells are often preserved as fossils within peat deposits

r-selected species those species that expand their populations rapidly but have limited ability to survive intense competition. They therefore cope well in unstable, disturbed conditions

salinity the concentration of salts in a solvent

salinization the increasing salt content in the wetlands of hot dry regions when they have no exit drainage. Salinization is a consequence of the evaporation of water leaving behind the salts contained in the incoming water

salt marsh coastal intertidal wetlands dominated by herbaceous plants

saltation the movement of a particle, such as a sand grain, by bouncing over a surface

saturated vapor pressure the total amount of water vapor that a volume of air can hold at a given temperature

secondary dispersal the additional transport of a seed following its initial dispersal, such as when a mouse or an ant picks up a wind-dispersed seed and takes it farther from the parent plant

sediment material that is deposited in an ecosystem, such as a lake or a peat land, and accumulates over the course of time. Sediments may be organic and/or mineral in nature

sedimentary rocks rocks formed by the gradual accumulation of eroded materials, either underwater or on land, eventually forming a compressed, stratified mass of material

sedimentation the process of SEDIMENT accumulation

snow patch an accumulation of snow that is sufficiently deep to survive well into the summer season and may even last for several years, in which case its edges melt each summer

solifluction the movement of soils down slopes in tundra conditions, which occurs because the surface of the soil melts while the lower layers remain frozen

solute a material that dissolves in a SOLVENT

solvent a liquid in which materials (SOLUTES) can dissolve

species richness the number of species of organisms within a given area; an important component of BIODIVERSITY

spores the dispersal propagules of algae, mosses, liverworts, ferns, and fungi

stadial a cold period in the history of the Earth that is less severe or shorter than a glacial episode

stochastic a chance event; one that cannot be predicted. *See also* DETERMINISTIC

stomata (singular stoma) the pores through which a plant exchanges gases with its environment and through which it loses water by transpiration

stone polygons patterns of stones caused by FROST HEAVING on level ground in tundra habitats

stone stripes lines of stones following the contours of a slope produced by FROST HEAVING in tundra soils

stratification the layering of sediments in a lake or of leaf layers in a canopy

stratigraphy the study of layering in rocks and sediments and the description of sediment profiles

stratosphere the part of the Earth's atmosphere lying above the TROPOSPHERE, from around nine to 30 miles (14–50 km)

structure the architecture of a vegetation canopy

subalpine the zone immediately below the alpine tundra zone and above the timberline on mountains

submerged aquatic plants freshwater plants rooted in soil that lies underwater and grow toward but not above the water surface. Some submerged aquatics have flowers that extend above the water surface

subduction zone a region where one tectonic plate pushes beneath another

sublimation the direct formation of a solid from a gas

succession the process of ecosystem development. The stages of succession are often predictable as they follow a directional sequence. The process usually involves an increase in the BIOMASS of the ecosystem, although the development of RAISED BOG from carr is an exception to this. Succession is driven by immigration of new species, FACILITATION by environmental alteration, competitive struggles, and eventually some degree of equilibration at the CLIMAX stage

tephra the glasslike dust particles emitted from erupting volcanoes. Layers of tephra in ice stratigraphy can serve as time markers because the dates of eruptions are well known and the chemistry of tephra particles often indicates the precise volcanic eruption involved

terrestrial occurring on land, as opposed to aquatic

terrestrialization the process of SUCCESSION whereby aquatic ecosystems gradually become infilled

texture refers to the proportions of different sized particles in soil. A soil containing a relatively even contribution from sand, silt, and clay is called a loam

thermohaline circulation the movement of water masses around the oceans of the world in a circulatory system that varies in the density of waters, caused by a combination of temperature and salt content. *See also* OCEANIC CONVEYOR BELT

till the detritus carried in and on the ice of a glacier, which is dumped as an unsorted mass when the glacier melts; sometimes called boulder clay

timberline the upper limit of forest on a mountain. Isolated trees, however, may survive beyond the timberline

topography the form of a landscape, including hills and valleys

transect a line along which vegetation is recorded

transpiration the loss of water vapor from the leaves of TERRESTRIAL plants through the stomata, or pores, in the leaf surface

tropopause the boundary between the TROPOSPHERE and the STRATOSPHERE

troposphere the lower layer of the Earth's atmosphere up to about nine miles (15 km)

tundra the open, low-stature vegetation of extremely cold conditions, found in the polar regions (ARCTIC TUNDRA) and on high mountains (ALPINE TUNDRA). Trees are absent, although dwarf shrubs and CHAMAEPHYTES are present

ultraviolet radiation short-wave radiation from the Sun that is largely absorbed by the ozone layer in the STRATOSPHERE. It is harmful to living organisms

uniformitarianism the theory that geological processes, such as erosion and sedimentation, took place in essentially the same way in the past as they do at present. The present thus acts as a clue to the past

vertebrate an animal with a backbone

vessels specialized cells in wood that are responsible for the transport of water up a stem or trunk

vulnerability the degree to which a site is threatened, such as when a wetland is in danger of drainage for alternative uses such as agriculture or forestry. *See also* FRAGILITY

water level the height of water above the surface of a soil

water table the level at which water is maintained within the soil of an ecosystem

watershed the region from which water drains into a particular stream or wetland (equivalent to CATCHMENT). The term is also used of the ridge separating two catchments, literally the region where water may be shed in either of two directions

wavelength the distance apart of crests in a wave motion

weathering the breakdown of a rock into MINERALS and chemical elements as a result of the effects of such factors as frost, solution, and biological activity

wetland a general term covering all shallow aquatic ecosystems (freshwater and marine) together with marshes, swamps, fens, and bogs

wildlife both the wild animals and the wild plants of a habitat

xeromorphic structural adaptations in plants associated with drought resistance

zonation the banding of vegetation along an environmental gradient, such as in the transition from lowland forest, through montane forest and cloud forest, to the timberline on a tropical mountain

Further Reading

GENERAL ENVIRONMENTAL REFERENCE

Archibold, O. W. *Ecology of World Vegetation.* New York: Chapman & Hall, 1995. A broad and useful introduction to all the major biomes of the world.

Bradbury, Ian K. *The Biosphere.* 2d ed. New York: Wiley, 1998. An introduction to global processes that link the various biomes and the human population of the planet.

Brown, J. H., and M. V. Lomolino. *Biogeography.* 3d ed. Sunderland, Mass.: Sinauer Associates, 2006. A very extensive and exhaustive coverage of the scientific principles that unite biology and geography in the study of the living world.

Cox, C. B., and P. D. Moore. *Biogeography: An Ecological and Evolutionary Approach.* 7th ed. Oxford: Blackwell, 2005. An introductory text dealing with the historical and modern factors that determine species distributions on Earth.

Gaston, K. J., and J. I. Spicer. *Biodiversity: An Introduction.* 2d ed. Oxford: Blackwell, 2004. An explanation of the concept of biodiversity, its meaning, and its importance in conservation.

Houghton, J. *Global Warming: The Complete Briefing.* 3d ed. Cambridge: Cambridge University Press, 2004. The authoritative account of the most recent research by the Intergovernmental Panel on Climate Change.

TUNDRA HISTORY

Allen, B. ed. *The Faber Book of Exploration.* London: Faber & Faber, 2002. A fascinating account of early travel in many remote parts of the world, including the tundra.

Imbrie, John, and K. P. Imbrie. *Ice Ages: Solving the Mystery.* Cambridge, Mass: Harvard University Press, 1979. A readable book concerned with the possible causes of ice ages.

John, Brian S. *The Winters of the World: Earth Under the Ice Ages.* London: David & Charles, 1979. The impact of ice ages on the history of the planet.

Pielou, E. C. *After the Ice Age: The Return of Life to Glaciated North America.* Chicago: University of Chicago Press, 1991. An excellent and very accessible account of the development of North American habitats following the last ice age.

Roberts, N. *The Holocene: An Environmental History.* 2d ed. Oxford: Blackwell, 1998. A general account of climatic, environmental, and biological change in the world over the past 10,000 years.

THE TUNDRA ECOSYSTEM AND ITS INHABITANTS

Barbour, M. G., and W. D. Billings. *North American Terrestrial Vegetation.* 2d ed. Cambridge: Cambridge University Press, 2000. An authoritative account of the vegetation of North America, including the tundra regions of Alaska and Canada.

Chabot, B. F., and H. A. Mooney. *Physiological Ecology of North American Plant Communities.* New York: Chapman & Hall, 1985. An examination of the vegetation of various North American ecosystems, including the tundra, with particular emphasis on the physiological adaptations of plants.

Crawford, R. M. M. *Studies in Plant Survival.* Oxford: Blackwell Scientific, 1989. The ways plants adapt to extreme environments, including very cold conditions.

Knystautas, A. *The Natural History of the U.S.S.R.* London: Century, 1987. A well-illustrated account of the habitats, flora, and fauna of Russia and surrounding areas, including the Siberian tundra.

Rydin, H., and J. Jeglum. *The Biology of Peatlands.* Oxford: Oxford University Press, 2006. An account of the ecology of the world's peatlands, including those of the tundra regions.

Sage, B. *The Arctic and Its Wildlife.* London: Croom Helm, 1986. A sound scientific account of the ecology and biology of the Arctic, with much detail concerning individual species.

Sparks, J. *Realms of the Russian Bear.* London: BBC, 1992. The ecology of the Russian boreal forests and their transition into the tundra regions.

POLAR TUNDRA

Aleksandrova, V. D. *Vegetation of the Soviet Polar Deserts.* Cambridge: Cambridge University Press, 1988. A very detailed account of the plant life of the Asian High Arctic.

Chernov, Y. I. *The Living Tundra.* Cambridge: Cambridge University Press, 1985. The ecology and biology of tundra regions, concentrating on the Russian region.

McGonigal, D., and L. Woodworth. *Antarctica: The Complete Story.* London: Frances Lincoln, 2003. Arguably the best book ever published on the geology, biology, and human history of Antarctica.

Ritchie, J. C. *Past and Present Vegetation of the Far Northwest of Canada.* Toronto: University of Toronto Press, 1984. A detailed account of the development of tundra vegetation in the Northwest Territories of Canada.

Young, S. B. *To the Arctic: An Introduction to the Far Northern World.* New York: Wiley, 1994. A popular and readable introduction to polar tundra studies.

ALPINE TUNDRA

Bowman, W. D., and T. R. Seastedt. *Structure and Function of an Alpine Ecosystem: Niwot Ridge, Colorado.* Oxford: Oxford University Press, 2001. A detailed research account of studies in the Rocky Mountains.

Hambrey, M., and J. Alean. *Glaciers.* Cambridge: Cambridge University Press, 1992. The geology and behavior of mountain ice masses and ice sheets.

Hedberg, O. *Features of Afroalpine Plant Ecology.* Västervik, Sweden: Acta Phytogeographica Suecica 49, 1995. A detailed survey of the remarkable mountain vegetation of East Africa.

Larson, D. W., U. Matthes, and P. E. Kelly. *Cliff Ecology.* Cambridge: Cambridge University Press, 2000. A unique review of the ecology and biology of cliff habitats, including those of mountain areas.

Matthews, J. A. *The Ecology of Recently Deglaciated Terrain.* Cambridge: Cambridge University Press, 1992. A detailed account of research on the recolonization of land laid bare by glacial retreat in Norway.

Rundel, P. W., A. P. Smith, and F. C. Meinzer. *Tropical Alpine Environments.* Cambridge: Cambridge University Press, 1994. A collection of reviews concerning the physiology and adaptations of tropical mountain plants.

Storer, Tracy I., and Robert L. Usinger. *Sierra Nevada Natural History.* Berkeley: University of California Press, 1997. The ecology and biology of the Sierra Nevada Mountains of California.

Web Sites

ANTARCTICAN
URL: http://www.antarctican.com
Accessed November 9, 2006. This Tasmanian Web site carries current news from Antarctica.

ARCTIC RESEARCH CONSORTIUM AUSTRIA
URL: http://www.arctic.at
Accessed November 9, 2006. This is a good site for a wide range of world links.

CONSERVATION INTERNATIONAL
URL: http://www.conservation.org
Accessed November 9, 2006. Particularly concerned with global biological conservation

EARTHWATCH INSTITUTE
URL: http://www.earthwatch.org
Accessed November 9, 2006. General environmental problems worldwide

GATEWAY ANTARCTICA
URL: http://www.anta.canterbury.ac.nz
Accessed November 9, 2006. Based at the University of Canterbury, New Zealand, this site has much information on wildlife conservation and management in Antarctica.

GLACIER
URL: http://www.glacier.rice.edu
Accessed November 9, 2006. Based at Rice University in Texas, this is a useful site for glacial geology information.

INTERNATIONAL UNION FOR THE CONSERVATION OF NATURE
URL: http://www.redlist.org
Accessed November 9, 2006. Many links to other sources of information on particular species, especially those currently endangered

NATIONAL AERONAUTICS AND SPACE ADMINISTRATION (NASA)
URL: http://earthobservatory.nasa.gov/Laboratory/Biome/biotundra.html
Accessed May 22, 2007. This has some basic information on the tundra environment and provides useful links to other relevant sites.

NATIONAL PARKS SERVICE OF THE UNITED STATES
URL: http://www.nps.gov
Accessed November 9, 2006. Information on specific conservation problems facing the National Parks

SCOTT POLAR RESEARCH INSTITUTE
URL: http://www.spri.cam.ac.uk
Accessed November 9, 2006. This is the world's leading database on the Antarctic.

SIERRA CLUB
URL: http://www.sierraclub.org
Accessed November 9, 2006. Covers general conservation issues in the United States and also covers issues relating to farming and land use

UNITED NATIONS ENVIRONMENTAL PROGRAM WORLD CONSERVATION MONITORING CENTER
URL: http://www.unep-wcmw.org
Accessed November 9, 2006. Good for global statistics on environmental problems

U.S. ANTARCTIC PROGRAM
URL: http://www.polar.org
Accessed November 9, 2006. This site includes information on the activities of U.S. vessels in the Antarctic region.

U.S. DEPARTMENT OF THE INTERIOR BUREAU OF LAND MANAGEMENT
URL: http://www.blm.gov/education/
Accessed May 22, 2007. This is a site that provides a wealth of environmental information, including much that is relevant to tundra habitats. It also includes ideas for practical work.

U.S. FISH AND WILDLIFE SERVICE
URL: http://www.nwi.fws.gov
Accessed November 9, 2006. A valuable resource for information on wildlife conservation

U.S. GEOLOGICAL SURVEY
URL: http://www.usgs.gov
Accessed November 9, 2006. Covers environmental problems affecting landscape conservation

U.S. NATIONAL SCIENCE FOUNDATION
URL: http://www.nsf.gov
Accessed November 9, 2006. A good site for satellite images

WORLD BIOMES
URL: http://worldbiomes.com/tiomes_tundra.htm
Accessed May 22, 2007. This site contains some basic information about tundra and also gives useful links to other sites.

Index

Note: *Italic* page numbers indicate illustrations; *m* refers to maps.

A

ablation 38, 46–47
acidity 53–54, 69. *See also* pH
acid rain 228
acids 56, 57
active layer 59, 61
Adélie penguin 225
adenosine triphosphate (ATP). *See* ATP
adiabatic cooling 22
adiabatic warming 25
Aëdes nigripes 152
aestivation 130
Agassiz, Louis 179, 180, 194
agriculture 203, 212
air 105
air pressure 122
Alaska 2, 197, 205, 215–216
Alaskan hare 169
Alaska willow 71
albatross *129*, 165, 174, 175, 230
albedo 8, 223
algae 78, 97, 143. *See also* cyanobacteria
alkalinity 54
Allen, Joel A. 114
Allen's rule 114–115, 158
Allerød 190
Alopex lagopus 108
alpine butterfly 119
alpine meadow *5*, 80–81, *81, 201, 231*
alpine pasque flower 150–151, *151*
alpine tundra *5, 24*, 79–80, *81*, 85, *209, 232*
  birds in 161–163
  climate of 21–26
  conservation of 231
  cows *201*
  early humans in 200–201

ecosystems of 102, 103
European exploration of 208–209, *209*
grazing *231*
location of 5–6
mammals of 170–173
rocks of 36–37
temperature cycle 23
tourism in 217
tropical alpine habitats 82–85, *83*
vascular plants of 149–152
wetlands *82*
altitude 122–123, 138
American golden plover 159
American mountain goat 171–172
amino acids 104, 110
ammonification 98
ammonium ions 98
amphi-Atlantic distribution 71
amphi-Beringian distribution 71
amphibians 155, 174
Amundsen, Roald 206–208, 210
anaerobic environment 121
Andean condor 163
Andes Mountains 83
angle, of Sun's rays 8–9
animal. *See also* mammals; *specific animals, e.g.:* caribou
  adaptation to cold 105
  of alpine tundra 80–81
  in anoxic environment 122
  of Antarctic tundra 79
  coloration 119
  of dwarf shrub tundra 69
  grazing by 129–130
  hibernation 135
  migration 131–133, 135
  of polar desert 66
  sexual reproduction 123–124, 128–129

animal cells 109
animal husbandry 203
animal reproduction 123–124, 128–129
anise swallowtail butterfly *153*, 154
annelid worms 147, 148
annuals 111–113
anoxia 121–122
anoxic environment 121
Anser albifrons 159
Antarctica *4m, 39*
  ecotourism in 217
  effects of climate change on 222, 225
  hunting in 212–213
  lichen in 144
  mineral exploitation of 215, 221
  polar tundra 85
  soil 148
  vascular plants on 149
Antarctic Peninsula 5
Antarctic Treaty System 215
Antarctic tundra xvi, *4m*, 78–79
  climate *17m*, 17–18
  European exploration of 206–208, *207m*
Anthelia juratzkana 145–146
anthocyanins 118, 119
anticyclone 11
antifreeze 110, 214
antlers 167
Aphodius holdereri 153
apomixis 124
Aptenodytes sp. 164
Aquila chrysaetos 161
Arachnida. *See* spiders
archaeology 202
Arctic 14
arctic-alpine 14
arctic fox 108
arctic hare 108, 169
Arctic National Wildlife Refuge 215–216

Arctic Ocean 2, 3, 13, 14
Arctic polygons *60, 67*
arctic poppies 69
Arctic small tool tradition 202
arctic tern 155*m*, 157, *157*
arctic tundra xvi, 3*m*, 14*m*, 85
　　aurora borealis 16–17
　　climate 14*m*, 14–16
　　European exploration of 203–206,
　　　204*m*
　　glacial history of 190
　　seasons in 16
Arctic wetland 74–78, *76*
arctic willow 66, 169
Arctostaphyllos uva-ursi 116, *116*
arête 41
Aristotle 206
art, early 200
arthropods 152–154
Asia 6
aspect 19, *19*
astronomic cycles 184. *See also*
　Milankovitch cycles
atmosphere
　　absorption of heat energy by 7
　　and carbon cycle *219*, 219–221
　　as carbon reservoir 220
　　circulation *9*, 14
　　and climate 10–11
　　greenhouse effect 7, *7*, 220
　　layers of *10*
　　as source of nutrients 94
ATP (adenosine triphosphate) 95, 97,
　109, 142
aurora borealis 16–17
Australia 197
autotrophic plants 86
autumn 119

B

Bacillariophyta 147
bacteria 98, 99, 174. *See also* microbes
Baffin, William 205
bald eagle 161
Barents, Willem 203–204
Barrow's goldeneye 160
basal ice 45
bats 169, 175
bearberry 116, *116,* 118
bearded vulture 163, 164, 231
bedrock 45, 52
bees 127
beetles 108, 153, 174, 187, 191
Bellingshausen, Thaddeus von 206, 210
beluga 213
Bergmann, Carl 113, 114

Bergmann's rule 113, 114, 153–154, 199
Bering, Vitus 205
Beringian invasion 197
Bering Strait 205
Betula sp. 69
bezoar stone 173
B horizon *56, 57*
Bible, as account of history of Earth
　179
bighorn ram *213,* 214
bighorn sheep *171,* 172–173, 214
biodiversity 136–175
　　arthropods 152–154
　　birds of alpine tundra 161–163
　　birds of northern tundra 154–161
　　birds of southern tundra 164–165
　　defined 136–137
　　at end of Wisconsinan glaciation
　　　193
　　lichens 143–145, *144*
　　mammals of alpine tundra 170–173,
　　　170–173
　　mammals of polar tundra
　　　165–170
　　microbes 141–143
　　mosses and liverworts 145–147
　　patterns 137–139, *139*
　　soil fauna 147–149
　　succession 139–141
　　value of tundra's contribution to
　　　211
　　vascular plants 149–152
biogeography 137
biological history. *See* history,
　geological/biological
biological spectrum 110, 113
biological weathering 54
biology, of tundra 104–135
　　adaptation to cold 104–109,
　　　106–108
　　body size 113–115
　　cold stress 109–110
　　coloration of plants/animals
　　　115–119
　　drought 119–121
　　grazing 129–130
　　hibernation 130–131
　　migration 131–133
　　oxygen shortage 121–123, *123*
　　plant forms 110–113
　　reproduction/dispersal 123–129
　　responses to climate change
　　　224–226
　　waterlogging 121–122
biomass 90–91, 100–103, 139, 228
biome 113
bioprospecting 214

biotechnology 214
birch 72, 73
birds
　　of alpine tundra 161–163
　　of Antarctic tundra 79
　　conservation of 231
　　cryptic coloration of 119
　　disjunct distribution patterns of
　　　193–194
　　insulation for 108–109
　　migration of 105, 131–133
　　of northern tundra 154–161
　　nutrient import by 96–97
　　and oxygen supply 123
　　and plant dispersal 118–119
　　of polar desert 66
　　reproductive strategies 135
　　seed dispersal by 128
　　for seed transport to young
　　　ecosystem 99–100
　　sexual reproduction 128
　　of southern tundra 164–165
　　use of lichens for nesting material
　　　145
bison 199, 200
biting midges 152–153
black-bellied plover 159
black brant goose 159
black-browed albatross *129*
black-capped marmot 171
blackflies 152
black-tailed godwit 133
Blanc, Mont *209*
blanket bog 75, 77
blood 122
blueberries 118–119
blue-green bacteria. *See* cyanobacteria
body size, of tundra life forms 113–115,
　134
bog mosses 85, 146–147
bomb calorimeter 89
boreal forest 225
boulder 50
boulder clay 46
bowhead whale 213
Branta bernicla 159
breeding 124, 132, 133, 168–169.
　See also sexual reproduction
British East India Company 208
bryophytes 145–147, 174. *See also*
　liverworts; mosses
Bryum argenteum 146
bumblebee 107
Buteo lagopus 161
butterflies 119, 127, *153,* 153–154
Buturlin, Sergei Alexandrovich 156

C

Cabot, John 203
Calcarius lapponicus 161
calcium carbonate 187
Calidris alba 158–159
California condor 163
Callorhinus ursinus 165
calories 89
calving 40
camels 193
camouflage 119, 134
Canada 2
capillarity 55
capillary water 55, 57
Capra hircus 173
carbon 88
carbon 14 180. *See also* radiocarbon dating
carbon budget *219,* 220–221, 236
carbon cycle *219,* 219–221
carbon dioxide
    and carbon loss from ecosystem 101
    and chemical weathering 53
    and climate cycles 186
    during glacial stages 186
    as greenhouse gas 220
    and late Carboniferous ice age 178
    recycling within ecosystem 88
    and transpiration 120
carbonic acid 53, 54
Carboniferous 178
carbon isotopes 96–97
carbon sink 220–221, 224, 228, 236
caribou *167*
    and conservation 230
    in dwarf shrub tundra 69
    early domestication of 202
    effect of grazing on biomass/productivity of ecosystem 93
    element cycling by 93
    grazing by 130
    in human diet 203
    lichen consumption by 145
    migration by 131
    migration of 93–94
    in polar tundra 166–168
    population cycle of 101
    as prey for Eurasian humans 202
    protection of herds 212
    and radioactive fallout 228
    range of *166m*
    survival after death of other megafauna 199
    thermoregulation by 115

carnivorous diet 198
carotene 118
carotenoids 118
carrying capacity 125
catastrophism 179
cave art 200
cell membranes 109, 134
cells 109
Cenozoic 178
cesium 137 145, 228
CFCs (chlorofluorocarbons) 227
chamaephytes 110–113, *111,* 115, 120–121, 134, 149
chamois 172, 214
Chancellor, Richard 203
chemically bound water 55
chemical reactions 104
chemical weathering 53
chemosynthetic autotrophs 98
Chernobyl nuclear accident 145, 228
chilling injury 109
China 51
china clay 53
chinstrap penguin 114, *114,* 165
chlorofluorocarbons (CFCs) 227
chlorophyll 87, 118, 119, 142
Chomonix, France *233*
C horizon *56,* 57
Chuckchi people 203
ciliates 147
cirque 41
Cladonia sp. 145
clay 50
climate 1–26, *11,* 137
    of alpine tundra 21–26, *24*
    of Antarctic 17–18
    of Arctic 14–16
    and atmosphere 10–11
    and aurora borealis 16–17
    global climate patterns 6–9
    greenhouse effect *7*
    humidity and 20
    microclimate 18–21
    oceans' effect on 11–14, *12*
    patterns of tundra distribution 1–6
climate change xviii, 199, 220–236, *223m, 224m*
climate cycles. *See* Milankovitch cycles
climatology 235
climax ecosystem 100, 103, 141
cloning 123
clothing 196, 200
cloudberry *115,* 118
cloud forest 84
clover 97, 126
Clovis people 197–198
clutch 128, 135, 160, 161

coastal tundra 74
cobble 50
coccolithophorids 187
coevolution 127, 128
Colaptes auratus 193
cold
    adaptation to 104–109, *106–108,* 134
    effect on metabolism 104
    hibernation as response 130–131
    migration as response to 131–133
    and psychrotrophic fungi 142
cold acclimation 109
cold glacier 42
cold starvation 114
cold stress 109–110
cold war 215
Coleoptera. *See* beetles
collared flycatcher 194
Collembola. *See* springtail
Colobanthus quitensis 149
coloration, of plants/animals *115–118,* 115–119, 127, 164, 167
Columbus, Christopher 203
comminution 148
community 86
competition 140
condors 163, 174
conifers 72
conservation, of tundra regions 137, 211–212, 229–233, *233*
conservative plate boundary 31
consumer animals 103
continental climate 12
continuous permafrost 58, 59
convection currents *32*
convergent evolution 163
Cook, James 206, 212
Coriolis, Gaspard Gustave de 13
Coriolis effect 13
cornified cells 105
corrie 41
cosmic rays 180
cottongrass 65, 67, 70, 151–152
coversands 52
cows *201*
crane flies 148
creeping stems 124
Cretaceous 178, 194
crevasses 44, 77
cruise ships 217
crustaceans 147
crustal plates *30*
crustose lichens 144
cryosols 58
cryptic coloration 119, 134
cryptophytes 112

cuticle 120–121
cyanobacteria 97, 99, 142, 143, 174
cycling, of elements 93–98, *94*
cyclone 11
cytoplasm 109–110

D

DALR (dry adiabatic lapse rate) 22
Darwin, Charles 123, 124
DDT 228
dead ice 49
decay 98, 219–220. *See also* decomposers; decomposition
deciduous plants 115, 134
deciduous shrubs 68–69
deciduous trees 72
decomposers 87–88, 91, 94, 103
decomposition 98–99, 149, 224
denatured protein 104
Dendrosenecio elgonensis 83, 84
Dendrosenecio sp. 149, 150
denitrification 99
Deschampsia antarctica 149
desiccation 121, 145
detergents 216
detritivore 66, 87–88, 99, 147–149, 174
dew point 20
diamicton 46
diamondback moth 154
diatoms 147, 187
diet 202–203
dimethyl sulfide 53
dioecious plants 127
Diomedea exulans 165
discontinuous permafrost 58–59
disjunct distribution 193, 195
dispersal 118–119
distribution patterns, of animals 123
distribution patterns, of tundra 1–6
diversity 141
DNA 142
domestication, of wild animals 202
Dondroica coronata 193
drainage 119, 134
Drake, Francis 206
drift 48
drought 119–121
drought avoiders 120
drought tolerators 120, 121
drumlin 48, *49*
dry adiabatic lapse rate (DALR) 22
Dryas sp. 66, 150
ducks 160
dung beetle 153
dwarf birch 69, 71
dwarf shrubs

in alpine tundra 80
and color 115
in dwarf shrub tundra 68–69
grazing by caribou 130
in tropical alpine habitats 84
dwarf shrub tundra 66–70, *68,* 85, 90–91

E

eagles 161
earthworms 148
East Africa 84, 85
ecosystem 86–103
alpine tundra 102
decomposition 98–99
defined 86–88
development/stability 99–101
element cycling 93–98
energy in 87
food web 90–93, *92*
nitrogen fixation 97
primary productivity in 88–91
productivity/population cycles 101–102
ecotone 22, 71, 72
ecotourism 216–219, *217, 221*
ecotypes 150, 226
ectoparasites 152
edema 122
element cycling 93–98, 103
elfin forest 84
Elizabeth I (queen of England) 203
elk *92, 132,* 229
eluviated horizon *56,* 57
emperor penguin 164, 225
endangered species 214
endemics 84
end moraine 47
energy balance 116–117
energy budget 100, 124, 128
energy consumption 131
energy fixation 102
energy input 138–139
energy loss 105
energy sink 14
enzymes 104, 109, 134, 142
enzyme-substrate complex 104
equator 9
equilibrium 100
Eremophila alpestris 162
Erica arborea 84
Eriophorum sp. 151–152
Eriophorum vaginatum 70
erosion 28, 94
erratic 48
esker 50

Eskimos 202
Espeletia 149–150
ethyl alcohol 122, 134
Eudyptula minor 164
eukaryote 141
Eurasia 202
Europe 6
Europeans, exploration by 203–209, *204m, 207m, 209*
eustatic sea level changes 183
Everest, George 208
evergreens 68, 72, 115–117, *116,* 134
evolution 123–124, 133, 137
exfoliation 52
exploration 203–209, *204m, 207m, 209*
extinction 193, 198

F

facilitation 139, 140
falcons 160
Falco sp. 160–161
farming 203, 229–230
fat 132, 171, 198, 214, 228
fauna 147–149
feathers 108–109, 164
feedback 223, 224
fellfield 79
fermentation 122
ferns 99, 149
fertilizer 96, 99, 228
filaments 126
fire, human use of 196
firn 38
fish 97, 109
fishing 212, 230
fission 123, 142
flavonoids 118
flavonols 118, 119
flies 107, 127
flowering plants 65–66, 99–100
flowers 89
flycatcher 194
flyways 133
föhn winds 25
folds, in protein 104
foliose lichens 144
follicles 105
food chain 92
food supply 131
food web 87, 90–93, *92,* 103, 227–228
foraminifera 187
forest limit 22
forest tundra 71–74, *73,* 85
fossil 114, 178, 187–190, 199
fossil fuels 222. *See also* petroleum
fourfold glaciation theory 180

freeze avoiders 110
freeze tolerators 110
freezing injury 109
Frobisher, Sir Martin 203
frost 58
frost boil 70
frost damage 109
frost hardy plants 109, 110
fruits 89, 128
fruticose lichens 144
fungi. *See also* microbes
    in Antarctic tundra 78
    and biodiversity 142–143
    as decomposers 91, 98
    and lichen 97, 143
    psychrophilic 109
    role in soil food web 174
    in young ecosystem 99
fur trade 205, 212
future, of tundra 222–233, *223m*, *224m*
    biological responses to climate
      change 224–226
    climate change 222–224
    conservation 229–232, *233*
    ozone hole 226, 227
    pollution 226–229, 233

G

Gama, Vasco da 206
gametophyte 145
gas pipeline *216*
Gavia sp. 158
geese 130, 159–160. *See also* snow geese
Geikie, Archibald 180
gene pool 136, 234
genetic diversity 123–124, 136, 226
genetic information xviii, 133
Gentiana verna 150
gentians 127
gentoo penguins 165
geological history. *See* history,
   geological/biological
geologists 235
geology of the tundra 28–63
    alpine rocks 36–37
    glacial debris/deposition 44–49
    ice caps and glaciers 40–41
    ice formations 38–44
    periglacial features 60–62
    plate tectonics 29–33, *30, 32, 34*
    polar rocks 35–36
    rock cycle 28–29, *29*
    rock weathering 52–54
    snow/ice 37–38
    soil formation/maturation 52–59

tectonic history of tundra lands
    33–35
    transport of debris 49–52
geomorphology 49
geophytes 111, *111*, 112, 120
geothermal energy 6
glacial 180–183
glacial debris 44–49
glacial erosion *41, 42*
glacial foreland 50
glacial history of Earth 176–183, *182m*
glacial ice 186
glacial lake 47
glacial milk 46
glacial retreat 139
glacial sole 46
glacial surge 43–44
glaciation
    causes of 183–186
    ice cover, Northern Hemisphere
      182m
    and sea level change 182–183, 194,
      195
    and succession studies 139
glacier *37, 41–44, 45, 47, 48, 51*
    current retreat of 222
    and evidence of climate cycles 186
    glacial history of Earth 176–183
    snow's role in formation of 38
    tropical 225
Glacier Bay, Alaska 205, 222, *223m*
glaciofluvial process 50
glaucous gull 156
gleysols 57
global climate patterns 6–9
global warming. *See* climate change
Glossopteris 34–35, 178
glycerol 110
goat 173, 175
gold 215, 221
golden eagle 161, *162*, 163, *232*
golden plover 159
granite 52, 53
grasses
    in Antarctic tundra 78
    grazing's effect on 129, 130
    in polar desert 65, 66
    replacement of mosses by 146
    in tropical alpine habitats 82–83
grass tussocks 84
gravel 50
gravitational water 55, 57
grazing 129–130, 135, *231*
    after Ice Age 192–193
    in alpine tundra 80
    by arctic hare 169

and biomass/productivity of ecosys-
    tem 93
    by caribou 167
    in coastal tundra 74
    and conservation 230
    and evergreens 117
    and mosses 146
    and transhumance 200
grazing animals 89–90
green algae 75
greenhouse effect 7, *7*, 220
greenhouse gas 7, 220
Greenland 1, 2, 69, 101, 210
griffon vulture 163, 164
gross primary productivity 89
guano 74, 97
Gulf Stream 13, 14, 192, 195
gulls 156–157
Gymnogyps californiarus 163
gymnosperms 126
Gypaetus barbatus. See bearded vulture
Gyps fulvus. See griffon vulture
gyrfalcon 160–161

H

hair 105–107, 122, 146, 168
hairy willow 193
Haliaetus sp. 161
halophyte 110
hanging valley 47–48, *48*
hares 128, 175
harp seal 165
hawks 161
hay 80–81
head (transported soil) 62
heat 92
heat conservation 114
heather 99, 121
heavy metals 227–228
height, of plants 110–111
heliotropism 151
hemicryptophytes 110–113, *111*, 134
herbivore-predator system 91, 169
herbivores 69, 87, 108, 129
herbs 71, 203
heterocyst 142
heterotrophic animals 87
hibernation 105, *130*, 130–131, 135,
    166, 171
High Arctic 65, 85
high latitude 1
high-pressure zone 11
Hillary, Edmund 209, 210
Himalayan snowcock 163
history, geological/biological 176–195
    causes of glaciation 183–186

glacial history of Earth 176–183
Ice Age, aftermath of 191*m*, 192–194
Ice Age, end of 190–192, 191*m*
Pleistocene glaciations 178–183
stratigraphy and recent history 186–190
histosol 58
hoary marmot 170
Holocene 181, 190, 192, 195
homiothermic animals 105
Homo sp. 196, 209
horned lark 163
horns 168
horses 114, 193
housefly 107
housing 203
Hubbard Glacier 43
Hudson, Henry 205
Hudson Bay 183, 205
humans 196–210
adaptation to cold by 105–106
alpine tundra exploration by Europeans 208–209, *209*
Antarctic exploration by Europeans 206–208, 207*m*
Arctic exploration by Europeans 203–206, 204*m*
and death of megafauna 199
early art by 200
early inhabitants of tundra 197–201, 210, 235
emergence of modern 196–197
and megafauna 197–201
modern tundra people *202*, 202–203
humidity 20
hummocky till 49
hunting 212–214, *213*, 221
by Clovis people 197–198
and conservation 229
by early humans 196–197, 199–200
and early tundra art 200
of endangered species 214
by Eurasians 202
Hutton, James 179
hydrogen ions 54
hydrophytes 111, *111*, 112
Hydrurga leptonyx 165

I

ibex 173, 214
ice 37–38, 106
Ice Age (Pleistocene). *See also* Pleistocene epoch
aftermath of 191*m*, 192–194

end of 190–192
human adaptation to 196–197
ice ages (various)
Louis Agassiz's theory of 179
in early history of earth 176–178, 194, 195
and failure of oceanic conveyor belt 13–14
periodicity of 183–186
iceberg 40, 45
ice cap 40–41, 62, 234
ice cores 186
ice cover, Northern Hemisphere 182*m*
ice fall 44
Iceland 203
"ice man" 201
ice sheet *4*
Antarctic 4, 5, 78
and climate change 222–224
and climate cycles 186
during glacials 183
and ice shelves 39–40
lead in 227
limiting effect on Greenland tundra 1
as source of atmospheric history 235
ice shelf 39–40
ice wedge 61
ICZ (Intertropical Convergence Zone) 10
igloo 203
igneous rocks 28
importance, of tundra. *See* value, of tundra
inbreeding 136
industry 220
inertia 100, 226
infill 61, 75
inflorescences 126, 127
insect 105, 107–108, 119, 122
insect pollination 126, 127, 135, 174, 188
insulation 105–109, 134, 164, 167, 203
interglacial 180, 183
interstadial 181
intertidal zone 74
Intertropical Convergence Zone (ICZ) 10
Inuit 202, *202*, 203, 212, 228
invertebrates 110, 114, 121, 123, 129. *See also* specific invertebrates, e.g.: insects
iridescence 119
iron 57
iron pan *56*, 57
isostasy 39

isostatic sea level changes 183
isotherm 72
isotope 181
Ivan IV the Terrible (czar of Russia) 203

J

jaegers 157–158
James' saxifrage *117*
jet stream 10–11
Jurassic 184

K

kairomones 152
kame delta 50
kame terrace 50
kaolinate 53
kea 163
keratin 105
kettle hole 49
kinetic energy 87
king penguin 164
Klondike gold rush 215
krill 213
krummholz 22, 72–74, 81, 82
K-selection 124, 125, 128, 134

L

lagomorphs 128
Lagoptus sp. 108–109, 119, 160
lake sediments 187
languages, Native American 197
Lapland longspur 161
lapse rate 21, 22
larch 72, 73
Larix laricina 72
larks 163
lateral moraine 47
latitude
and angle of Sun's rays 8–9
and biodiversity 137, *138*
and climate change 225
defined 1, *2*
and lichen 144
and temperature 9
leaching 28, 55, 56
lead 227
leaf 88
leatherjackets. *See* crane flies
leaves 120
legumes 97
lemmings
grazing by 129
in polar desert 66
in polar tundra 168–169, 175

population cycle of 101
predation by snowy owls 161
reproductive strategy of 124, 128
Leo Africanus 208
leopard seal 165
Lepidoptera. *See* butterflies
leptosol 58
Lepus sp. 108, 169
lichens 143–145, *144*
  in Antarctic tundra 78
  biological weathering by 54
  in coastal tundra 74
  in dwarf shrub tundra 69
  grazing of, by caribou 130, 167
  hair on 106
  nitrogen fixation by 97
  as poikilohydric plants 121
  in polar desert 65
  in tropical alpine habitats 82
  in young ecosystem 99
life-form 112
Limosa limosa 133
lipids 109
litters 128–129
little penguin 164
liverworts 65, 78, 145–147, 174
lobelia 84
lodgement till 46, 48
loess 51–52
logarithmic growth curve 125
longitude 1
long-lining 230
loon 158
Low Arctic 56, 66
low latitude 1
Lyell, Charles 178–179

M

macaroni penguin 165
Mackenzie, Alexander 205
macroclimate 18
Magellan, Ferdinand 206
magnesium 94, 95
magnetic field, of Earth 133, 135
Mallory, George Leigh 208–210
mammals
  of alpine tundra 80, 170–173,
    170–173
  and arrival of humans in tundra
    regions 198
  body size of 113–115, 134
  grazing by 129
  hair 105
  need for insulation by 108–109
  of polar tundra 165–170
  reproductive strategy of 128–129

mammoths 198, 199
Mammuthus primigenius 198
marmot 105, 131, 170–171, 175, 214
Marmota caligata 170
Marmota camtschatica 171
marsh 90
mass balance 39
meat 202–203
medial moraines 44, 47
Medvezhiy Glacier 43
megafauna 197–201, 209–210
melanism 119
meltwater 43
merino sheep 230, *231*
metabolism 104, 131
metamorphic rocks 28
methane 186
Mexico 82
microbes 141–143, 235
  and climate change 224
  need for nitrogen 99
  nitrogen fixation by 97
  and ozone hole 226
  role in decomposition 98, 103
  use by biotech industry 214
microclimate xvii, 18–21, *19*, 74,
  98–99, 139
microhabitat xvi, 136–137
midges 152–153
migration 93–94, 96–97, 105, 131–133,
  135
Milankovitch, Milutin 185–186
Milankovitch cycles *184–185,* 184–186,
  192
minerals xviii. *See also* rock
  mining, quarrying, and prospecting
    214–216, *216,* 228–229
  and plant growth 53
  in precipitation near ocean 94–95
mining 215, 226–229
Miocene 189
Mirounga leonine 165
mites 148, 152
mitochondria 109
mollusks 148
molt 164
Montifringilla nivalis 162–163
Montreal Protocol 227
moose 95
moraine 47
mor humus 56
mosquito 107–108, 152
mosses 145–147, 174
  of alpine tundra 80
  of Antarctic tundra 78
  of dwarf shrub tundra 69
  hair on 106

as poikilohydric plants 121
of polar desert 65
sexual reproduction by 126
of tropical alpine habitats 82
in young ecosystem 99
mountain xvi–xvii, 5–6, 138, 208–209.
  See also alpine tundra
mountain alder 71, 72
mountain aven 66, 150, 193
mountain goat 231
mountain sheep 173
Mount Everest 208–210
Muir, John 205, 230
muskeg 75, 77
musk ox 66, 69, 114, 168, 202, 229
mycelium 142, 143
mycobiont 143
mycorrhizae 98

N

Native American languages 197
navigation 133, 135
Neanderthal man 196
negative feedback 224
negative interaction 139
nematode worms 147–148
Nepali woman *123*
Nestor notabilis 163
Netherlands, explorers/traders from
  203–204, 206, 210
net primary production 103
net primary productivity 89
Newfoundland 203, 210
Newton's law of cooling 105, 106
New World 197
New World primates 196
nidifugous birds 108–109
nitrate ion 98
nitrates 228
nitrification 98, 99, 102
Nitrobacter 98
nitrogen
  in alpine tundra 102
  conversion to radioactive carbon
    180
  and decomposition 99
  and evergreens 117
  as limiting factor in ecosystem
    development 95–96
nitrogenase 104
nitrogen cycle 99
nitrogen fixation 97, 100, 142, 145
nitrogen mineralization 98
nitrogen oxides 53, 228
Nitrosomonas 98
Niwot Ridge, Colorado 102

Norgay, Tenzing 209, 210
Norse people 203
North Africa 6
North America
    alpine tundra 6
    arrival of humans in 197, 209
    atmospheric circulation patterns 11
    loess deposits in 51–52
    Viking discovery of 203
North Atlantic Ocean 13
northeast passage, search for a 203–204
northern flicker 193
northern fur seal 165
Northern Hemisphere 5, 8, 13
northern tundra 154–161
North Pole 3, 205
northwest passage 206, 210
Norwegian lemming 168
nunatak 38, 39, 193
nutrient 100
nutrient availability 98, 226
nutrient balance 100
nutrient cycle 88, 93, *94*, 143
Nyctea scandiaca 109, 161

O

oceanic circulation 13–14, 186
oceanic climate 12
oceanic conveyor belt *12*, 13, 192,
    223–224. *See also* thermohaline
    circulation
oceanic currents *11, 12*
oceans
    effect on climate *11,* 11–14, *12*
    effect on nutrient content of precipi-
        tation 94–95
    role in guano production 97
    as sink for atmospheric carbon 220
Odobenus rosmarus 165
Oeneis uhleri 118
O horizon 56, 57
oil drilling. *See* petroleum
oil spills 216, 229
Older Dryas 190
Old World primates 196
Oligocene epoch 178, 189
ombrotrophic mire 75
ombrotrophic wetland 77
open-pit mining 215
Ordovician 178, 194
Oreamnos americanus 171–172
organic pollutants 228
outbreeding 136
overgrazing 229, 230
Ovis canadensis. See bighorn sheep
Ovis moschatus 66, 168

Ovis nivicola 173
owl 161
oxygen
    isotopes of 181, 182, 187, 189, 195
    shortage of 121–123, *123,* 134
    and soil formation 57
ozone hole xviii, 106, 226, 227,
    232–234

P

pachycaul 83
pack ice 74
paintbrush 82
paleomagnetism 31
palsa 62, *76, 77*
palsa mire 77
palynology. *See* pollen analysis
Pangaea 178
Papaver sp. 69
Papilio zelicaon 153, 154
parabolic flowers 151
páramo 83
parrots 163
Parry, William Edward 205
Peary, Robert 205, 210
peat
    blanket bog 77
    in blanket bog 77
    as carbon sink 221
    and climate change 224
    development from moss 147
    in dwarf shrub tundra 69
    and palsa mound 62
    and soil formation 58, 85
penguins
    in Antarctic tundra 164–165,
        174–175
    effect of ecotourism on 217
    effects of climate change on 225
    hunting of 212–213
    reproductive strategy of 124
    sexual reproduction of 128
peregrine falcon 160
periglacial features 60–62
periglacial regions 59, 63
permafrost 58*m*
    and cryptophytes 111
    defined xvi, 14
    discontinuous/continuous 58–59
    in dwarf shrub tundra 67–68
    in tall shrub tundra 71
permanent wilt 120
pesticides 228
Peter I the Great (czar of Russia) 205
petroleum 215–216, *216,* 221
pH 53, *53,* 54, 228

phanerophytes *111,* 111–113
Phippsia algida 66
phloem 149
Phoca sp. 165
phosphorus 95, 96
photobiont 143
photosynthesis
    in bacteria 142
    carotenoids' role in 118
    in cyanobacteria 142
    in ecosystem 86
    energy trapping by 88, 102–103
    in evergreens 116
    in lichen 143–145
    oxygen generation by 121
    transpiration during 120
    in tundra wetlands 75
photosynthetic organisms 54
phycocyanin 142
physical weathering 52
physiognomy 112
Picea sp. 72
pied flycatcher 194
piedmont glacier 41–42
pigment 105, 119. *See also* coloration,
    of plants/animals
pika 80
pillow lava 32
pine 73, 83
pingo *61, 62*
pioneer 140, 141
pipelines 216, *216*
plant 110–113. *See also specific plants,*
  e.g.: grasses
    of alpine tundra 79–81
    of Antarctic tundra 78
    and capillary water 55
    classification of *111,* 112
    cold's effect on 104–105, 134
    drought's effect on 119
    energy trapping by 88
    freezing protection mechanisms in
        110
    frost's effect on cells 109
    grazing's effect on 129–130
    hair on 105
    and oxygen starvation 121
    of polar desert 65–66
    resource allocation by 88, 89
    seed size 127–128
    sexual reproduction 125–126
    of tropical alpine habitats 82–83
    vascular plants 149–152
    vegetative reproduction 123–125
plant hairs. *See* trichomes
plant reproduction/dispersal 123–129

plate tectonics 29–35, *30, 32, 34,* 62, 235
Plectrophenax nivalis 161
Pleistocene epoch
    evolution of *Espeletia* during 150
    glaciations during 178–183, 194, 195
    mammoths of 198
    shortage of fossil material from 187, 189
Pliny 171
Pliocene 189–190
Plutella xylostella 154
Pluvialis sp. 159
podzol 55, 56, *56*
poikilohydric plants 121, 145
poikilothermic animals 105, 123
polar bear *96,* 97, 165–166, 175, 203
polar desert 14–15, 65–66, 78, 85, 90, 225
polar front 10, 11
polar rocks 35–36
polar semidesert 14–15
polar steppe 66
polar tundra 1–5, 23, 81, 85, 165–170
pollen analysis 187–190, 195
pollen grains 126
pollination 117–118, 126–127, 134–135, 151
pollution xviii, 226–229, 233
    in alpine tundra 102
    fertilization from atmosphere 96
    from gold mining 215
    from oil extraction 216
polygon *59, 60,* 60–63, *67*
    in dwarf shrub tundra 67–68
    in polar desert 65
    in tundra wetlands 75
polygon mire 75
Polytrichum sexangulare 146
polyuronic acids 75
pools 74–75
population 86, 125
population cycles 101–103, 168
population ecology 124
positive feedback 223, 224
positive interaction 139
potential energy 87
Prasiola crispa 145
Precambrian shield 36
precipitation 12, 43, 94, 224, 227. *See also* snow
predators 132, 168, 196–198
prevailing wind 25
primary consumers 87
primary producer 86
primary productivity 88–91

in alpine tundra 102
and biodiversity 137
and grazing 129
and lichen 144
and ozone hole 226
and population cycles 101
and rate of element cycling 93
primates 196
productivity. *See* primary productivity
prokaryote 141. *See also* microbes
proline 110
protein 98, 99, 104, 131
Proterozoic era 176
protist 79
psychrophilic bacteria 142
psychrophilic fungi 109
psychrotrophic bacteria/fungi 99
psychrotrophic fungi 142
ptarmigan 108–109, 119, 160, 163
Pulsatilla alpina 150–151, *151*
purple saxifrage 66, 149, 151
Pygoscelis antarctica 114, *114*
pyramid of biomass 93

Q

Queen Charlotte Islands 193
Quelccaya ice cap 225

R

rabbits 128
Racomitrium lanuginosum 146
radioactive fallout 228
radiocarbon dating 180, 190–191, 194–195
ragwort 84
rain shadow 25
rainwater 53
Ramonda myconi 121
Rangifer tarandus. See caribou
Raunkiaer, Christen 110, 112, 113
recreation 216–219
red deer 229
refugia 193, 195
reindeer. *See* caribou
reindeer moss 145
relative humidity 20
relict species 193
reproduction, biological 123–129, 142, 143
reproductive strategies 124
reptiles 155
resilience 100
resource allocation 88, 89, 102
respiration 87, 88, 92, 109, 122
rheotrophic wetland 75
rhizobia 97

rhizomes 124
Rhizopoda 147
ribbon seal 165
riparian distribution 72
rock
    of alpine tundra 36–37
    classification by particle size 50
    exfoliation 52
    fragments carried by glacier 44–48, 62–63
    polar 35–36
    as source of nutrients 94
    weathering of 52–54
    in young ecosystem 99
rock cycle 28–29, *29*
rock detritus *47*
rock flour 44
rock platform *36*
rock ptarmigan 160
root systems
    in alpine tundra 102
    and drought stress 121
    hair on 105
    and oxygen starvation 121, 122
    in tundra plants 97–98
    and water potential 120
Ross, James Clark 156, 206
Ross, John 205
Ross Ice Shelf 206
Ross's gull 156–157
rotifers 147
rough-legged hawks 161
r-selection 124, 125, 128, 134
Rubus chamaemorus 69
Rumex acetosa 126
Rupicapra rupicapra 172
Russia 3, 12, 205

S

Salix alaxensis 71
Salix arctica 66, 169
SALR (saturated adiabatic lapse rate) 22
saltation 49
salt marsh 74
Sami people 212
sand 50, 52, 53
sand dunes 74
sanderling 158–159
saturated adiabatic lapse rate (SALR) 22
Saxifraga caepitosa 149
Saxifraga oppositifolia 66, 149
saxifrage 66, *117,* 149, 151
Scandinavia 2–3, 6
scent 127
scientific research 235
Sclerotinia borealis 142

Scotland 6
Scott, Robert Falcon 207–208, 210
sea birds 156–158, 174, 216
sea cliff 74
sea ice 40
sea level 182–183, 194, 195, 222–223
seals 97, 165, 203, 212, 213, 217
sea mayweed 127
seasons 8–9, 16, 101, 119, 127
seawater, density of 13
secondary consumer 87
second law of thermodynamics 87
sedge marsh 90
sedge meadows 90
sedges 129, 174
sedimentary rocks 28
seeds 99–100
seed size 127–128
senecio 83, 84
sérac 44
setae 105
sexual reproduction 123–129
sheep 175, 229, 230, 231
shoots 125
shorebirds 158–159, 174
Siberia 69
Sierra Nevada 24
size, of tundra life-forms 113–115
skiing 217–218, 221
snout, glacial 42, 44, 45
snow 37–38, 131, 145–146
snow bunting 161
snowcock 163
snow cover 65, 71
snowfall 65
snow finch 162–163
snow geese 74, 93, 130, 130
snow leopard 173, 174
snow patch 38
snow sheep 173
snowshoe hare 101, 107, 108
snowy owl 108, 109, 128, 161, 168
sodium 94, 95
soil xvi, 52
    acidity of 53–54
    alkaline 54
    in alpine tundra 80
    fauna of 147–149
    formation/maturation 52–59, 63
    from loess deposits 51–52
    microbes in 141–142
    nitrogen depletion in 99
    podzol 55, 56, 56
    of polar desert 66
    stabilization by bryophytes 146
    water potential of 120
soil profile 56, 57

solar angle 91
solar cycle 185
solar radiant energy 6–9
solar wind 16–17
solifluction 62, 66
solute 109–110
sorrel 126
South America 52
southern elephant seal 165
Southern Hemisphere 5, 8, 11, 13
southern tundra 164–165
South Pole 3, 207–208, 227
species richness 141
Sphagnum sp. 69, 75, 85, 146–147, 154
spiders 106, 107, 148, 154
spikelets 127
spodosol. See podzol
sporangia 126
spores 99, 142–143
sporophyte 145
spring gentian 150, 150
springtail 148, 152
spruce 72, 73, 225
stability 100–101
stadial 181
starch 87
Steller's eider 160
Steller's sea lion 212
stems, creeping 124
Stercorarius sp. 157–158
Sterna paradisea 157
stolons 124
stomata 120
stone 60, 65
stone field 65
stone polygons 61
stone stripes 61–62
stratigraphy 60, 180, 186–190, 195
stratosphere 11
stratospheric ozone. See ozone hole
strip mining 215
stromatolites 142
structure (vegetation) 64–65, 85
subarctic 14
subduction zone 31
substrate 104
succession 139–141, 140
sugar 87, 110
sugar alcohol 110
sulfides 53
sulfur 53
Sun 6, 91
sunflower 83
sunlight 64, 86–88, 133
survival 124
Svalbard, Norway 43, 142, 143, 148
symbiosis 127, 144

T

taiga 225
tall shrub tundra 71, 85, 90–91
tardigrades 148
Tasman, Abel Janszoon 206
tectonic history 33–35
temperature 9, 12, 98–99, 103. See also
    cold
temperature cycle 23
temperature inversion 25
tephra 38, 186
Teratornis merriami 163
terminal moraine 47, 180
terns 157
Terra Australis Incognita 206
tertiary consumer 87
Tetraogallus tibetanus 162
thallus 143, 145
thermodynamics 87
thermohaline circulation 13–14
thermoregulation 115
thermosphere 16
therophytes 111–113, 120
Tierra del Fuego 149
till 46, 48, 179–180
tillites 46
timberline 14, 22, 73, 81–82, 84
tools 197–198, 202
torpid state 131
tourism xviii, 211, 216–219, 218
trade routes, search for 203, 210
Trans-Alaska Pipeline 216
transhumance 200
transpiration 117, 120
trapping 212
tree 72, 73, 81–84, 192
tree heather 84
tree limit 22
trichomes 105
Trifolium sp. 126
trimline 38
Trimmer, Joseph 179–180
Tripleurospermum maritmum 127
trophic levels 92–93
tropical alpine habitats 82–85, 83
Tropics 137
tropopause 11
troposphere 11
tufted saxifrage 149
tundra (defined) xv–xvi
tundra (map) xviim
tundra ecosystem. See ecosystem
tundra wetlands. See wetlands
turbulence 10
tussock 65, 70, 78, 79, 82–83, 225
tussock-heath tundra 70, 70

Tyrrell, Joseph Burr 183
Tyrrell Sea 183

U

Uhler's arctic butterfly *118*
ultraviolet radiation 106, 119, 143, 226, 234
uniformitarianism 178, 179, 194
Ursus maritimus. See polar bear
U-shaped valley 41

V

Vaccinium sp. 118–119
vacuole 109
valley, hanging 47–48, *48*
valley glacier *43, 46,* 50
value, of tundra xviii, 211–221
    carbon cycle *219,* 219–221
    ecotourism and recreation 216–219, *217, 218*
    evaluation issues 211–212
    hunting, fishing, and trapping 212–214, *213, 214*
    mineral reserves 214–216, *216*
Vancouver, George 205
vapor pressure 20
vapor pressure deficit 20
varve 51
vascular bundles 149
vascular plants 149–152, 174
vegetation. *See* plant
vegetation structure. *See* structure
vegetative reproduction 123–125
Venezuela 83
Vespertilionidae 169
Vikings 203, 210
Vinland 203

vivipary 125
volcanoes 186
voles 129, 175
vulnerability, of tundra ecosystems 101, 103
vultures 163–164, 174
Vultur gryphus 163

W

wading birds 96–97, 105, 158
walleye pollock 212
walrus 165
wandering albatross 165
warm glacier 42
water
    and chemical weathering 53
    effect on cells when frozen 109
    effect on climate 11–12
    in lichen 144
    loss during photosynthesis 120
    oxygen distribution in 121
    and oxygen isotopes 181
    and physical weathering 52
    and pingo creation 62
    role in soil formation 54–59
    sorting of debris by 49, 50
water bears 148
waterfall *48*
water lily 122
waterlogging 57, 121–122, 134
water potential 120
water sedge 67
weathering 28, 52–54, 63, 94, 143
wetlands 74–78, *76, 82,* 85
whaling 213
white-fronted goose 159, 159*m*
whitewater rafting 218

Whymper, Edward 208
wildfowl 159–160, 174
willow 71, 72, 106
willow grouse 163
willow ptarmigan 119, 160
wilt 120
wind 51–52, 188
wind chill effect 15
wind pollination 126–127, 134
Wisconsinian glaciation 182, 183, 190, 193
wolf *92, 132, 172, 236*
    and caribou 101, 167
    and conservation 229
    early humans' cooperative hunting with 197
woodrat 114
woolly mammoth 198, 199

X

Xanthoria parietina 145
xylem 149

Y

yak *230*
yeast 122
yellow-bellied marmot *170,* 214, *214*
yellow-billed loon 158
yellow-eyed penguin 165
yellow-rumped warbler 193
Younger Dryas 190, 191, 195

Z

Zemlya Aleksandra Island 65